Fortressing Pixels

Other related titles:

You may also like

- PBPC082 | Dalal | Generative AI in Multimedia Content Processing: Security and privacy perspectives | contracted 30 Sept 2024 (Multimedia book series)
- PBSE029 | Rani | Machine Learning and Deep Learning Driven Techniques for Multimodal Data Security in the Internet of Multimedia Things | contracted May 2024 (Multimedia book series)
- PBPC066 | Srivastava | Federated Learning for Multimedia Data Processing and Security in Industry 5.0 | contracted May 2022 (Multimedia Book Series)
- PBPC061 | Lv | Access Control and Security Monitoring of Multimedia Information Transmission | pub date Dec 2023 (Multimedia book series)
- PBPC064 | Rajput | Intelligent Multimedia Processing and Computer Vision: Techniques and applications | pub date Nov 2023 (Multimedia Book Series)

We also publish a wide range of books on the following topics:
Computing and Networks
Control, Robotics and Sensors
Electrical Regulations
Electromagnetics and Radar
Energy Engineering
Healthcare Technologies
History and Management of Technology
IET Codes and Guidance
Materials, Circuits and Devices
Model Forms
Nanomaterials and Nanotechnologies
Optics, Photonics and Lasers
Production, Design and Manufacturing
Security
Telecommunications
Transportation

All books are available in print via https://shop.theiet.org or as eBooks via our Digital Library https://digital-library.theiet.org.

IET SECURITY SERIES 030

Fortressing Pixels

Information security for images, videos, audio and beyond

Edited by
Subhrajyoti Deb, Adnan Abdul-Aziz Gutub and
Aditya Kumar Sahu

The Institution of Engineering and Technology

About the IET

This book is published by the Institution of Engineering and Technology (The IET).

We inspire, inform and influence the global engineering community to engineer a better world. As a diverse home across engineering and technology, we share knowledge that helps make better sense of the world, to accelerate innovation and solve the global challenges that matter.

The IET is a not-for-profit organisation. The surplus we make from our books is used to support activities and products for the engineering community and promote the positive role of science, engineering and technology in the world. This includes education resources and outreach, scholarships and awards, events and courses, publications, professional development and mentoring, and advocacy to governments.

To discover more about the IET, please visit https://www.theiet.org/.

About IET books

The IET publishes books across many engineering and technology disciplines. Our authors and editors offer fresh perspectives from universities and industry. Within our subject areas, we have several book series steered by editorial boards made up of leading subject experts.

We peer review each book at the proposal stage to ensure the quality and relevance of our publications.

Get involved

If you are interested in becoming an author, editor, series advisor, or peer reviewer please visit https://www.theiet.org/publishing/publishing-with-iet-books/ or contact author_support@theiet.org.

Discovering our electronic content

All of our books are available online via the IET's Digital Library. Our Digital Library is the home of technical documents, eBooks, conference publications, real-life case studies and journal articles. To find out more, please visit https://digital-library.theiet.org.

In collaboration with the United Nations and the International Publishers Association, the IET is a Signatory member of the SDG Publishers Compact. The Compact aims to accelerate progress to achieve the Sustainable Development Goals (SDGs) by 2030. Signatories aspire to develop sustainable practices and act as champions of the SDGs during the Decade of Action (2020-2030), publishing books and journals that will help inform, develop, and inspire action in that direction.

In line with our sustainable goals, our UK printing partner has FSC accreditation, which is reducing our environmental impact to the planet. We use a print-on-demand model to further reduce our carbon footprint.

British Library Cataloguing in Publication Data

A catalogue record for this product is available from the British Library

978-1-83724-163-7 (hardback)
978-1-83724-164-4 (PDF)
978-1-80705-149-5 (epub3)

Typeset in India by MPS Limited

Printed in the UK by CPI Group (UK) Ltd

Cover image: Donald Iain Smith/Tetra images via Getty Images

Contents

Foreword   xv
About the editors   xvii
Introduction   xix

1   A perceptual image hashing using a CSLBP-based method   1
    Moumita Roy and Irshed Hussain
    1.1   Introduction   1
    1.2   Literature survey   3
    1.3   Proposed system   5
        1.3.1   Center-symmetric local binary pattern   5
        1.3.2   Detailed design   5
        1.3.3   Distance measure and performance evaluation   7
    1.4   Experimental results and discussion   8
        1.4.1   Perceptual robustness   8
        1.4.2   Analysis of discrimination   10
        1.4.3   Performance comparison   11
    1.5   Conclusion and future work   12
    References   12

2   ECG signal protection using redundant discrete wavelet
    transform-based data hiding   15
    Moad Moad Med Sayah, Zermi Narima, Khaldi Amine, Kafi Med
    Redouane and Aditya Kumar Sahu
    2.1   Introduction   16
    2.2   Related works   17
    2.3   Proposed watermarking scheme   18
        2.3.1   Watermark generation   19
        2.3.2   RDWT-Schur watermark integration process   19
        2.3.3   RDWT-Schur watermark extraction process   21
    2.4   Experimental results and discussions   21
        2.4.1   Capacity test   22
        2.4.2   Imperceptibility test   23
        2.4.3   Robustness test   26
    2.5   Conclusion   27
    2.6   Compliance with ethical standards   27
    References   27

3   **A blind watermarking framework for authentication of digital images**   31
    Samrah Mehraj, Subreena Mushtaq and Shabir A. Parah
    3.1   Introduction   31
          3.1.1   Critical requirements for an efficient digital watermarking
                  scheme   33
          3.1.2   Classification of digital watermarking methods   35
    3.2   Literature survey   38
    3.3   Proposed work   40
          3.3.1   Image pre-processing and block division   40
          3.3.2   LSB modification   41
          3.3.3   DCT transformation and DC coefficient extraction   41
          3.3.4   LSB embedding of bits   41
          3.3.5   Watermark extraction   42
    3.4   Experimental outcomes   42
          3.4.1   Imperceptibility analysis   43
          3.4.2   Fragility analysis   43
          3.4.3   Timing analysis   49
    3.5   Conclusion   49
    References   50

4   **Exploring technological trend and collaboration analysis in reversible data hiding in encrypted images**   53
    Ankur, Sonal Gandhi, Rajeev Kumar and Ki-Hyun Jung
    4.1   Introduction   53
    4.2   Publication trend and productivity analysis   55
          4.2.1   Data collection source and strategy   56
          4.2.2   Publication structure analysis   56
    4.3   Co-authorship analysis   60
          4.3.1   Author coauthor linkages   60
          4.3.2   Organizational coauthor linkages   64
          4.3.3   Country coauthor linkages   65
    4.4   Citation analysis   66
          4.4.1   Author citation linkages   66
          4.4.2   Country citation linkages   68
          4.4.3   Document citation linkages   69
          4.4.4   Organizational citation linkages   71
    4.5   Conclusion   72
    References   73

5   **Blockchain architecture, consensus mechanisms, and their applications in watermarking**   77
    Gandharba Swain, Anita Pradhan, Satish Muppidi,
    Pramoda Patro and Monalisa Sahu
    5.1   Introduction   77

5.2 Blockchain architecture   79
5.3 Consensus mechanisms   82
5.4 Types of blockchains   84
5.5 Applications of blockchain in image watermarking   86
5.6 Conclusion   88
References   89

6 **Cloud-based analysis with quantum cryptography-based cloud
   security model (QC-CSM) for enhanced data security in storage
   and access**   93
   *Amit Kumar Chandanan, Vivek Kumar Sarathe, Akhilesh Dwivedi,
   Raja Chandrasekaran, Vandana Roy and Adıtya Kumar Sahu*
   6.1 Introduction   94
   6.2 Related work   97
   6.3 Objective of the chapter   99
   6.4 Motivation   100
   6.5 Proposed work   100
        6.5.1 Quantum key distribution (QKD)   101
        6.5.2 Quantum no-cloning theorem   102
        6.5.3 Secure key distribution using QKD   102
        6.5.4 Attribute-based encryption (ABE) for secure data storage 103
        6.5.5 Quantum authentication mechanism   104
        6.5.6 Quantum certificates and digital signatures for integrity
               and non-repudiation   105
        6.5.7 Secure key distribution using QKD   105
        6.5.8 Quantum key distribution protocol process flow   106
   6.6 Results and analysis   106
   6.7 Conclusion   112
   References   113

7 **An improved deep reinforcement learning approach for
   security-aware virtual network embedding algorithm**   117
   G. Yogarajan and G. Rajasekaran
   7.1 Introduction   117
   7.2 Backgrounds   119
        7.2.1 Virtual network embedding (VNE)   119
        7.2.2 Security-based virtual network embedding algorithms 120
        7.2.3 Machine learning-based virtual network embedding
               algorithms   121
        7.2.4 Graph convolutional network (GCN)   121
   7.3 Network models and evaluation indicators   122
        7.3.1 System model   122
        7.3.2 Evaluation metrics   124
   7.4 Introduction to a deep reinforcement learning algorithm for the
        policy-based network   125

	7.4.1	Feature extraction	125
	7.4.2	Policy network	126
	7.4.3	Graph convolutional network	127
	7.4.4	Training and testing	127
	7.4.5	GCN–VNE algorithm	128
	7.4.6	Analysis of computational complexity	130
7.5	Analysis and evaluation of performance		130
	7.5.1	Setting parameter	130
	7.5.2	Comparison algorithm	131
	7.5.3	Result analysis	132
7.6	Conclusion		134
References			134

**8 Demystifying IoT, cloud, and blockchain for multimedia
security – a scientific study   139**
*P.K. Paul, Tatayya Bommali, Abhijit Bandyopadhyay,
Sanjukta Chakraborty, Mustafa Kayyali and S.K. Sharma*

8.1	Introduction		140
	8.1.1	Objective of the chapter	140
	8.1.2	Methods adopted	141
	8.1.3	Existing works and research gap	141
	8.1.4	Internet of things and multimedia security	144
	8.1.5	Challenges and considerations in IoT multimedia security	149
	8.1.6	Emerging trends and research pathways in IoT multimedia security	152
	8.1.7	Cloud computing and multimedia security	155
	8.1.8	Techniques for securing multimedia in the cloud	156
	8.1.9	Blockchain and multimedia security	157
	8.1.10	IoT, cloud, and blockchain in multimedia security: Trends, direction, and future aspects	160
	8.1.11	Concluding remarks	162
References			164

**9 Bibliometric analysis and research trends in image encryption:
securing sensitive visual content (2020–2024)   171**
*Roseline Oluwaseun Ogundokun, Pius Adewale Owolawi, Elizabeth
Mkoba, Abdulwasiu Bolakale Adelodun, Akinyomade O. Owolabi,
Monalisa Sahu and Gandharba Swain*

9.1	Introduction		172
9.2	Background and motivation		173
9.3	Materials and methods		174
	9.3.1	Research methodology	175
	9.3.2	Data collection	175
	9.3.3	Data wrangling	176
	9.3.4	Data analyses	176
	9.3.5	Visualization	177

9.4   Result   178
    9.4.1   Experimental setup   179
9.5   Discussion   186
    9.5.1   Interpretation of the findings   186
    9.5.2   Evaluating research goals   190
9.6   Conclusion   191
References   192

**10 Privacy protection of medical data using NTRU-based
post-quantum cryptography   197**
*B.V.S.S. Praneeth, Rupa Ch, Ch.N. Manikanta, D. Pavan Kumar,
Monalisa Sahu and Aditya Kumar Sahu*
10.1   Introduction   197
    10.1.1   Problem statement   198
    10.1.2   Motivation   198
    10.1.3   Research contributions   199
10.2   Literature survey   199
10.3   Preliminaries   201
    10.3.1   Polynomial ring unit   202
    10.3.2 Nth degree truncation   202
    10.3.3   Cyclic convolution function   202
10.4   Proposed methodology   202
    10.4.1   Key generation   202
    10.4.2   NTRU encryption   204
    10.4.3   NTRU decryption   206
10.5   Results and analysis   206
    10.5.1   Performance evaluation   206
10.6   Conclusion   209
References   209

**11 Advances in image steganography: a survey of methods, metrics,
and emerging trends across domains   213**
*Biswajit Patwari, De Rosal Ignatius Moses Setiadi, Debosree Ghosh,
Srishti Dey, Utpal Nandi, Sudipta Kr Ghosal and Monalisa Sahu*
11.1   Introduction   214
11.2   Techniques in digital steganography   216
    11.2.1   Spatial domain techniques   217
    11.2.2   Transform domain techniques   221
11.3   Evaluation metrics   223
    11.3.1   Capacity   224
    11.3.2   Imperceptibility   225
    11.3.3   Robustness   226
11.4   Recent trends and advances in steganography   226
    11.4.1   Adaptive and hybrid techniques   226
    11.4.2   Machine learning and AI in steganography   228

11.4.3 Quantum steganography   229
11.4.4 Cross-domain and cross-modal steganography   231
11.4.5 Linguistic steganography   232
11.5 Challenges and open issues   234
11.5.1 Trade-off among evaluation metrics   234
11.5.2 Robustness against attacks in steganography   235
11.5.3 Detection and steganalysis resistance   236
11.6 Applications of steganography   238
11.7 Conclusion   240
References   241

**12 Video tampering detection: deep learning techniques for
     temporal and motion artifacts   251**
Ajantha Devi Vairamani
12.1 Introduction   252
12.1.1 Rise of video manipulation technologies   252
12.1.2 Threats posed by deepfakes and video tampering   252
12.1.3 Role of AI and deep learning in video security   253
12.2 Types of video tampering   254
12.2.1 Spatial forgery through intra-frame manipulations   254
12.2.2 Temporal forgery (inter-frame manipulation)   256
12.2.3 Spatio-temporal forgery (combination of spatial and
        temporal manipulation)   258
12.2.4 More manipulation techniques and their detection   259
12.3 Detection of video tampering in deep learning   260
12.3.1 Convolutional neural networks (CNNs) for frame-level
        analysis   261
12.3.2 Recurrent neural networks (RNNs) for temporal
        consistency verification   261
12.3.3 Vision transformers (ViTs) for motion-based analysis   262
12.3.4 Hybrid models for spatiotemporal analysis   263
12.4 Feature extraction for forged video detection   263
12.4.1 Optical flow and anomalies of motion   264
12.4.2 Frame transition inconsistencies   265
12.4.3 Temporal attention mechanisms   265
12.5 Datasets from video-tampering detection   266
12.5.1 FaceForensics++   266
12.5.2 Deepfake detection challenge (DFDC)   267
12.5.3 Celebrity deepfake dataset (Celeb-DF)   267
12.5.4 UCF-crime dataset   268
12.5.5 Custom dataset creation for video integrity   268
12.6 Hands-on: detecting video deepfakes using a pretrained
      transformer model   268
12.6.1 Loading the pre-trained video deepfake detector   270
12.6.2 Extracting frames and features   271

12.6.3 Using optical flow for temporal analysis   272
12.6.4 Running the model and interpreting predictions   273
12.6.5 Fine-tuning on a custom dataset   274
12.7 Conclusion   277
References   278

Conclusion   **281**
Index   **283**

Foreword

Multimedia (and more generally multimodal data) stands as one of the most demanding and exciting aspects of the information era. The processing of multimedia has been an active research area with applications in secure multimedia contents on social networks, digital forensics, digital cinema, education, secured e-voting systems, smart health care, automotive applications, the military, finance, insurance, and more. The advent of the Internet of Things (IoT), cyber-physical systems (CPSs), robotics, as well as personal and wearable devices, now provides many opportunities for the multimedia community to reach out and develop synergies.

Our book series comprehensively defines the current trends and technological aspects of multimedia research with a particular emphasis on interdisciplinary approaches. The authors will review a broad scope to identify challenges, solutions, and new directions. The published books can be used as references by practicing engineers, scientists, researchers, practitioners, and technology professionals from academia, government, and Industry working on state-of-the-art multimedia processing and security in Industry 5.0 applications. It will also be useful to senior undergraduate and graduate students, as well as PhD students and postdoc researchers.

This book, entitled "Fortressing Pixels: Information Security for Images, Videos, Audio, and Beyond," provides a detailed study of the rising concern of securing multimedia content against growing threats like tampering, surveillance, and unauthorized access. It brings together cutting-edge research, practical frameworks, and diverse perspectives from computer science, signal processing, cryptography, and AI. At first, it introduces techniques related to image authentication and tamper detection. Also, it focuses on protecting ECG signals through wavelet-based signal processing. Next, it explores image watermarking and reversible data hiding in encrypted domains, offering both practical solutions and a technical survey. Further, it dives into blockchain-based multimedia security. Finally, the book balances theoretical explanation with practical applications, offering clear mathematical derivations and experimental analysis. It is intended for researchers, cybersecurity experts, engineers, and graduate students working in multimedia processing and the security domain.

We hope the readers will find this book of great value in its visionary words.

Dr. Amit Kumar Singh, Book Series Editor
Department of Computer Science and Engineering,
National Institute of Technology, Patna 800005, India

Prof. Stefano Berretti, Book Series Editor
Department of Information Engineering,
University of Florence, Florence 50139, Italy

About the editors

Subhrajyoti Deb is an academician and researcher currently serving as an assistant professor in the Department of Information Technology at the Indian Institute of Information Technology Bhopal (IIIT Bhopal), a premier institute under the Ministry of Education, Government of India. A Visvesvaraya PhD Fellow from NEHU with postdoctoral experience at the Indian Statistical Institute (ISI) Kolkata, he specializes in lightweight cryptography, multimedia security, and IoT authentication, with an impressive research portfolio including 20+ SCI-indexed publications in prestigious journals (*IEEE Transactions, ACM TOMM, Elsevier*), 2 authored books, 2 patents, and supervision of 4 PhD scholars (3 awarded, 1 ongoing). His international research credentials include an invited visit to Nanyang Technological University (NTU) Singapore (2024), IRIS Program Fellowship from the Embassy of Japan (2019–2020), and collaborations with researchers across six countries. He served as the organizing chair for MICA 2023. He has served as a co-editor for the "Special Issue on Recent Advances in IoT and Its Applications" in *IEEE MMTC Communication Frontier*. He is a member of IEEE and CRSI. An active contributor to the academic community, he serves as a reviewer for top-tier journals (IEEE, ACM, Elsevier) and a technical program committee member for international conferences.

Adnan Abdul-Aziz Gutub is ranked as Professor (since 2012) currently affiliated with CyberSecurity Department at College of Computing within Umm Al-Qura University (UQU), Makkah - Saudi Arabia. Adnan's academic experience in CyberSecurity was gained from his previous long-time work as Associate Professor, Assistant Professor, Lecturer, and Graduate Assistant, all in cybersecurity professionalism linked with Computer Engineering department at King Fahd University of Petroleum and Minerals (KFUPM) in Dhahran, Saudi Arabia. He received his Ph.D. degree (2002) in Electrical & Computer Engineering from Oregon State University, USA. He had his BS in Electrical Engineering and MS in Computer Engineering both from KFUPM, Saudi Arabia. Adnan's research work can be observed through his 200+ publications (journals and conferences) as well as his 5 US patents registered officially by USPTO. His main research interests involved optimizing, modeling, simulating, and synthesizing Very Large Scale Integrated-circuits (VLSI) hardware for crypto and security computer arithmetic operations. He worked on designing efficient integrated circuits for the Montgomery inverse computation in different finite fields. He has some work in modeling architectures for RSA and elliptic curve crypto operations. His interest in xviii

computer security also involved steganography focusing on image based steganography and Arabic text steganography as well as watermarking and secret sharing focusing on counting-based secret-sharing.

Aditya Kumar Sahu is working in the Department of Computer Science and Engineering, SRM University-AP, Andhra Pradesh 522502, India. His research area includes multimedia forensics, digital image watermarking, image tamper detection and localization, image steganography, reversible data hiding, and convolution neural network based data hiding. He has authored 70+ publications. He is an associate editor of Journal of Electronic Imaging, Journal of Human-centric Computing and Information Sciences (HCIS), Journal of Information Processing Systems (JIPS), and International Journal of Neuroscience and Neuroinformatics (IJNN). He is also an editorial board member of Int. J. of Blockchains and Cryptocurrencies, Int. J. of Systems, Control and Communications, International Journal of Creative Computing and Discover Internet of Things, and an editorial review board member of International Journal of Digital Crime and Forensics (IJDCF) and International Journal of Fog Computing (IJFC). He holds a PhD degree in Digital Image Steganography and Steganalysis.

Introduction

The exponential rise in multimedia content – images, videos, audio, and streaming data – has changed communication, learning, entertainment, and scientific cooperation. With this rise comes a parallel escalation of security threats: unauthorized access, manipulation, surveillance, and data breaches. Confirming multimedia information's authenticity, confidentiality, and integrity has become a significant research problem across domains from computer science and electrical engineering to healthcare and forensics. The book *Fortressing Pixels: Information Security for Images, Videos, Audio and Beyond* brings together current advancements, critical understandings, and creative frameworks that address this growing challenge through diverse, multi-perspective lenses.

This book is structured to present foundational techniques and advanced security models, as well as imaging multimedia security's interdisciplinary and developing nature. The chapters represent a mix of algorithmic development, empirical studies, and system-oriented techniques from leading researchers and domain experts.

The opening chapter introduces a perceptual image hashing technique using center symmetric local binary pattern (CSLBP), presenting a compact and robust representation of image content for tamper detection and authentication. It sets the technique for adapting traditional image processing techniques for security purposes.

Chapter 2 expands the content to physiological signal protection, specifically ECG data, demonstrating how signal processing combined with repetitive discrete wavelet transforms can serve both diagnostic commitment and data confidentiality.

Chapters 3 and 4 present image watermarking in random frameworks and reversible data hiding within encrypted fields. The technical survey in Chapter 4 offers an exhilarating collaboration research and highlights how security paradigms develop through academic and industrial subsidies.

Blockchain technology, which has appeared as a powerful tool for decentralized security, is explored in Chapters 5 and 8. This assistance dissects concurrence mechanisms and their alignment with watermarking, IoT, and cloud security. In a complementary direction, Chapter 6 presents a quantum cryptography-based cloud security model, indicating post-quantum enthusiasm in multimedia environments.

Artificial intelligence is increasingly essential in enabling adaptive and intelligent security mechanisms. Chapter 7 presents this narrative by incorporating deep reinforcement learning with virtual network implanting for security-aware multimedia services.

Bibliometric analyses in Chapters 4 and 9 offer valuable perspectives on how research in image encryption and reversible data hiding is shaping. In contrast, Chapter 10 highlights medical data, particularly NTRU lattice-based encryption, for safeguarding sensitive health information.

Arising directions in image steganography are comprehensively covered in Chapter 11, looking at algorithms and metrics while determining future trends. The book concludes with Chapter 12, which explores deep learning-based methods for witnessing video tampering, a pressing problem in the age of artificial media and misinformation.

Prevailing, *Fortressing Pixels* is more than a collection of tactics – it is a curated map of continuous creations and directions in multimedia security. As editors, we intend to bridge technical stringency with application materiality, offering this book as a reference for researchers, professionals, and students seeking to extend the boundaries of secure multimedia systems. We expect this book to contribute meaningfully to the evolving digital trust and resilience conversation in a progressively visual and connected world.

Chapter 1

A perceptual image hashing using a CSLBP-based method

Moumita Roy[1] and Irshed Hussain[2]

Abstract

With the recent advancements in image editing tools, digital images can be easily manipulated. To address this, research efforts are focusing on developing methods for image authentication. Perceptual image hashing (PIH) is essential for various applications, including image retrieval, copyright protection, image authentication, etc., where a compact and unique hash is generated that captures the perceptual content of an image. In this work, a center-symmetric local binary pattern (CSLBP)-based hashing algorithm is designed to process an image and produce a compact and discriminative hash code. CSLBP captures texture information by comparing center-symmetric pairs of pixels, or we can say it considers circular symmetric neighborhoods around each pixel instead of rectangular ones. This method is highly efficient in representing local image structures and is less sensitive to noise and small image variations, making it an ideal candidate for perceptual hashing. The proposed algorithm is evaluated through a series of experiments to test its robustness against various image manipulations, including rotation, scaling, and noise addition. The results demonstrate that the CSLBP-based perceptual hashing algorithm achieves a high degree of robustness, maintaining consistent hash values for perceptually similar images while producing significantly different hashes for perceptually distinct images.

Keywords: Perceptual image hashing; CSBP; Image hashing; Copyright protection; Rotation-invariant feature

1.1 Introduction

There has been a surge of digital content in the last decade. With it, a huge amount of editing software and different technologies are introduced that have made it easy to create manipulated content on the web. While it has its good side, its dark side

[1]Computer Science and Information Technology, SOA University, India
[2]Department of Information Technology, Sri Sivasubramaniya Nadar College of Engineering, Kalavakkam, Tamil Nadu, India

has also taken a toll on people who become victims of the malicious content being circulated for harmful purposes. Every now and then, reports of cybercrimes and adulterated content are available, and with the strong hardware and software availability, it becomes a hard task to provide the authenticity [1] of the original and fake content.

But these developments have also brought up serious issues with digital multimedia data security and integrity. With the increasing accessibility of multimedia content comes an increased risk of illicit distribution and illegal access. Malicious actors can readily get and misuse private and sensitive media, and the ease of sharing digital content can result in unauthorized distribution and possible intellectual property theft as well as financial loss for content providers.

Over the last ten years, a lot of research has been done on ways to identify forgeries and safeguard digital picture intellectual property. Image watermarking [2] allows for the detection of illicit usage by encoding secret markers in photos to confirm ownership and authenticity. In order to detect manipulation, such as discrepancies in texture and lighting, digital image forensics examines the attributes of images. Perceptual image hashing (PIH) [3,4] is useful for identifying duplication and confirming authenticity since it produces distinct hashes based on visual information that are resilient to changes like scaling and compression. When combined, watermarking, forensics, and perceptual hashing provide reliable defenses against unapproved use and manipulation of digital images.

PIH has emerged as a key technique for securing and authenticating multimedia data, serving as a complement to traditional cryptosystems. Traditional cryptographic hash functions play a crucial role in ensuring data integrity and supporting data retrieval processes. However, it is essential to recognize the fundamental differences between cryptographic hash functions and perceptual image hash functions.

The hash functions generated through cryptographic methods are extremely sensitive to changes. Even a one-bit change results in a severely different hash vector output. These kinds of changes are essential for security and integrity-based applications where 1-bit information is too crucial to ignore.

However, in the case of digital images, there exists frequent editing that falls under the content-preserving manipulations like blurring, denoising, reducing quality, resizing the image, etc. These changes do not change the appearance of the images that meet the eye, though they alter a few information in the image. For the said reason, the need for a system that can withstand these common operations on the images and does not significantly change the hash vector after the manipulation is a requirement of the time. And perceptual hash functions do the same. It generates the hash vector from the content of the data and is robust and discriminative at the same time. However, despite the attempt of several researchers, it still requires a better balance between detecting the similar images as the same and different images as different from the original image.

To tackle this issue, this work incorporates center-symmetric local binary pattern (CSLBP) for generating the hash vector for similarity purposes. The notable contributions of the work can be summarized as follows:

1. Local binary pattern's (LBP) rotation invariance property is being utilized to create a system capable of resisting geometric attacks.
2. The algorithm incorporates CSLBP to extract the robust feature vector for creating the stable features pertaining to the image.
3. Experimental results conducted upon the images from standard datasets demonstrated that our approach is indeed more resistant than state-of-the-art methods. Furthermore, it exhibits a significant ability to classify discrimination across different images.

1.2   Literature survey

The last decade presented a crucial amount of research on developing robust image-hashing systems. This section details a few of the important ones.

Ouyang *et al.* [5] presented a method utilizing the combination of log polar transform (LPT) and quaternion discrete Fourier transform (QDFT). In this approach, a secondary image is generated using LPT, and QDFT is applied to each of the three color channels. The method demonstrated strong performance, particularly in handling large-angle rotations.

Tang *et al.* [6] presented an approach that used ring partition and invariant vector distance techniques. They created a four-dimensional representation of each ring's visual content using statistical measures, namely, mean, variance, skewness, and kurtosis. These distances are robust against a few common image processing operations (CPOs).

Yan *et al.* [7] described a method based on a quaternion that leverages information (color and structural) to detect a broader range of image modifications. The geometric hash is calculated using the quaternion Fourier–Mellin transform (QFMT), and the feature image hash is derived from the QFT. To further enhance the accuracy of detection, they also developed an adaptive technique for tamper localization.

Abbas *et al.* [8] introduced a PIH method based on a noise-resistant local binary pattern (NRLBP) utilizing singular value decomposition (SVD). In this approach, SVD is applied to each sub-block of non-overlapping blocks, followed by the application of the NRLBP method to these blocks. The final image hash is generated by concatenating the NRLBP histograms.

Qin *et al.* [9] presented a PIH technique that utilizes a selective sampling approach to outdraw structural features. They employed dual-threshold edge detection using a Canny operator to carefully sample non-overlapping blocks from the input image that contain the most structurally significant perceptual features. The structural feature was constructed by extracting the two dominant discrete cosine transform (DCT) coefficients from these sampled blocks, along with the corresponding position data. The final image hash was generated using two feature matrices: the location information matrix and the dominant DCT coefficient matrix. These concatenated features were then compressed using principal component analysis (PCA) to produce the final hash.

Srivastava *et al.* [10] introduced a hashing method focused on copy detection by leveraging statistical image attributes. Following preprocessing, they generated a row vector by combining the DCT with the Radon transformation. For hash generation, they selected four statistical variables – mean, variance, skewness, and kurtosis – that represent central value, dispersion, asymmetry, and peakedness. The similarity between hash pairs was then measured using the hamming distance (HD).

Hira *et al.* [11] first proposed using Laplacian pyramids to create hash codes. This method involves generating two Laplacian pyramids through multi-scale processing that extracts filters at two different levels. Then, its difference is calculated to produce a hash vector. The hash vector exhibits good robustness and is discriminative. While this technique is effective in detecting even minor changes in images, it has limitations when dealing with rotated images.

Sajjad *et al.* [12] proposed a highly secure method to generate the hash of images for protecting the distribution of data in surveillance networks in smart industries. Their work incorporates the usage of the salient edge detection technique and selective block sampling to identify blocks with high and low information content. This process, at the end, results in generating both a hash code and a key, ensuring data security in the network.

In Li [13], vital and differentiating capable features are extracted from the image signals. These extracted features are represented using a sparse set of coefficients through sparse coding techniques. A content fingerprint or hash is then generated from these sparse coefficients. The resulting hash is stored in databases for fast indexing and retrieval. The technique ensures better hash matching for the multimedia content, thereby leading to good real-time processing. However, scaling to large databases remains a challenge for sparse coding.

Ambeth Kumar [14] introduced a novel technique of saliency detection utilizing global and local feature extraction with an analyzed color space model based on image color distribution. This method aims to enhance the detection of image content for retrieval purposes. More so, it calculates several distance measures for feature identification for a large dataset of test images. The technique is capable of identifying similar images effectively.

Kanaparthi and Raju [15] utilized the MapReduce technique to handle large chunks of images, to improve the image retrieval works. Their work enhances the search retrieval rate by considering six different methods to extract the features of an image. The method provides a good content retrieval rate.

Roy *et al.* [16] proposed a method where the KAZE features are extracted on the strongest keypoints within an image. These local features are then used to create a hash function, resulting in a robust and discriminative value effective against various attack combinations, including double manipulations. However, the method is moderately rotation-invariant. To address this limitation, another proposed algorithm [17] combines KAZE features with those of the LBP histogram Fourier (LBP-HF), enhancing rotation invariance.

1.3 Proposed system

1.3.1 Center-symmetric local binary pattern

Traditional LBP techniques have gained significant popularity in image hashing due to their simplicity and effectiveness in characterizing textures. However, these methods often face challenges in maintaining robustness against common image alterations, namely, scaling, lighting conditions, and large-angle rotations.

To reduce the effects, several versions of LBP have been developed, the CSLBP version of which works well, as it takes advantage of the symmetry of pixel combinations. CSLBP offers a smaller descriptor size with enhanced resistance to rotations and noise effects. This results in lesser execution time and is better suited for PIH applications.

1.3.2 Detailed design

The section delves into explaining the steps of the proposed system.

1. Preprocessing: Firstly, the input color (RGB) image is resized to a standard fixed size of M × M pixels to ensure the same hash length for varying image sizes. This step is achieved via bilinear interpolation, which is an efficient image quality-preserving technique. This step also enables to resist scaling changes in the attacked image.

2. CSLBP feature extraction: Followed by preprocessing, in the second step, the input parameters of the CSLBP are defined along with the extraction of the CSLBP patterns.
   - Here, the radius is considered as 1 and neighbors as 8. The radius specifically determines the distance between the center pixel and the surrounding pixel, and neighbors decide the count of the surrounding pixels considered for the pattern.
   - The individual channels of the resized image, viz. red, green, and blue channels, are separately considered for CSLBP pattern calculation. This ensures that the texture patterns pertaining to each color channel are included in the feature.
   - While comparing with the center pixel, if the value of one neighbor in the pair is greater than or equal to its counterpart, a binary "1" is assigned; otherwise, a "0" is assigned.
   - This comparison generates an 8-bit binary pattern for each pixel, representing its local texture information.

3. CSLBP pattern histogram:
   - After computing the CSLBP patterns for the red, green, and blue channels, histograms of these patterns are created for each channel.
   - These histograms, $cslbp_{hR}$, $cslbp_{hG}$, and $cslbp_{hB}$, count the frequency of each possible CSLBP pattern across the entire channel.
   - The histograms from the red, green, and blue channels are then combined into a single, unified histogram.

4. Hash generation: To generate the final binary hash, each bin in the combined histogram is compared with the value of the next bin. If the current bin's value is greater than or equal to the value of the subsequent bin, the corresponding bit in the final hash is set to "1." Otherwise, the bit is set to '0'. This comparison-based method translates the histogram information into a binary sequence, forming the final hash.

The algorithm for the CSLBP pattern histogram is given in Algorithm 1.
The overall proposed system is given in algorithmic format in Algorithm 2.
The algorithm for CSLBP feature extraction is given in Algorithm 3.

Algorithm 1 Algorithm for CSLBP_Histogram(cslbp_image)

Require: CSLBP image *cslbp_image*
Ensure: Histogram of CSLBP patterns *histogram*
1: $max\_value \leftarrow max(cslbp\_image(:))$ ▷ maximum value of the CSLBP pattern
2: $histogram \leftarrow zeros(1, max_{value} + 1)$
3: $histogram \leftarrow zeros(max\_value + 1, 1)$
4: **for** $value \leftarrow 0$ to max_value **do**
5:    $histogram(value + 1) \leftarrow \sum(cslbp\_image == value)$
6: **end for**
7: **return** *histogram*

Algorithm 2 Algorithm for Image Hash Generation

Require: RGB image *I*
Ensure: Binary hash *hash*
1: $I_{rez} \leftarrow imresize(M \times M), radius \leftarrow 1, neighbors \leftarrow 8$
2: $cslbp\_R \leftarrow CSLBP(I_{rez}(:,:,1), 1, 8)$
3: $cslbp\_G \leftarrow CSLBP(I_{rez}(:,:,2), 1, 8)$
4: $cslbp\_B \leftarrow CSLBP(I_{rez}(:,:,3), 1, 8)$
5: $cslbp_{hR} \leftarrow CSLBP\_hist(cslbp\_R)$
6: $cslbp_{hG} \leftarrow CSLBP\_hist(cslbp\_G)$
7: $cslbp_{hB} \leftarrow CSLBP\_hist(cslbp\_B)$
8: $hash \leftarrow min(cslbp_{hR}, min(cslbp_{hG}, cslbp_{hB}))$
9: **for** $i \leftarrow 1: length(hash) - 1$ **do**
10:   **if** hash(1,i) >= hash(1,i+1) **then**
11:      $hash(1,i) \leftarrow 1$
12:   **else**
13:      $hash(1,i) \leftarrow 0$

14:   **end if**
15: **end for**
16: **return** *hash*

Algorithm 3 Algorithm for CSLBP(I, radius, neighbors)

Require: *I, radius, neighbors*
Ensure: CSLBP image *cslbp_I*
1: $[rows, cols] \leftarrow size(I)$
2: $cslbp_I \leftarrow zeros(rows, cols)$
3: **for** $i \leftarrow 1 + radius : rows - radius$ **do**
4:    **for** each pixel at position (i, j) **do**
5:        $center\_pixel \leftarrow I(i, j)$
6:        $binary\_pattern \leftarrow zeros(1, neighbors)$
7:        **for** $k \leftarrow 1 : neighbors$ do
8:            $x \leftarrow \text{round}(i + radius \cdot \cos(2\pi(k - 1)/neighbors))$
9:            $y \leftarrow \text{round}(j - radius \cdot \sin(2\pi(k - 1)/neighbors))$
10:           **if** $x \geq 1$ && $x \leq rows$ && $y \geq 1$ && $y \leq cols$ **then**
11:               $binary\_pattern(k) \leftarrow I(x, y) \geq center\_pixel$
12:           **end if**
13:       **end for**
14:       $cslbp\_value = \text{bi2de}(binary\_pattern,' left - msb')$
15:       $cslbp\_I(i, j) \leftarrow cslbp_value;$
16:   **end for**
17: **end for**
18: **return** *cslbp_I*

1.3.3   Distance measure and performance evaluation

We have utilized the normalized hamming distance (NHD) for distance measures. The NHD provides the difference in bits calculated between the hash vectors. In an NHD calculation, the closer the value is to 0, the greater the similarity between the hashes, and vice versa. This can be calculated using (1.1).

$$NHD = 1/Lt(k, 2) \sum_{i=1}^{L(k,2)} |hash_1(i) - hash_2(i)| \qquad (1.1)$$

where $Lt(k, 2)$ is the length of the hash vector and $hash_1(i)$ and $hash_2(i)$ are the *m*th values of the image hashes.

For the overall performance analysis, the receiver operating characteristics (ROC) curve is exploited, which assesses the robustness and discrimination capabilities of the system. This curve is generated by examining the true positive rate (TPR) of the system for different thresholds against the false positive rate (FPR).

Moreover, the area under the curve (AUC) value is computed, which provides a quantitative value for checking the system's ability to classify between similar and dissimilar images. The nearer the AUC value to 1, the better the classification. For the same purpose, we employ (1.2 and 1.3) for calculating the TPR and the FPR values.

$$P_{TPR} = n_s/N_S \qquad\qquad (1.2)$$

$$P_{FPR} = n_d/N_D \qquad\qquad (1.3)$$

where n_s represents the number of similar images classified as visually identical, N_S denotes the total number of identical images, n_d denotes the number of different images judged as perceptually similar, and N_D represents the total number of different images.

The ROC curve illustrates the performance tradeoff between robustness and discrimination capability.

1.4   Experimental results and discussion

This section discusses the experimental results concerning the efficacy of the hashing method with respect to the robustness offered for similarity, and the capacity to discriminate dissimilarity.

1.4.1   *Perceptual robustness*

To enable the robust nature of the system, different standard dataset images are considered, and visual attacks (manipulations) are produced using MATLAB tools. Different types of single manipulations are used in this work. The experimental databases for conducting the overall evaluation include databases, such as USC-SIPI [18], Petitcolas Photo Database [19], CEI Image Database [20], McGill Calibrated Color Image Database [21], and Ground Truth Database [22].

Here, we have used nine manipulations: rotation with crop (RCrp), salt and pepper noise (SPN), scaling (SCL), Gaussian filter (GAUF), circular blur (CIRB), average blur (AVGB), Gaussian noise (GAN), speckle noise (SPKN), and motion blur (MNB). The attack parameters are mentioned in Table 1.1.

For assessment of the robustness pertaining to each of the 9 manipulations, we considered 40 images from [18–20] and subjected them to the attacks with parameters in Table 1.1, resulting in 3160 visually similar images. Figure 1.1 demonstrates the distribution of the hash distances in the bin range for the original image and its attacked images.

The x-axis in Figure 1.1 provides the hash distance range, and the y-axis shows the count of the hash distances lying within the given range. As can be seen in the figure, the system with λ values of 0.15, 0.2, 0.25, and 0.3 can correctly classify 68.20%, 84.72%, 92.47%, and 96.30% of images as perceptually similar, respectively.

Table 1.1   Attack parameters with step size (SZ)

Attack types	Parameter description	Parameter value	No. of attacked images
RCrp	Angle	5–45, SZ: 5	9
SPN	Noise density	0.001–0.01, SZ: 0.001	10
SCL	Scale	0.5–5, SZ: 0.5	10
GAUF	Standard deviation	0.3–3, SZ: 0.3	10
CIRB	Radius	0.5–3, SZ: 0.5	6
AVGB	Filter size	1×1–5×5, SZ: 1×1	5
GAN	Variance	0.01–0.1, SZ: 0.01	10
SPKN	Variance	0.001–0.01, SZ: 0.001	10
MNB	(Len, Theta)	(1, 2, 3) (0, 5, 10), SZ: (1, 5)	9
Total			79

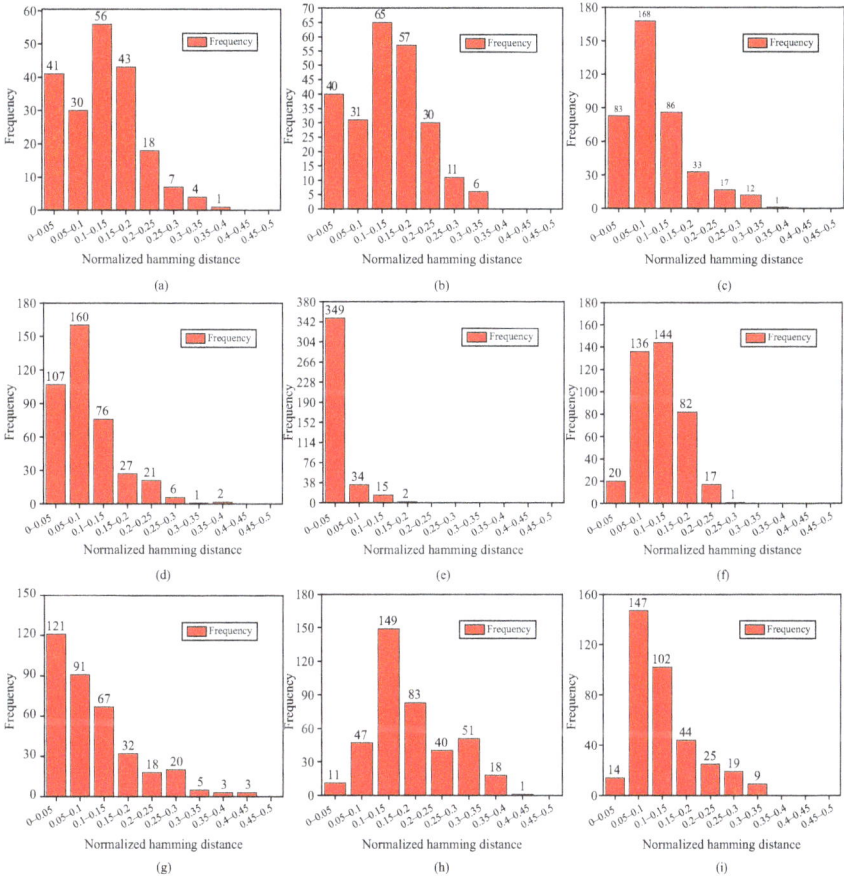

Figure 1.1   Distribution of hash distances of the 40 images and their attacked versions. (a) AVGB, (b) CIRB, (c) GAUF, (d) SCL, (e) SPN, (f) SPKN, (g) MNB, (h) GAN, and (i) RCrp.

1.4.2   Analysis of discrimination

To evaluate the ability of the system to discriminate images of different content, we formed a dataset of 200 RGB images. This dataset includes images from standard datasets [18–22] and additionally images that are chosen from the internet. Sample examples of those images are provided in Figure 1.2.

The frequency of distribution for the 39,800 hash distances computed by comparing the hashes of 200 unique images with each other is illustrated in Figure 1.3. Setting the threshold (λ) value to 0.05 would prevent any images from being incorrectly classified. However, it is to be noted that with this high threshold,

Figure 1.2   Sample images from the Internet

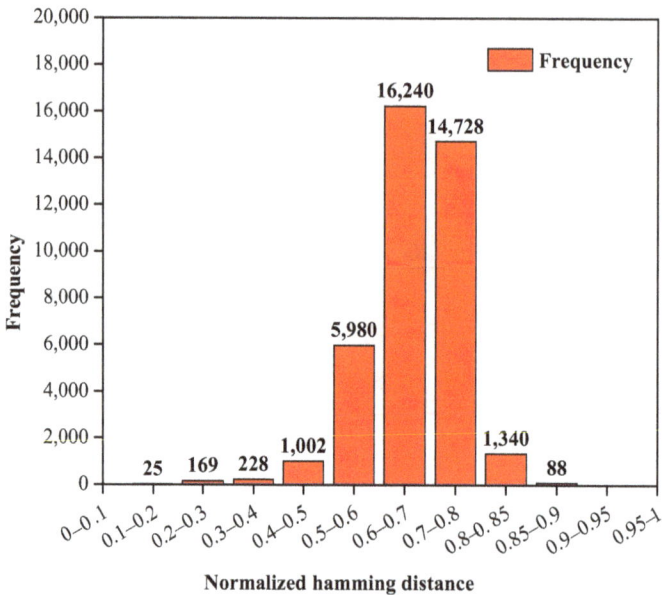

Figure 1.3   Dissimilar image hash distance distribution

the robustness of the system shall reduce owing to the misclassification of perceptually similar images as a different one.

To retain a balance of discrimination against robustness, a threshold (λ) of 0.2 is found to be suitable. With $\lambda = 0.2$, approximately 0.06% of images may be incorrectly classified. This choice is intended to maintain strong discrimination and, at the same time, ensure a reasonable level of robustness in the system.

1.4.3 Performance comparison

The effectiveness of PIH algorithms is defined by robustness and discrimination. These factors, though separately, can be adjusted by modifying a range of thresholds, but finding a good tradeoff between them is a difficult task. To thoroughly evaluate the performance metrics of a variety of algorithms, the ROC curve is used. The AUC value is then evaluated to give a quantitative performance measure.

The comparative analysis of the system includes several state-of-the-art algorithms, namely *RINGINV* [6], *SIFTSVD* [23], *STRUCTURE* [24], and *KAZE* [16]. This comparison assessed how our system performs with respect to established systems. Furthermore, for fair comparison among the algorithms, we used the same datasets as before, along with the corresponding thresholds specified in their respective works. The comparison was conducted based on the similarity measures provided by each algorithm. In addition to this, the hash distances are calculated based on their corresponding similarity or distance measures.

The ROC curve presents a set of points where the abscissa (*x*-axis) represents the FPR and the ordinate (*y*-axis) represents the TPR. These values are extracted for individual thresholds and for every combination of attacks to generate the final ROC curve as shown in Figure 1.4.

It is evident from Figure 1.4 that the RINGINV and STRUCTURE methods demonstrate lower efficacy. This inefficiency arises because RINGINV has a high rate of misclassification, while STRUCTURE-based methods lack both robust and discriminating capacity. The SIFTSVD system performs moderately due to its limited robustness against blurring and geometric attacks. Similarly, the KAZE-based system shows poor robustness to rotation attacks. However, the computation time of our system is higher in comparison to some systems; nested loops are running in the algorithm. But, our proposed algorithm outperforms the others, as reflected by its AUC value of 0.9881 as shown in Table 1.2, which surpasses the AUC values with respect to the state-of-the-art algorithms.

Table 1.2 Performance comparison among different algorithms

Systems	AUC	Length of hash	Time(s)
SIFTSVD [23]	0.7469	121 bits	4.5
RINGINV [6]	0.6724	40 digits	0.3
STRUCTURE [24]	0.6519	294 bits	0.29
KAZE [16]	0.8700	64 bits	0.35
PROPOSED	0.9881	255 bits	1.9

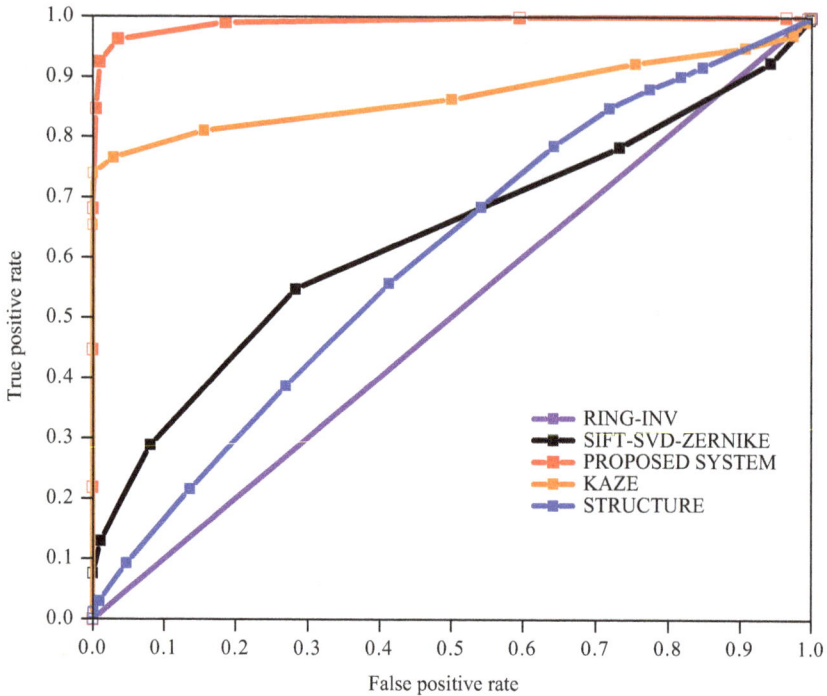

Figure 1.4   Performance comparison in terms of ROC

1.5   Conclusion and future work

This work introduces a perceptual image-hashing algorithm utilizing CSLBP feature extraction, an extension of the LBP. Our approach involved working with standard datasets for deciding specific thresholds for differentiating between visually similar images and images having different content. Experimental analysis has been thoroughly validated for both robustness and discrimination. Comparisons involving ROC curves and AUC values indicate that our algorithm's performance is better than the state-of-the-art techniques, alongside a comparable balance between robustness and discrimination.

Further work may involve enhancing the algorithm's performance with deep learning techniques and extending its applicability to other types of visual data.

References

[1]   Huang, Y., Zhang, J., Pan, L. and Xiang, Y.: 'Privacy protection in interactive content based image retrieval', *IEEE Transactions on Dependable and Secure Computing*, 2018, **17**, (3), pp. 595–607.

[2] Meesala, P., Roy, M. and Thounaojam, D.M.: 'A robust medical image zero-watermarking algorithm using Collatz and Fresnelet transforms', *Journal of Information Security and Applications*, 2024, **85**, pp. 103855.

[3] Roy, M., Thounaojam, D.M. and Pal, S.: 'Various approaches to perceptual image hashing systems – a survey'. *2023 International Conference on Intelligent Systems, Advanced Computing and Communication (ISACC)*, 2023. pp. 1–9.

[4] Neog, P.S., Roy, M., Sangale, T., *et al.*: 'Pyram: a robust and attack-resistant perceptual image hashing using pyramid histogram of gradients', *International Journal of Information Technology*, 2024, pp. 1–19.

[5] Ouyang, J., Coatrieux, G. and Shu, H.: 'Robust hashing for image authentication using quaternion discrete Fourier transform and log-polar transform', *Digital Signal Processing*, 2015, **41**, pp. 98–109

[6] Tang, Z., Zhang, X., Li, X. and Zhang, S.: 'Robust image hashing with ring partition and invariant vector distance', *IEEE Transactions on Information Forensics and Security*, 2015, **11**, (1), pp. 200–214.

[7] Yan, C.P., Pun, C.M. and Yuan, X.C.: 'Quaternion-based image hashing for adaptive tampering localization', *IEEE Transactions on Information Forensics and Security*, 2016, **11**, (12), pp. 2664–2677.

[8] Abbas, S.Q., Ahmed, F., Živić, N. and Ur. Rehman, O.: 'Perceptual image hashing using SVD based noise resistant local binary pattern'. *2016 8th International Congress on Ultra Modern Telecommunications and Control Systems and Workshops (ICUMT)*, 2016, pp. 401–407.

[9] Qin, C., Chen, X., Dong, J. and Zhang, X.: 'Perceptual image hashing with selective sampling for salient structure features', *Displays*, 2016, **45**, pp. 26–37.

[10] Srivastava, M., Siddiqui, J. and Ali, M.A.: 'Robust image hashing based on statistical features for copy detection'. *2016 IEEE Uttar Pradesh Section International Conference on Electrical, Computer and Electronics Engineering (UPCON)*, 2016, pp. 490–495.

[11] Hamid, H., Ahmed, F. and Ahmad, J.: 'Robust image hashing scheme using Laplacian pyramids', *Computers & Electrical Engineering*, 2020, **84**, pp. 106648.

[12] Sajjad, M., Haq, I.U., Lloret, J., Ding, W. and Muhammad, K.: 'Robust image hashing based efficient authentication for smart industrial environment', *IEEE Transactions on Industrial Informatics*, 2019, **15**, (12), pp. 6541–6550.

[13] Li, Y.N.: 'Robust content fingerprinting algorithm based on sparse coding'. *IEEE Signal Processing Letters*, 2015, **22**, (9), pp. 1254–1258.

[14] Ambeth Kumar, V.: 'Coalesced global and local feature discrimination for content-based image retrieval', *International Journal of Information Technology*, 2017, **9**, (4), pp. 431–446.

[15] Kanaparthi, S.K. and Raju, U.: 'Content based image retrieval on big image data using local and global features', *International Journal of Information Technology*, 2022, **14**, (1), pp. 49–68.

[16] Roy, M., Thounaojam, D.M. and Pal, S.: 'Perceptual hashing scheme using KAZE feature descriptors for combinatorial manipulations', *Multimedia Tools and Applications*, 2022, **81**, (20), pp. 29045–29073.

[17] Roy, M., Thounaojam, D.M. and Pal, S.: 'A perceptual hash based blind-watermarking scheme for image authentication', *Expert Systems with Applications*, 2023, **227**, pp. 120237.

[18] Weber, A.G.: 'The USC-SIPI image database version 6', *USC-SIPI Report*, 2018, **432**, pp. 1–24.

[19] Petitcolas, F. Accessed: 2025 January. https://www.petitcolas.net/watermarking/image_database/

[20] Dataset, C.E.I. Accessed: 2025 January. https://web.archive.org/web/20210114012713/https://homepages.cae.wisc.edu/ ece533/images/

[21] Olmos, A. and Kingdom, F.A.: 'A biologically inspired algorithm for the recovery of shading and reflectance images', *Perception*, 2004, **33**, (12), pp. 1463–1473.

[22] Shapiro, L.G. Accessed: 2025 February. https://imagedatabase.cs.washington.edu/groundtruth/

[23] Singh, K.M., Neelima, A., Tuithung, T. and Singh, K.M.: 'Robust perceptual image hashing using SIFT and SVD', *Current Science*, 2019, **117**, (8), pp. 1340.

[24] Khan, M.F., Monir, S.M. and Naseem, I.: 'Robust image hashing based on structural and perceptual features for authentication of color images', *Turkish Journal of Electrical Engineering and Computer Sciences*, 2021, **29**, (2), pp. 648–662.

Chapter 2

ECG signal protection using redundant discrete wavelet transform-based data hiding

Moad Med Sayah[1], Zermi Narima[2], Khaldi Amine[3], Kafi Med Redouane[1] and Aditya Kumar Sahu[4]

Abstract

Telemedicine provides a variety of products and services aimed at expediting the flow of digitized patient records. For efficient consultations, it is essential to centralize patient data, enabling easy access to the patient's medical history during consultations. Consequently, it is crucial for patients to utilize a secure tool with comprehensive security solutions to safeguard their information. In our effort to enhance the security of electrocardiogram (ECG) signals exchanged in telemedicine, we propose a frequency-domain watermarking approach in this chapter. This method involves concealing electronic patient records within corresponding ECG signals. The signal is initially transformed into a two-dimensional (2D) image, and the frequency content is extracted using a redundant discrete wavelet transform (RDWT). The resulting coefficients undergo Schur decomposition, and the watermark bits are incorporated by adjusting the least significant bit of the eigenvalues. Imperceptibility tests demonstrate that this approach generates a watermarked signal closely resembling the original, thereby preserving the diagnostic content. Robustness tests further indicate that the watermark can withstand commonly employed attacks in watermarking.

Keywords: Telemedicine; watermarking; electrocardiogram signals; quick response code; Schur decomposition; redundant discrete wavelet transform

[1]Department of Electronics, Faculty of Sciences and Technology, Electrical Engineering Laboratory, University of Kasdi Merbah, Algeria
[2]Electronics Department, Faculty of Engineering Sciences, Laboratories of Automation and Signals of Annaba (LASA), Badji Mokhtar Annaba University, Algeria
[3]Computer Science Department, Faculty of Sciences and Technology, Artificial Intelligence and Information Technology Laboratory (LINATI), University of Kasdi Merbah, Algeria
[4]Department of Computer Science and Engineering, SRM University-AP, Andhra Pradesh, India

2.1   Introduction

Telemedicine applications have evolved somewhat organically, responding to the needs of healthcare practitioners or, more frequently, aligning with initiatives from the political, pharmaceutical, or IT sectors [1]. These applications often amalgamate telemedicine services and products with specific telecommunications infrastructures, either Internet-based or within closed networks [2]. Ensuring secure exchanges is needed for medical practitioners to communicate and transmit medical data while protecting the privacy of doctor–patient interactions [3]. Given that medical information is typically personally identifiable, it necessitates robust security measures [4]. Upholding medical secrecy dictates that, during electronic exchanges, these data must be protected and secured to prevent interception and unauthorized access [5]. Cryptography appears to be an effective means of safeguarding patient data, yet its protection is contingent upon file encryption; once decrypted, any user can access the patient's data [6]. On the contrary, watermarking allows for concealing confidential patient information within multimedia files, ensuring continuous protection [7]. Access to patient data remains exclusive to authorized individuals. Furthermore, since the information is embedded in the file, patient authentication remains possible even after files have been transferred between medical practitioners [8]. Watermarking approaches are categorized based on the spatial or frequency domain used. In spatial techniques, the host signal's values are directly modulated to integrate the watermark [9]. With the frequency domain technique, the signal's frequency content must be extracted via a transform, and the resulting coefficients must then be modulated to incorporate the watermark [10]. Typically, the image's frequency content is extracted using the discrete wavelet transform (DWT), as it allows for effective spatiotemporal localization [11]. Nevertheless, the varying shifts in the DWT result in significant alterations to DWT coefficients, leading to data loss during watermark extraction [12]. Recognizing the limitations of DWT, particularly its susceptibility to data loss due to shift variations, the redundant discrete wavelet transform (RDWT) has been suggested as an alternative owing to its shift-invariant properties and directionality [13]. In this context, our work introduces a frequency-domain watermarking approach to embed electronic patient records within their corresponding electrocardiogram (ECG) signals. In order to retrieve the image's frequency content, the RDWT is applied after the signal was converted into a two-dimensional (2D) image. Subsequently, Schur decomposition is applied to the resulting coefficients, and the watermark bits are incorporated by adjusting the least significant bit of the eigenvalues. The watermarked image is then generated through inverse Schur decomposition and an inverse RDWT. This 2D image is further converted into a one-dimensional (1D) signal representing the watermarked ECG signal. To assess the proposed approach, ECG signals from the MIT-BIH Arrhythmia Database were employed. Distortion measurements were calculated to evaluate the obtained watermarked signals, and to assess robustness, various commonly used watermarking attacks were performed to gauge the watermark strength. The structure of this chapter is organized as follows: Section 2.2 presents a review of recent-related works, while Section 2.3 details the proposed approach. Section 2.4 outlines the experiments and evaluation of imperceptibility and robustness. Finally, Section 2.5 concludes the work and presents future perspectives.

2.2   Related works

Bhalerao recommends employing a deep neural network for estimating the center of the ECG signal to embed information through watermarking [14]. Artificial neural networks are utilized to estimate the midpoint of each block once the signal is divided into non-overlapping segments (see Figure 2.1). The dissimilarity between the estimated and actual values is then measured, and these estimated values are combined with the watermark to align the two sets of values. In a watermarking method proposed by Sanivarapu *et al.*, the initial step involves transforming the signal into a 2D image [15]. This image then undergoes discrete wavelet decomposition to create frequency sub-bands. Subsequently, the water-mark and detail components are subjected to quick response (QR) decomposition, and the watermark integration is carried out by manipulating the resulting Q and R matrices. In the approach presented by Natgunanathan *et al.*, the optimal regions for watermarking are initially identified, corresponding to areas with minimal fluctuations [16]. These regions are then subjected to a discrete cosine transform (DCT) to decompose the signal into high and low frequencies. To enhance the security of the watermark, a private key is employed to encrypt the binary sequence corresponding to the original watermark. The encrypted watermark bits are ulti-mately integrated by modulating the acquired DCT coefficients.

Yang *et al.* propose an approach based on the integer wavelet transform for watermarking ECG signals [17]. In this method, a 1D integer wavelet transform is employed to segregate the ECG signal into high and low-frequency sub-bands. The resulting coefficients from both low and high frequencies are subjected to

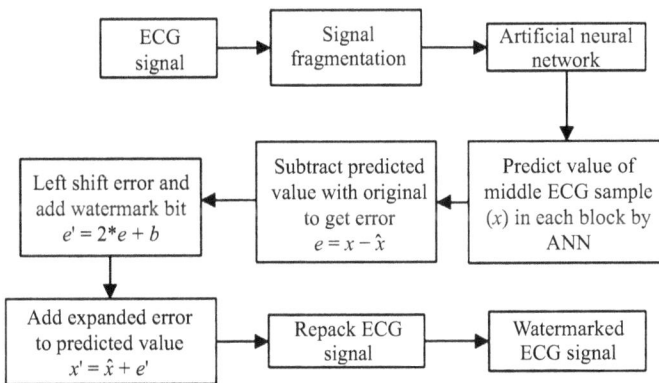

Figure 2.1   Bhalerao watermarking scheme

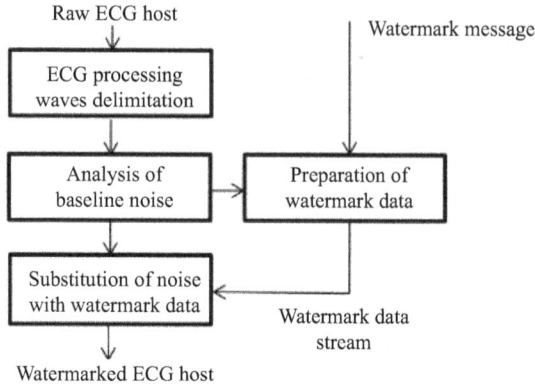

Figure 2.2   Augustyniak watermarking scheme

modulation, and the watermark bits are subsequently integrated. Augustyniak introduces a technique for watermarking ECG signals [18]. In this method, the baseline is defined using Kirchhoff's voltage law to estimate the tracks necessary for watermarking (see Figure 2.2). The disparity between the anticipated and actual values is then measured and treated as noise. Following this, the watermark is modulated to align with the measured noise.

Mathivanan *et al.* [19] introduce an innovative method for watermarking ECG data through the application of singular value decomposition and a configurable Q-factor wavelet modification. Employing a configurable Q-factor wavelet transformation, the host ECG signal is partitioned into low and high frequencies. The high-frequency components undergo singular value decomposition, and the watermark bits are embedded into the resulting singular values. In the approach presented by Mathivanan and Ganesh [20], the patient data forming the watermark is transformed into a QR code. Subsequently, a Daubechies 4 wavelet is employed to decompose the ECG signal into a 1D wavelet. The non-zero coefficients are then selectively modified to choose the high-frequency coefficients for watermark integration. Goyal proposes a method for watermarking ECG signals utilizing curvelet transformations [21]. A rapid discrete curvelet transform is applied to resize the image, decompose it under various angles, and enhance the representation of curvilinear objects and edges. The optimal coefficients for the decomposition are selected following a clustering process. To embed the watermark bits, the chosen coefficients are subsequently modulated.

2.3   Proposed watermarking scheme

Generally, the watermarking process involves three primary phases: the initial phase is dedicated to generating the watermark, followed by the integration of watermark bits into the host signal. Subsequently, the watermark is extracted through the extraction process [22].

2.3.1 Watermark generation

In the proposed method, for robust patient authentication, the watermark comprises both the patient's data and the capture parameters of the electrocardiogram. After concatenating this information and converting it into binary form, it is transformed into a QR code before integration. A QR code is a 2D matrix-style barcode composed of black modules on a white background, arranged in a square configuration [23]. The arrangement of these dots defines the information encoded in the code. One of its main benefits is that it can hold more data than a conventional barcode, particularly data that programs can recognize right away. A QR code can actually hold up to 7,089 numeric characters and 4,296 alphanumeric characters, which is a substantial improvement over a barcode's 10–13 character limit. QR codes utilize the Reed–Solomon method to correct errors, allowing for up to 30% redundancy [24]. Additionally, QR codes can incorporate graphics like logos or drawings without compromising the code's readability, thanks to Reed–Solomon error-correcting technology. As a result, this code encapsulates all the essential data for the patient's identity and information regarding signal capture (see Figure 2.3).

2.3.2 RDWT-Schur watermark integration process

Given that the host ECG signal in this chapter is a 1D signal, it undergoes conversion into a 2D image before the initiation of the watermarking process. Recognizing the diagnostic significance of P and T waves in ECG data [25], the Tompkins method is employed for their extraction through bandpass filtering of the signal. Averaging the signals using a window function aids in noise reduction, and a mask is generated by focusing on the QRS properties within a specific interval. Removal of duplicate detections facilitates obtaining the R wave and the QRS. To convert the signal into an image, 64 values are retained for each sample from the computed points. Although the resulting image size is influenced by the produced signal, the minimal data loss seems inconsequential to the diagnostic capabilities of the signal. In the proposed method, the frequency content of the image is extracted, and RDWT is employed for image decomposition. Similar to DWT, this transformation breaks down the image into four frequency bands (LL, LH, HL, and HH) at each level, using low-pass and high-pass filters. The primary limitation of using DWT lies in the shift variation induced by the down-sampling operation. However, employing the RDWT helps maintain a uniform sampling rate in the time domain. Translational invariance is guaranteed by eliminating undersampling, thereby generating an overcomplete representation of the frequency coefficients. The RDWT coefficients c_j and d_j are calculated using low- and high-pass filters $h[-k]$

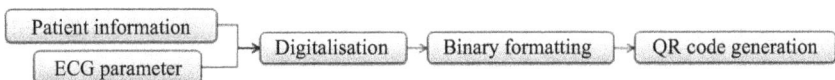

Figure 2.3 *Watermark generation process*

and $g[-k]$ as follow:

$$\begin{cases} c_j[k] = c_{j+1}[k] * h_j[-k], \\ d_j[k] = c_{j+1}[k] * g_j[-k]. \end{cases} \tag{2.1}$$

The LL sub-band coefficients obtained prior to the integration process undergo Schur decomposition to enhance the robustness of the watermark (see Figure 2.4). Leveraging the eigenvalues for watermark integration contributes to increased robustness due to their inherent stability. Consequently, the diagonal coefficients of matrix S represent the eigenvalues of matrix M. Schur decomposition involves splitting a square matrix M into a unitary matrix and an upper triangular matrix. The Schur decomposition of a matrix M can be determined using the following formula [26]:

$$M = U * S * U', \tag{2.2}$$

where U is a unitary matrix ($U * U' = I$) and S the upper triangular matrix.

The watermark bits W_1 and W_2 are then integrated by modulating the successive eigenvalues (S_1, S_2, and S_3) obtained by the following formula:

$$\begin{cases} S2 + +, \mathit{If}((W1 \neq (S1\%2) \oplus (S2\%2)) \text{ and } (W2 \neq (S3\%2) \oplus (S2\%2))), \\ S1 + +, \mathit{If}((W1 \neq (S1\%2) \oplus (S2\%2)) \text{ and } (W2 = (S3\%2) \oplus (S2\%2))), \\ S3 + +, \mathit{If}((W1 = (S1\%2) \oplus (S2\%2)) \text{ and } (W2 \neq (S3\%2) \oplus (S2\%2))). \end{cases} \tag{2.3}$$

To achieve this modulation, the least significant bits of the eigenvalues undergo modification. Given that this alteration is minimal and exclusively impacts the final least significant bits, it results in fewer distortions compared to the original signal, ensuring the maintenance of satisfactory imperceptibility. The revised coefficients are subsequently acquired through an inverse Schur decomposition, and the watermarked image is generated using an inverse RDWT:

$$c_{j+1}[k] = \frac{1}{2} \left(c_j[k] * h_j[k] + d_j[k] * g_j[k] \right), \tag{2.4}$$

where c_j and d_j represent the RDWT coefficients, $h[-k]$ and $g[-k]$ represent the low- and high-pass filters, respectively.

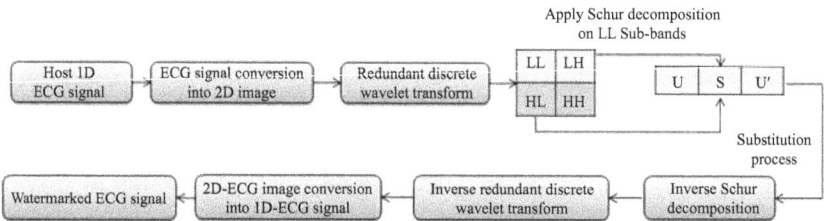

Figure 2.4   Watermark integration process

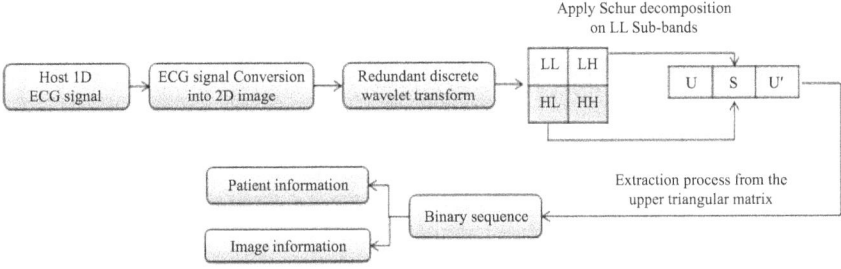

Figure 2.5   Watermark extraction process

Finally, the watermarked ECG signal is obtained by converting the 2D-watermarked image into a 1D-ECG signal.

2.3.3   RDWT-Schur watermark extraction process

Watermark extraction is commonly carried out in a non-blind manner, necessitating both the watermark and the original image, or in a semi-blind manner, requiring only the watermark, or in a blind manner, where neither the original file nor the watermark is needed [27]. The proposed watermark in this chapter is blind, signifying that the extraction process does not require either the original file or the watermark itself. The watermark can be extracted directly from the watermarked file. For watermark extraction and integration procedures, the ECG signal is initially transformed into a 2D image (see Figure 2.5). The RDWT is employed to extract the image's frequency content. The resulting coefficients then undergo Schur decomposition, and the watermark bits are retrieved from the eigenvalues using the following formula on the least significant bits:

$$\{W1 = (S1\%2) \oplus (S2\%2),$$
$$\{W2 = (S3\%2) \oplus (S2\%2), \tag{2.5}$$

where W_1 and W_2 represent two watermark bits and S_1, S_2, and S_3 represent three successive eigenvalues.

The retrieved bits form the distinct QR code corresponding to the patient information. The patient's clear identification and the acquisition of the ECG capture parameters are made by this QR code.

2.4   Experimental results and discussions

The proposed method was evaluated by watermarking ECG signals obtained from the MIT-BIH Arrhythmia Database. This database comprises 48 ECG recordings, totaling half an hour, collected from a diverse group of 47 individuals at Boston's Beth Israel Hospital. The experiments were conducted on a system with a 2.27 GHz

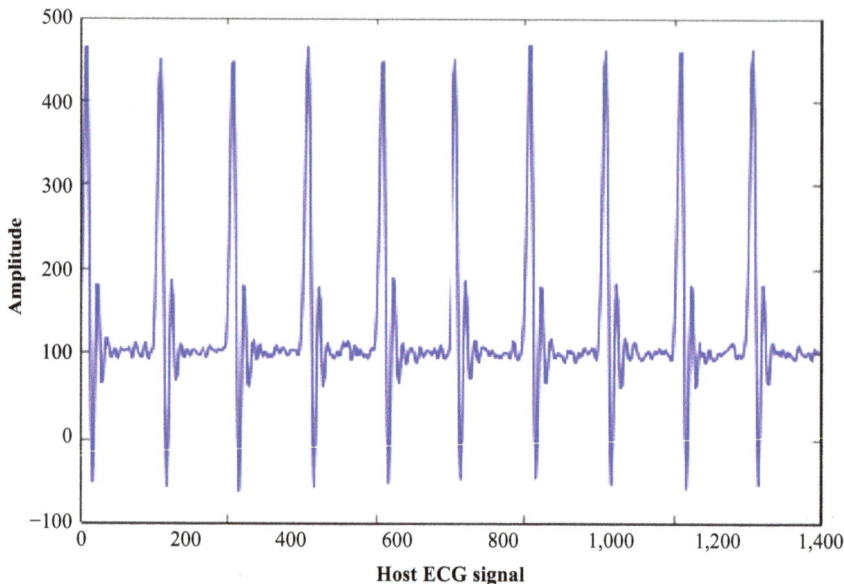

Figure 2.6   Host ECG signal example

Table 2.1   Capacity results

	Yang and Wang	Mathivanan and Ganesh	Mathivanan *et al.*	Goyal *et al.*	Augustyniak	Proposed approach
Payload (bits)	19,556	64,000	21,056	4,489	645	33,000

CPU, 2.00 GB RAM, Windows 7, and MATLAB® 8.5.0 (R2015a). An example of an ECG signal from this database is illustrated in Figure 2.6.

2.4.1   Capacity test

Capacity refers to the amount of information that can be seamlessly integrated into a medium without noticeable degradation [28]. In our approach, following the conversion of the signal into one dimension and applying RDWT and Schur decomposition for each color plane, we achieve 96,000 embeddable coefficients. Our embedding process alters only 1 coefficient out of 3, resulting in an integration capacity of 33,000 bits. Table 2.1 demonstrates that the method proposed by Mathivanan and Ganesh allows for the integration of approximately twice as much information as our method. Nevertheless, our approach's capacity remains substantial and reasonable when compared to numerous related works.

2.4.2   Imperceptibility test

The imperceptibility constraint, mandating that the distortions introduced in the original image should remain visually imperceptible, must be adhered to during watermark insertion [29]. This constraint ensures that the quality of the water-marked image closely matches that of the original image. Figure 2.7 illustrates a watermarked ECG signal, with the host signal corresponding to the one shown in Figure 2.6. Notably, the signal remains unaffected by the integration of the watermark, as evident in the figure. The distortions resulting from the integration process are visually undetectable, and the two figures appear to be visually identical.

Various objective evaluation metrics have been computed to evaluate the imperceptibility of the proposed approach:

- Peak signal-to-noise ratio (PSNR) is a distortion metric commonly employed in digital images, particularly in image compression [30]. It assesses the reconstruction quality of a compressed image by comparing it to the original image, providing a means to evaluate the performance of encoders. The PSNR is defined by the following formula:

$$(PSNR)_{dB} = 10 \log_{10} \left[M * N \; \frac{\mathrm{max}I^2(i,j)}{\sum i,j \, [I(i,j) - J(i,j)]^2} \right] \quad (2.6)$$

where $I(i,j)$ is the value of the luminance of the reference pixel (i,j) and $J(i,j)$ that of the image to be tested, the two images being of size $[M * N]$.

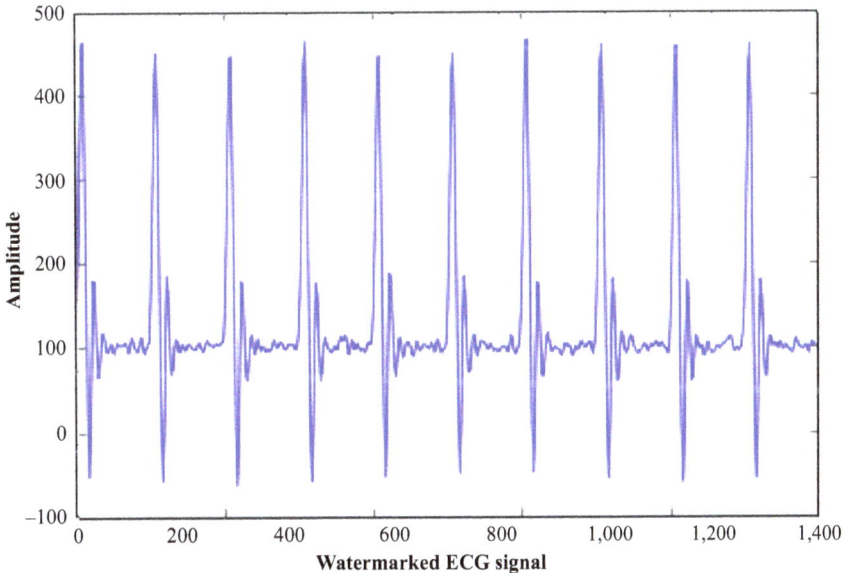

Figure 2.7   Watermarked ECG signal example

- The term "bit error rate" (BER) pertains to the error rate calculated between two signals, I and J, each having dimensions i and j, considering the amount of attenuation and/or disturbance of a transmitted signal [31]. This rate is measured upon the reception of a digital transmission:

$$\text{BER} = \frac{\sum_{i,j}(I(i,j) \oplus J\,(i,j))}{M*N}.$$
(2.7)

- The percentage residual difference (PRD) represents the square difference between the original ECG signal I and the watermarked signal J, and a smaller value indicates higher quality in the watermarked signal. This metric is calculated as follows [32]:

$$\text{PRD} = \sqrt{\frac{\sum i,j[I(i,j) - J(i,j)]^2}{\sum i,j \ \sum_{i,j}I(i,j)^{2^2}}}.$$
(2.8)

- Structural similarity (SSIM) serves as a measure of image similarity. It was designed to evaluate the visual quality of a compressed image in comparison to the original image. Unlike PSNR, which focuses on pixel-to-pixel differences, SSIM aims to assess the SSIM between the two images [33]. The SSIM is calculated as follows:

$$\text{SSIM}(x,y) = \frac{\left(2\mu_x\mu_y + C_1\right)\left(2\delta_{xy} + C_2\right)}{\left(\mu_x^2 + \mu_y^2 + C_1\right)\left(\delta_x^2 + \delta_y^2 + C_2\right)},$$
(2.9)

where μ_x is an average of x; μ_y is an average of y. δ_x^2 is the variance of x. δ_y^2 is the variance of y. δ_{xy} is the covariance of x and y. $C_1 = k_1L_2$ is the constant to avoid instability when $\mu_x^2 + \mu_y^2$ is close to zero; $k_1 = 0.01$, $=255$.
$C_2 = k_2L_2$ is the constant to avoid instability when $\delta_x^2 + \delta_y^2$ is the close to zero; $k_2 = 0.03$, $=255$.

The 48 ECG signals from our database underwent watermarking, and the resulting signals were compared to the host signals to quantify the distortion introduced by the integration process. As depicted in Table 2.2, the average PSNR obtained is 47.21 dB. While certain approaches achieve higher PSNR values than ours (such as those by Mathivanan, Ganesh, and Goyal), the results from our approach remain reasonably satisfactory when compared to other recent relevant works. The minimal alteration of information during the integration process contributes to generating fewer distortions, thereby having an insignificant impact on the host signal. This results in a signal that is structurally similar to the original signal, with an average SSIM rate of 0.9994. The BER and PRD rates obtained affirm the integrity of the watermark. The integration of watermark bits into the eigenvalue of the Schur decomposition provides reasonable stability for the watermark. Consequently, the extraction of the watermark is executed with minimal errors during the extraction process.

Table 2.2 *Imperceptibility results*

	Yang and Wang	Sanivarapu et al.	Natgunananathan et al.	Mathivanan and Ganesh	Mathivanan et al.	Goyal et al.	Bhalerao et al.	Proposed approach
PSNR	40.67	44.12	49.81	58.60	38.75	66.82	42.96	47.21
BER	/	/	1.42	0.4522	/	0.24	/	0.8984
PRD	0.0093	2.1197	/	0.029	0.09	0.0967	0.7935	0.7831
SSIM	/	/	/	0.9995	/	1	/	0.9994

2.4.3 Robustness test

The robustness assesses the watermark's resilience against various attacks. To measure the resistance of the watermark, the original watermark (I) and the extracted watermark (J) are compared using the normalized cross correlation (NCC). This metric serves as a reliable indicator of the watermark's ability to withstand attacks and is calculated as follows [34]:

$$\text{NCC} = \frac{\sum_{i,j}(I(i,j) * J(i,j))}{\sum_{i,j}I(i,j)^2}. \tag{2.10}$$

We calculated the PRD and the NCC to evaluate the robustness of the watermark and to make comparisons with related works. This approach allows for a more comprehensive assessment of our results, enabling a more accurate comparison with recent related works. Nonetheless, it is important to note that many related works lack robustness tests. The majority of watermarking studies on ECG signals found in the literature primarily focus on evaluating the imperceptibility of the proposed approaches. This emphasis is understandable, given that imperceptibility is a key requirement in medical file watermarking, ensuring the preservation of diagnostic content to avoid compromising the file's validity.

Based on the results presented in Table 2.3, we can infer that the watermark maintains its resilience against commonly employed attacks in watermarking. The proposed approach ensures a watermark that withstands rotation, compression, cropping, and scaling attacks. The achieved values are reasonably satisfactory, highlighting the significance of this resistance. Preserving the watermark during extraction is crucial to accurately identifying the patient and preventing confusion between different files.

Table 2.3 Robustness results

	Percentage residual difference			Normalized cross correlation	
	Yang and Wang	Proposed		Goyal et al.	Proposed
Truncation	0.9096	0.9762	Gaussian Noise (0.01)	0.9992	0.9996
Translation	0	0.7924	Salt and pepper (0.01)	0.9853	0.9915
Gaussian noise	0.9924	0.9912	Rotation (5°)	0.9798	0.9841
Cropping	1.2364	1.2494	Compression (5%)	0.9832	0.9891
Inversion	1.6054	1.7203	Median filter (3*3)	0.9783	0.9901
Scaling (*0.95)	1.0796	1.1294	Cropping (5%)	0.9212	0.9391

2.5 Conclusion

Sharing medical data are often a crucial necessity for healthcare professionals engaged in telemedicine. Consequently, security has evolved into an essential criterion, and ensuring the safeguarding of this data are now a vital requirement for healthcare practitioners. Security standards are guided by ethical and legislative rules, encompassing aspects like confidentiality, availability, and reliability, as reflected in the integrity and authenticity of medical information. Trustworthy information instills confidence in healthcare professionals. The digital watermarking technique emerges as a valuable tool for securing biosignal content, ensuring reliability, and, depending on the application context, enabling traceability. In our work, we propose a frequency-domain watermarking approach to enhance the security of exchanged ECG signals in telemedicine, concealing electronic patient records within the corresponding ECG signals. Imperceptibility tests confirm that this approach generates a watermarked signal akin to the originals, thereby preserving the diagnostic content of the signal. Robustness tests demonstrate that the watermark exhibits resilience against commonly used attacks in watermarking. However, it is important to note that since the watermark is not encrypted, someone with knowledge of the integration process could extract the watermark and access patient data. An encryption scheme can be added during the watermark creation process to strengthen security and address this problem. Subsequently, the watermark extractor would necessitate a decryption key to access the extracted patient data.

2.6 Compliance with ethical standards

Data availability statements: The data used to support the findings of this chapter are available at: https://archive.physionet.org/physiobank/database/mitdb/.

Conflict of interest: The authors declare that they have no known competing financial interests or personal relationships that could have appeared to influence the work reported in this chapter.

Ethical approval: This article does not contain any studies with human participants or animals performed by any of the authors.

References

[1] P. S. Sneha, S. Sankar, and A. S. Kumar, "A chaotic colour image encryption scheme combining Walsh–Hadamard transform and Arnold–Tent maps," *J Ambient Intell Human Comput*, vol. 11, no. 3, pp. 1289–1308, 2020, doi:10.1007/s12652-019-01385-0.
[2] S. B. B. Ahmadi, G. Zhang, M. Rabbani, L. Boukela, and H. Jelodar, "An intelligent and blind dual color image watermarking for authentication and copyright protection," *Appl Intell*, vol. 51, no. 3, pp. 1701–1732, 2021, doi:10.1007/s10489-020-01903-0.

[3]   N. Zermi, A. Khaldi, M. R. Kafi, F. Kahlessenane, and S. Euschi, "Robust SVD-based schemes for medical image watermarking," *Microprocess Microsyst*, vol. 84, p. 104134, 2021, doi:10.1016/j.micpro.2021.104134.

[4]   M. Jafari Barani, M. Yousefi Valandar, and P. Ayubi, "A new digital image tamper detection algorithm based on integer wavelet transform and secured by encrypted authentication sequence with 3D quantum map," *Optik*, vol. 187, pp. 205–222, 2019, doi:10.1016/j.ijleo.2019.04.074.

[5]   A. K. Sahu and M. Sahu, "Digital image steganography and steganalysis: A journey of the past three decades," *Open Comput Sci*, vol. 10, no. 1, pp. 296–342, 2020, doi:10.1515/comp-2020-0136.

[6]   A. Khaldi, M. R. Kafi, and M. S. Moad, "Wrapping based curvelet transform approach for ECG watermarking in telemedicine application," *Biomed Sig Process Control*, vol. 75, p. 103540, 2022, doi:10.1016/j.bspc.2022.103540.

[7]   V. M. Manikandan, A. Amrutham, and A. A. Bini, "An improved reversible data hiding through encryption scheme with block prechecking," *Proc Comput Sci*, vol. 171, pp. 951–958, 2020, doi:10.1016/j.procs.2020.04.103.

[8]   S. B. B. Ahmadi, G. Zhang and H. Jelodar, "A robust hybrid SVD-based image watermarking scheme for color images," *2019 IEEE 10th Annual Information Technology, Electronics and Mobile Communication Conference (IEMCON)*, Vancouver, BC, Canada, 2019, pp. 0682–0688, doi:10.1109/IEMCON.2019.8936229.

[9]   P. Ayubi, M. Jafari Barani, M. Yousefi Valandar, B. Yosefnezhad Irani, and R. Sedagheh Maskan Sadigh, "A new chaotic complex map for robust video watermarking," *Artif Intell Rev*, vol. 54, pp. 1237–1280, 2020, doi:10.1007/s10462-020-09877-8.

[10]   S. Euschi, K. Amine, R. Kafi, and F. Kahlessenane, "A Fourier transform based audio watermarking algorithm," *Appl Acoust*, vol. 172, p. 107652, 2021, doi:10.1016/j.apacoust.2020.107652.

[11]   A. K. Sahu and G. Swain, "An optimal information hiding approach based on pixel value differencing and modulus function," *Wirel Pers Commun*, vol. 108, no. 1, pp. 159–174, 2019, doi:10.1007/s11277-019-06393-z.

[12]   A. Alarifi, S. Sankar, T. Altameem, K. C. Jithin, M. Amoon, and W. El-Shafai, "A novel hybrid cryptosystem for secure streaming of high efficiency H.265 compressed videos in IoT multimedia applications," *IEEE Access*, vol. 8, pp. 128548–128573, 2020, doi:10.1109/ACCESS.2020.3008644.

[13]   Z. Narima, A. Khaldi, K. Redouane, K. Fares, and E. Salah, "A DWT-SVD based robust digital watermarking for medical image security," *Forensic Sci Int*, vol. 320, p. 110691, 2021, doi:10.1016/j.forsciint.2021.110691.

[14]   S. Bhalerao, I. A. Ansari, A. Kumar, and D. K. Jain, "A reversible and multipurpose ECG data hiding technique for telemedicine applications," *Pattern Recognit Lett*, vol. 125, pp. 463–473, 2019, doi:10.1016/j.patrec.2019.06.004.

[15]   P. V. Sanivarapu, K. N. V. P. S. Rajesh, N. V. R. Reddy, and N. C. S. Reddy, "Patient data hiding into ECG signal using watermarking in transform

domain," *Phys Eng Sci Med*, vol. 43, no. 1, pp. 213–226, 2020, doi:10.1007/s13246-019-00838-2.

[16]   I. Natgunanathan, C. Karmakar, S. Rajasegarar, T. Zong, and A. Habib, "Robust patient information embedding and retrieval mechanism for ECG signals," *IEEE Access*, vol. 8, pp. 181233–181245, 2020, doi:10.1109/ACCESS.2020.3025533.

[17]   C. Yang, W. Wang, and C. Lai, "Adaptive data-hiding in electrocardiogram based on integer wavelet transform domain and incremental approach," in *2020 9th International Conference on Industrial Technology and Management (ICITM)*, Feb. 2020, pp. 285–290, doi:10.1109/ICITM48982.2020.9080382.

[18]   P. Augustyniak, "Differential watermarking of multilead ECG baseline," in *2019 41st Annual International Conference of the IEEE Engineering in Medicine and Biology Society (EMBC)*, July 2019, pp. 5681–5684, doi:10.1109/EMBC.2019.8856684.

[19]   P. Mathivanan, S. Edward Jero, and A. Balaji Ganesh, "QR code-based highly secure ECG steganography," in *International Conference on Intelligent Computing and Applications*, Singapore, 2019, pp. 171–178, doi:10.1007/978-981-13-2182-5_18.

[20]   P. Mathivanan and A. B. Ganesh, "ECG steganography based on tunable Q-factor wavelet transform and singular value decomposition," *Int J Imag Syst Technol*, vol. 31, no. 1, pp. 270–287, 2021, doi:10.1002/ima.22477.

[21]   L. M. Goyal, "Improved ECG watermarking technique using curvelet transform," *Sensors (Basel)*, vol. 20, no. 10, pp. 1–15, 2020, doi:10.3390/s20102941.

[22]   F. Kahlessenane, A. Khaldi, M. R. Kafi, N. Zermi, and S. Euschi, "A value parity combination based scheme for retinal images watermarking," *Opt Quant Electron*, vol. 53, no. 3, p. 161, 2021, doi:10.1007/s11082-021-02793-3.

[23]   V. M. Manikandan and V. Masilamani "A novel bit-plane compression based reversible data hiding scheme with Arnold transform," *Int J Eng Adv Technol*, vol. 9, no. 5, pp. 417–423, 2020, doi:10.35940/ijeat.E9517.069520.

[24]   S. Bagheri Baba Ahmadi, G. Zhang, S. Wei, and L. Boukela, "An intelligent and blind image watermarking scheme based on hybrid SVD transforms using human visual system characteristics," *Vis Comput*, vol. 37, no. 2, pp. 385–409, 2021, doi:10.1007/s00371-020-01808-6.

[25]   N. Tsafack, S. Sankar, B. Abd-El-Atty, *et al.* A new chaotic map with dynamic analysis and encryption application in Internet of Health Things. *IEEE Acc* 2020, 8, 137731–137744, doi:10.1109/ACCESS.2020.3010794.

[26]   M. Y. Valandar, M. J. Barani, P. Ayubi, and M. Aghazadeh, "An integer wavelet transform image steganography method based on 3D sine chaotic map," *Multimed Tools Appl*, vol. 78, no. 8, pp. 9971–9989, 2019, doi:10.1007/s11042-018-6584-2.

[27]   F. Kahlessenane, A. Khaldi, M. R. Kafi, and S. Euschi, "A color value differentiation scheme for blind digital image watermarking," *Multimed Tools Appl*, vol. 80, pp. 19827–19844, 2021, doi:10.1007/s11042-021-10713-6.

[28] V. M. Manikandan, and V. Masilamani, "A novel image scaling based reversible watermarking scheme for secure medical image transmission," *ISA Trans*, vol. 108, pp. 269–281, 2021, doi:10.1016/j.isatra.2020.08.019.

[29] M. Sahu, N. Padhy, S. S. Gantayat, and A. K. Sahu, "Shadow image based reversible data hiding using addition and subtraction logic on the LSB planes," *Sens Imag*, vol. 22, no. 1, p. 7, 2021, doi:10.1007/s11220-020-00328-w.

[30] S. B. B. Ahmadi, G. Zhang, and S. Wei, "Robust and hybrid SVD-based image watermarking schemes," *Multimed Tools Appl*, vol. 79, no. 1, pp. 1075–1117, 2020, doi:10.1007/s11042-019-08197-6.

[31] E. Salah, K. Amine, K. M. Redouane, and K. Fares, "Spatial and frequency approaches for audio file protection," *J Circuit Syst Comp*, vol. 30, no. 12, p. 2150210, 2021, doi:10.1142/S0218126621502108.

[32] K. C. Jithin and S. Sankar, "Colour image encryption algorithm combining Arnold map, DNA sequence operation, and a Mandelbrot set," *J Inform Sec Appl*, vol. 50, p. 102428, 2020, doi:10.1016/j.jisa.2019.102428.

[33] V. M. Manikandan and V. Masilamani, "A novel reversible data hiding scheme that provides image encryption," *JOIG*, vol. 6, no. 1, pp. 64–68, 2018, doi:10.18178/joig.6.1.64-68.

[34] A. K. Sahu and G. Swain, "Reversible image steganography using dual-layer LSB matching," *Sens Imag*, vol. 21, no. 1, p. 1, 2019, doi:10.1007/s11220-019-0262-y.

Chapter 3

A blind watermarking framework for authentication of digital images

Samrah Mehraj[1], Subreena Mushtaq[1] and Shabir A. Parah[1]

Abstract

Digital watermarking is important for verifying the integrity and authenticity of digital images. A fragile watermarking method based on least significant bit (LSB) manipulation and discrete cosine transform (DCT)-based embedding for detection of tampering and authentication is proposed in this chapter. The technique starts with dividing the image into 8×8 non-overlapping blocks to have a systematic embedding process. Before applying DCT, the LSBs of all pixel values are initialized to zero, forming a stable embedding situation. The DC coefficient of every block is coded into the first 64 bits of DCT coefficients, which are inserted into the LSBs of the pixel values. Being extremely sensitive to changes, any change in the image causes it to be destroyed, thus being best suited for authentication and integrity checking. Experimental outcomes prove that the suggested method efficiently identifies even slight changes without sacrificing much imperceptibility. This study validates the efficacy of the suggested fragile watermarking method in applications where content authentication and tamper detection are essential.

Keywords: Medical images; authentication; integrity; watermarking; embedding

3.1 Introduction

Over the past few years, the ubiquitous expansion of digital media, especially medical imaging, and the unhindered flow of sensitive medical information over the internet have posed important issues of security. With medical images being transmitted over unsecured networks regularly for telemedicine, remote diagnosis, and electronic health records (EHRs), the necessity for strong security measures to safeguard patient confidentiality and avoid data breaches has become more important than ever [1,2]. Medical image security solutions can typically be

[1]Department of Electronics and Instrumentation Technology, University of Kashmir, India

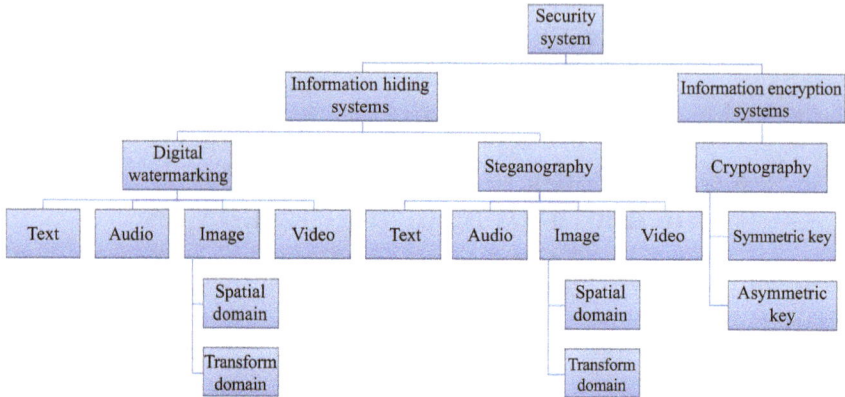

Figure 3.1   The detailed categorization of different types of security systems

divided into two major categories: information encryption systems, which address the security of data by using cryptographic methods, and information-hiding systems (e.g., digital watermarking), which hide information inside medical images to ensure authentication, integrity checking, and ownership [3]. A more specific categorization of these security systems is demonstrated in Figure 3.1.

Encryption is a core method for protecting digital information, such as extremely sensitive medical images employed in telemedicine, EHRs, and remote diagnosis. Encryption is the process of encoding data into an unintelligible state that is unreadable to anyone other than the authorized user without decryption through an assigned cryptographic key [4–6]. There are three main categories of information encryption systems:

(a) *Symmetric encryption*: It uses the same key for encryption and decryption. Symmetric encryption is effective, but it necessitates a secure distribution of keys to avoid unauthorized use, which makes it a reasonable choice for the protection of medical image transmissions on controlled networks.

(b) *Asymmetric encryption*: Also referred to as public-key cryptography, this method uses a pair of keys – a public key for encryption and a private key for decryption. Asymmetric encryption is most useful in secure telemedicine applications, where medical images need to be exchanged between healthcare professionals without violating patient confidentiality.

(c) *Cryptography*: Cryptography consists of encrypting sensitive data into an unreadable bit sequence to make it unintelligible to unauthorized persons. It is used extensively in medical data transmission and storage, such that patient histories and diagnostic images are not available to cyber hackers and unauthorized amendments.

Information-hiding methods add another level of protection by embedding sensitive information into other information in a way that makes it hard for unauthorized individuals to identify or extract hidden information [7,8]. Such methods are especially beneficial for medical images, providing safe image authentication, patient record safeguarding, and tampering detection in healthcare environments. The three main information-hiding methods are:

(a) *Digital watermarking*: Digital watermarking is the most efficient way to secure medical images. It inserts invisible but recoverable data into an image, providing authentication, copyright, and integrity verification. In medical imaging, watermarking assists in inserting critical metadata like patient ID, diagnosis information, and timestamps into the image, so that critical information is retained and securely associated with the right patient.

The two important features of watermarking are:

- *Imperceptibility*: The watermark should not be perceptible to the naked eye to preserve the image's diagnostic quality.
- *Robustness*: The watermark should be resistant to image processing operations like compression, filtering, and adding noise so that secure retrieval can be ensured.

(b) *Steganography*: Steganography is the complement to information hiding, where sensitive information is embedded within a digital image without affecting its perceptual quality. Steganography can be used to safely hide confidential patient information within diagnostic images so that unauthorized access is minimized, and safe transmission is facilitated between healthcare centers. As opposed to encryption, which informs possible attackers that data are being hidden, steganography hides the fact that data are being embedded.

(c) *Cryptography in information hiding*: Although cryptography is most directly related to encryption, it is also involved with covert data transmission in medical images. By transforming critical information into an unrevealing form before inserting it into a cover image, cryptographic methods increase the security of encoded medical information.

Of the many security systems, digital watermarking is perhaps the most reliable method of guarding medical images. By embedding security attributes in medical images, watermarking makes it possible to verify ownership, transmit images securely, and defend against forgery or unauthorized tampering. In contemporary studies, medical image watermarking is preferred over other types of watermarking because it can harmonize imperceptibility, robustness, and security to ensure that patient information is kept confidential while the diagnostic quality of medical images is preserved.

3.1.1 Critical requirements for an efficient digital watermarking scheme

The nature of the application dictates the inherent requirements of any digital watermarking scheme. These needs tend to involve a trade-off, such that enhancing one could compromise another. An efficient watermarking scheme must therefore reconcile such conflicting factors toward optimal performance. The critical requirements for creating an effective and robust watermarking system are as follows:

(a) *Robustness*

Robustness is the capability of a watermarking scheme to endure changes in the watermark and the cover media under different image processing attacks [9]. According to the type of robustness needed, watermarking methods can be classified into the following categories:

- *Fragile*: A fragile watermark is very sensitive and is damaged even with minor changes to the cover media. It is mostly applied for authentication and integrity checking, as it can identify tampered regions in the content.
- *Semi-fragile*: This watermark is robust against minor, accidental changes (like compression or format conversion) but is annihilated by malicious attacks.
- *Robust*: A strong watermark is created to survive common image processing procedures and attacks such as compression, filtering, cropping, and geometric transformation. It is often utilized for copyright protection, broadcast monitoring, and copy control. The robustness of the watermark in resisting such attacks is a key measure of robustness.

(b) *Imperceptibility*

Imperceptibility guarantees that the watermarked image is visually indistinguishable from the original cover image. This is essential in digital watermarking, where the embedded watermark must be imperceptible to the naked eye to avoid any distortion that could be perceived. One of the biggest challenges in watermarking is the interdependence of imperceptibility, robustness, and payload capacity. Increasing one of these features tends to decrease the others. For example, robustness may decrease imperceptibility, and payload capacity may impact both robustness and imperceptibility. Thus, a good trade-off has to be achieved to preserve image quality while guaranteeing watermark security.

(c) *Payload capacity (embedding capacity [EC])*

The payload of a watermarking scheme is used to denote how many bits can be embedded within the cover image while not visibly altering its visual appearance. An increase in capacity or payload sometimes coincides with trade-offs from a decrease in the robustness of the image, as well as a lack of detectability. As is evident from Figure 3.2, these three properties create a conflict triangle such that enhancing one property requires compromising the other two. For example, if the requirement is for high imperceptibility, either payload capacity or robustness has to be compromised.

(d) *Security*

Security is an essential requirement in any watermarking system. The watermark embedded must be retrievable only by authentic users who know the precise embedding algorithm. The system must be robust enough to resist malicious attacks intended for the erasure or modification of the watermark. To provide added security, watermarking methods can include encryption before embedding the watermark into the image. Various encryption algorithms, including Arnold scrambling, chaotic encryption, DNA encryption, and other

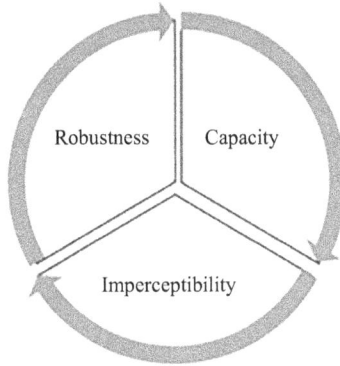

Figure 3.2   Conflict triangle between robustness, imperceptibility, and capacity

cryptographic techniques, can be used to render the watermark immune to unauthorized extraction or tampering.

(e) *Computational complexity*

Computational complexity is defined as the time and computational resources needed for embedding and extracting the watermark. Watermarking systems for real-time applications need to have low computational complexity for quick and efficient processing. High-security applications, on the other hand, need more sophisticated algorithms, which can translate into higher computational overhead.

A good digital watermarking scheme should balance robustness, imperceptibility, payload, security, and computational complexity based on the target application.

3.1.2   Classification of digital watermarking methods

Digital watermarking methods can be classified according to several factors, including robustness, availability of reference data, visibility, human perception, medium used, and embedding domain. The most important classifications are presented below:

(a) *Domain-based watermarking methods*

Watermarking methods can be categorized according to the domain where the watermark is inserted. These belong to two broad categories:

• *Spatial domain watermarking*:

In this method, the watermark is inserted directly by altering the pixel values of the cover image according to the watermark bits [10]. Spatial domain methods provide low computational complexity and are thus ideal for applications that need rapid processing. The least significant bit (LSB) substitution method is the most popular method, in which the LSB of pixel values of the cover image is modified according to watermark bits. Other spatial domain techniques are histogram adjustment, spread spectrum, and code division multiple access (CDMA)-based methods [11]. These methods are primarily used for embedding fragile or semi-fragile watermarks for content authentication.

- *Transform domain watermarking*:
  Here, the cover image is mathematically transformed, and the watermark is embedded by adjusting the transformed coefficients. Typical transformations are:

  o *Discrete cosine transform (DCT)*: The DCT is one of the most common transformations utilized in image processing and compression, especially in standards such as JPEG. The DCT is used to transform an image from the spatial to the frequency domain, expressing the image as a sum of cosines oscillating at various frequencies. The most important benefit of DCT is that it can focus the majority of the energy of the image in a small number of low-frequency coefficients, which is beneficial for compression and watermarking. By ignoring higher frequency terms, considerable data reduction can be obtained without degrading visual quality. DCT also offers resilience against small distortions, so it is an ideal candidate for digital watermarking applications.

  o *Discrete wavelet transform (DWT)*: The DWT is a robust multiresolution analysis technique that breaks down an image into various frequency subbands without losing spatial information. In contrast to DCT, which works on fixed-frequency components, DWT uses localized wavelet functions to examine both high and low-frequency details of an image. The decomposition divides an image into four subbands: low-low (LL), low-high (LH), high-low (HL), and high-high (HH), to facilitate improved edge detection and hierarchical representation. This characteristic renders DWT extremely efficient in image compression and watermarking because the watermark can be inserted into chosen subbands to be robust and imperceptible.

  o *Singular value decomposition (SVD)*: SVD is a matrix factorization method that transforms an image into three matrices, U, Σ (Sigma), and V. Matrix Σ is the diagonal matrix, which holds the singular values; these are intrinsic properties of the image and are less vulnerable to the usual distortions. This is why SVD is a resilient transformation for digital watermarking: changes in singular values do not have a tremendous impact on the quality of an image. SVD is usually used in conjunction with DCT or DWT to enhance robustness and imperceptibility, such that the embedded watermarks are resistant to different attacks and have high image fidelity.

  Transform domain methods are widely employed for embedding robust watermarks for copyright protection and verification [12,13].

(b) *Perception-based watermarking techniques*
Depending on the perception of the watermark by the human visual system, watermarking is categorized into two types:

- *Visible watermarking*:
  The watermark can be easily seen in the media and does not need to be detected with the help of any special equipment. Visible watermarking is utilized frequently for brand recognition and ownership confirmation.
- *Invisible watermarking*:
  The watermark is not visible and can only be identified or extracted by a specialized extraction algorithm. Invisible watermarking is commonly utilized for copyright protection and security purposes.

(c) *Reference-based watermarking techniques*
Watermarking techniques may also be differentiated based on whether the original watermark or cover media is needed in extraction:
- *Blind watermarking*:
  The watermark can be recovered without requiring the original watermark or cover image. This is the most difficult and sophisticated watermarking method.
- *Non-blind watermarking*:
  In non-blind watermarking, both the original cover image and the original watermark are required for extraction.
- *Semi-blind watermarking*:
  In semi-blind watermarking, specific features of the original cover image are required for watermark extraction, not the entire image.

(d) *Cover medium-based watermarking techniques*
Watermarking methods can also be classified according to the media type on which the watermark is embedded:
- *Text watermarking*:
  This technique is employed to secure text-based digital data like e-books, bank statements, digital libraries, and research papers. Watermarks are inserted to track unauthorized copying, forgery, or modification. Text watermarking is generally employed only for formatted documents.
- *Image watermarking*:
  The most common method used is image watermarking because there are such immense amounts of digital images derived from various sources, such as photographs, medical images, and satellite images. Watermark insertion can be carried out by adjusting pixel values, frequency coefficients, boundaries, or textures of an image. Imperceptibility and robustness are the major aims of image watermarking.
- *Video watermarking*:
  Video watermarking consists of inserting a watermark into a series of frames in a digital video. Because videos consume more bandwidth than images, more watermark data can be inserted. Most image watermarking methods have been applied to video watermarking purposes [14].
- *Audio watermarking*:
  Audio watermarking is employed to avoid unauthorized copying and distribution of audio files. With the convenience of downloading and sharing

digital audio, watermarking protects content by inserting special signals into the audio waveform. Methods utilize human auditory perception to insert watermarks without compromising the audio quality [15].

Watermarking methods are categorized according to their embedding domain, perceptibility, reference demands, strength, and cover medium. The different categories are employed for different purposes, from authentication and tamper detection to copyright protection and content security. An appropriate watermarking method is to be selected depending on the target application and security level.

3.2   Literature survey

Researchers across the globe are constantly looking for new ways to improve security and neutralize possible threats. Among several methods of securing transmitted digital data, digital watermarking stands out as one of the most promising ones. Digital watermarking has emerged as a key technique for authenticating images and detecting tampering. Various watermarking approaches have been suggested till now for improving the security and integrity of digital images. Cao *et al.* [16] presented a self-embedding watermarking scheme that employs an adaptive selection strategy for data embedding. A random binary array formed from most significant bits (MSB) layers of pixel values is encoded so that the system can identify the tampered area precisely and recover the modified segments efficiently. Qin *et al.* [17] designed a fragile watermarking scheme for the detection of tampering, with data from non-overlapping blocks to form a watermark. The pixel values also function as watermarks for the recovery of content so that tampered areas are efficiently recovered. Swaraja *et al.* [18] introduced a double watermarking approach for securing clinical images, embedding two watermarks via DWT, Schur transforms, and a hybrid optimization method based on Particle Swarm and Bacterial Foraging algorithms. The watermarks are Lempel–Ziv–Welch (LZW) lossless compressed before embedding, improving payload capacity with visual perception maintenance. Lin *et al.* [19] proposed a hybrid authentication scheme based on the segment truncation method for compressed images. Their scheme adaptively chooses embedding methods according to distinct image block features, exhibiting 96.6% efficiency in identifying tampered areas and collage attack resistance. In addition, Sarreshtedari and Akhaee [20] investigated image protection and recovery through source-channel encoding. The watermark is encoded using the Reed–Solomon method, and extracted data are 1 bpp coded. The erasure position identification is employed to recover the actual coded image by the system, but its effectiveness is reduced when the tampered regions are more than a certain threshold. Gul and Ozturk [21] proposed a fragile watermarking technique for the detection of tamper. The host image is separated into 32×32 non-intersecting segments and further sub-disintegrated into four sub-segments, in each of which a hash-based watermark is inserted into the LSBs of the fourth sub-segment. This method offers high watermarked image quality (57 dB peak signal-to-noise ratio [PSNR]) and resistance to linear and non-linear attacks but, because of using large block sizes, offers the least-resolution localization of tamper. Feng *et al.* [22] suggested a semi-fragile watermarking technique that embeds an authentication and recovery watermark into the last two LSBs of the original image. Pixel value compression and parity checks are utilized for tamper detection, whereas image restoration is achieved through torus-image-block averaging so even scrambled images are authenticable. Finally, an image integrity verification and authentication is provided by a vector quantization (VQ)-based watermarking scheme [23]. This method utilizes a pixel neighborhood clustering method to define intentional and unintentional attacks so that it can be immune to data-preserving operations but is susceptible to malignant alterations. Rajput and Ansari [24] suggested a multi-median authentication scheme to enable tamper detection of images. In this approach, copies of the cover photographs are embedded in their LSBs through pseudo-random codes. However with LSB embedding, the image clarity degrades, and the model can recover up to 20% of the tampering regions only. A hybrid and blind watermarking technique based on discrete Gould transform (DGT) and integer wavelet transform (IWT) was presented by Selvan *et al.* [25]. The host image is divided into regions with IWT, and less-energy elements are opted for implanting embedding. The sub-bands are segmented into blocks, on which DGT is performed preceding watermark embedding followed by inverse DGT for rebuilding. This compound technique outshines single-transformation techniques in performance but bears the drawback of high distortion with low payload capability. Prasad and Pal [26] also presented a watermarking approach whose watermark bits were created through a logistic map and Hamming code. The watermarking is done by taking advantage of the difference between two successive pixels, providing a fair trade-off between image perceptual quality and EC. Gull *et al.* [27] suggested an approach for clinical image tamper localization. In this method, the detection of temper information is hidden in the bottom half of an image segment, and the localization of temper information in the upper section so that modifications can be identified correctly. Sahu [28] suggested a fragile watermarking scheme using a logistic map for tamper localization. The watermark is created through the logistic map and inserted in the intermediate significant bit of the image. Azeroual and Karim [29] proposed a real-time watermarking scheme that utilizes Faber–Schauder-based DWT for the detection and location of tamper. The method has a quicker embedding process despite its suboptimal perceptual image quality. Gull *et al.* [30] also suggested a self-embedding fragile watermarking method for medical image authentication and tamper localization. The system uses two watermarks, namely the authentication watermark and localization watermark, which are derived from 4×4 non-overlapping blocks of the image. Chaotic encryption is applied to each block prior to watermark generation, and the 4-MSBs of every pixel decide the authentication watermark, which is also DNA-encoded to improve security. The literature identifies various watermarking methods that seek to provide content authentication, tamper detection, and integrity protection while compromising between imperceptibility and payload capacity. Nevertheless, the increasing number of security attacks point to the critical necessity for new security algorithms with improved imperceptibility, strength, and payload efficiency.

In summary, although fragile image watermarking has undergone significant improvement, issues relating to computational complexity, imperceptibility, and adaptability across a wide range of situations remain a motivation for continued research in this area. The structure of this chapter is as follows: Section 3.3 describes the proposed work, Section 3.4 presents the experimental results, and Section 3.5 provides the concluding remarks on the proposed scheme.

3.3 Proposed work

The suggested fragile image watermarking model is based on a hybrid approach and utilizes LSB adjustment and DCT for the watermark insertion. This section explains each step of the proposed methodology in a systematic manner, such as image partitioning, LSB modification, DCT transformation, DC coefficient embedding, and watermark extraction. The systematic approach maintains an equilibrium between imperceptibility, fragility, and computational complexity. The block diagram of the proposed method is depicted in Figure 3.3.

3.3.1 Image pre-processing and block division

The grayscale image is described as a two-dimensional intensity matrix where each pixel has a value in the range of 0–255. Suppose $I(x,y)$ is a grayscale image of dimension $N \times N$. Preprocessing aims at making the watermarking operation localized and spread evenly over the image. To manipulate the image effectively in

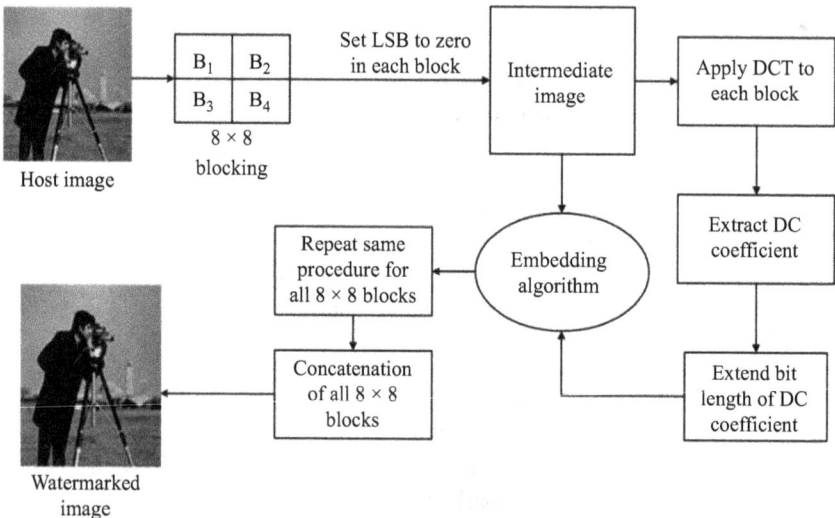

Figure 3.3 Block diagram of the proposed scheme

the frequency domain, it is split into non-overlapping 8×8 blocks. The reason behind taking 8×8 blocks lies in the JPEG compression standard, where DCT is applied on small blocks to perform efficient compression. Mathematically, the splitting process can be represented as:

$$B(k,l) = I(i : i + 7, j : j + 7) \quad \text{for } i,j = 8x, \ 8y \tag{3.1}$$

where x and y are indices that indicate block positions in the image.

3.3.2   LSB modification

For every pixel "X" within a block, the LSB is reset with a bitwise AND operation:

$$X' = X \wedge 254 \tag{3.2}$$

X' is the altered pixel with zeroed LSB. The binary form of 254 is 11111110, where all bits are kept constant except for the LSB. The embedding within the LSB maintains the visual quality of the image.

3.3.3   DCT transformation and DC coefficient extraction

The DCT is commonly applied in image processing because it can concentrate energy into a smaller number of coefficients. It converts an image block from the spatial into the frequency domain, enabling modifications that are less noticeable to the human eye. The 2D DCT transformation of an 8×8 block is expressed as follows:

$$T(u,v) = \frac{1}{4} D(u)D(v) \sum_{i=0}^{7} \sum_{j=0}^{7} B(i,j) \cos\left(\frac{(2i+1)u\pi}{16}\right) \cos\left(\frac{(2j+1)v\pi}{16}\right) \tag{3.3}$$

where,

$$D(k) = \begin{cases} \dfrac{1}{\sqrt{2}}, & k = 0 \\ 1, & k \neq 0 \end{cases}$$

$B(i,j)$ is the pixel intensity at position (i,j). $T(u,v)$ are the DCT coefficients at frequency indices (u,v). Also, the DC coefficient $T(0,0)$ is the value in the top-left corner of the DCT matrix, and is the average intensity of the block.

3.3.4   LSB embedding of bits

Once the DC coefficient is obtained, it is converted to a 64-bit binary string for embedding purposes. If the binary string contains less than 64 bits, it is padded with zeros. If it is more than 64 bits, only the first 64 bits are used. Now these 64 bits are embedded in the LSBs of the 64 elements of the block. This ensures that imperceptibility is achieved, as a human's perception is less sensitive toward changes in the LSB.

3.3.5   *Watermark extraction*

The watermark extraction process initiates with performing DCT on the attacked or altered watermarked image to gain the frequency domain coefficients. For every 8×8 block of the watermarked image, LSB extraction is performed, where LSBs from pixel values are extracted and reserved separately. To improve the precision of extraction, the LSBs of every block are cleared to zero, and DCT is again applied to calculate the DC coefficients. These extracted DC coefficients are compared with the originally embedded DC coefficient matrix to check for watermark integrity. If the extracted coefficients are the same as the original, the watermark is said to be successfully recovered; otherwise, differences suggest tampering or alteration in the image.

3.4   Experimental outcomes

The performance of the proposed algorithm has been assessed through simulations using MATLAB® R2017a with an Intel Core i7 processor operating at 3.6 GHz. The experiments were conducted on test images – "Image1," "Image2," "Image3," and "Image4" taken from the Openi database and "Aerial," "Boat," "Couple," and "Tank" – sourced from the USC SIPI database, each having a resolution of 256×256 pixels. Figure 3.4 presents the test images used for experimentation.

The various parameters used to evaluate the performance include the Structural Similarity Index Measure (SSIM), PSNR, normalized cross correlation (NCC), and bit error rate (BER). The PSNR and SSIM reveal the imperceptibility, while the NCC and BER give the fragility of the algorithm.

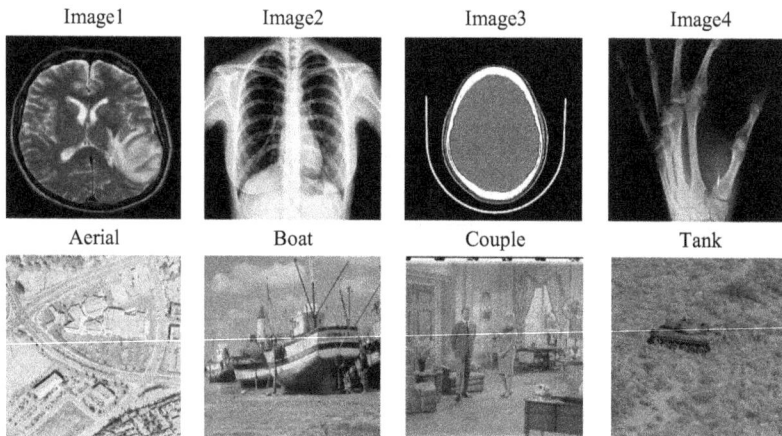

Figure 3.4   Test images used

3.4.1   Imperceptibility analysis

Imperceptibility is a basic requirement in strong watermarking, which guarantees that the embedded watermark is not visually perceptible while maintaining the host image's perceptual quality [31]. A watermarking scheme is said to have high imperceptibility when the changes made during embedding do not result in perceivable distortions. This is generally quantified via PSNR and SSIM, with greater values of PSNR (more than 40 dB) and SSIM close to 1 stating the absence of severe perceptual distortion. In the proposed method, LSB-based embedding coupled with frequency-domain transformations (DCT) preserves the subtlety of the changes while taking advantage of the resilience of the DC coefficient. The watermarked image thus maintains its original visual features without compromising the watermark integrity during different image processing operations. Table 3.1 presents the PSNR and SSIM values of the watermarked test images.

3.4.2   Fragility analysis

Fragility is an important requirement in digital watermarking, where the embedded watermark should not be recoverable even after the image has been subjected to several intentional or unintentional manipulations [32]. A fragile watermarking scheme does not withstand standard image processing operations, geometric attacks, and compression methods, so that unauthorized removal or degradation of the embedded watermark is detected. The suggested watermarking method employs a hybrid strategy that incorporates LSB-based embedding with DCT-domain transformations. To test the fragility of the suggested method, rigorous experiments were performed by applying various attack modes to the watermarked images, such as JPEG compression, Gaussian noise, salt-and-pepper noise, median filtering, histogram equalization, rotation, and hybrid attacks. Figure 3.5 shows the subjective analysis of watermarked images after undergoing various signal processing operations. The watermark extracted was tested using BER, and NCC to

Table 3.1   PSNR and SSIM of watermarked test images

Test images	PSNR	SSIM
Image1	51.1273	0.9965
Image2	50.5832	0.9807
Image3	51.1552	0.9978
Image4	51.1284	0.9980
Aerial	51.1427	0.9991
Boat	51.1374	0.9983
Couple	51.1537	0.9984
Tank	51.1275	0.9988
Average	51.0700	0.9959

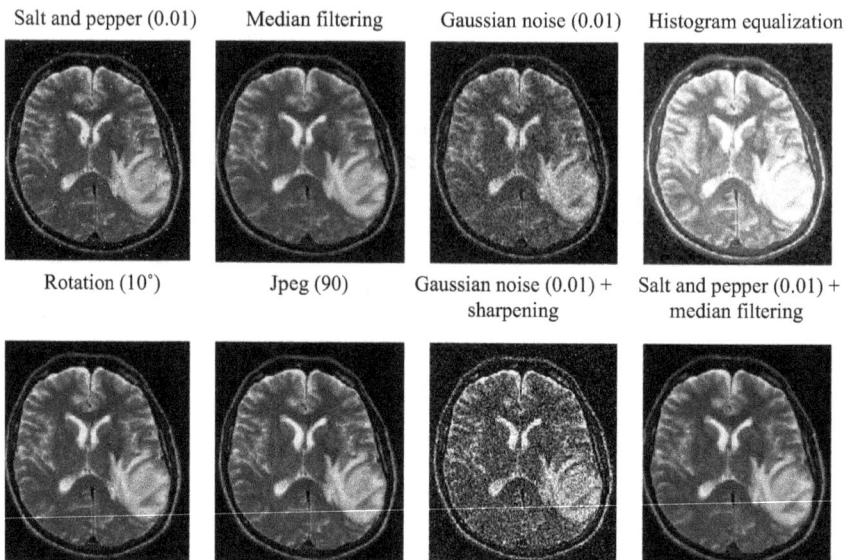

(a)

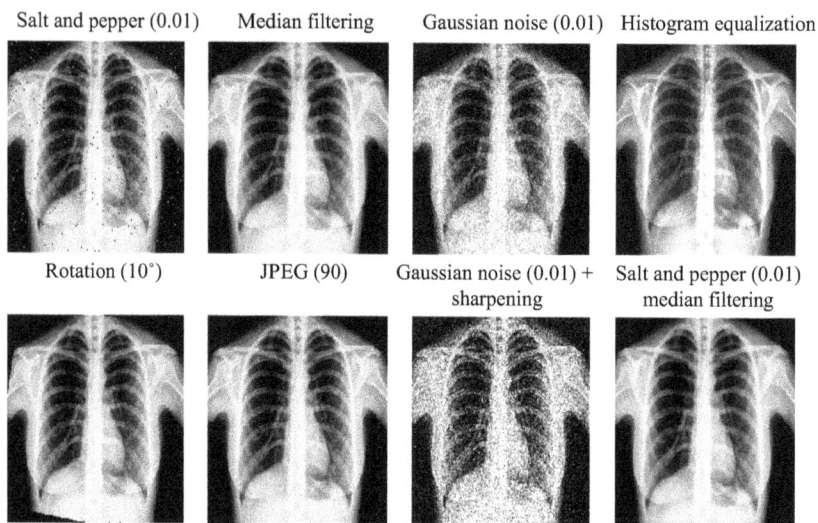

(b)

Figure 3.5   (a) Watermarked images after subjugating under various signal processing operations. (b) Watermarked images after subjugating under various signal processing operations. (c) Watermarked images after subjugating under various signal processing operations. (d) Watermarked images after subjugating under various signal processing operations. (e) Watermarked images after subjugating under various signal processing operations. (f) Watermarked images after subjugating under various signal processing operations. (g) Watermarked images after subjugating under various signal processing operations. (h) Watermarked images after subjugating under various signal processing operations.

Salt and pepper (0.01)   Median filtering   Gaussian noise (0.01)   Histogram equalization

Rotation (10°)   JPEG (90)   Gaussian noise (0.01) +   Salt and pepper (0.01) +
sharpening   median filtering

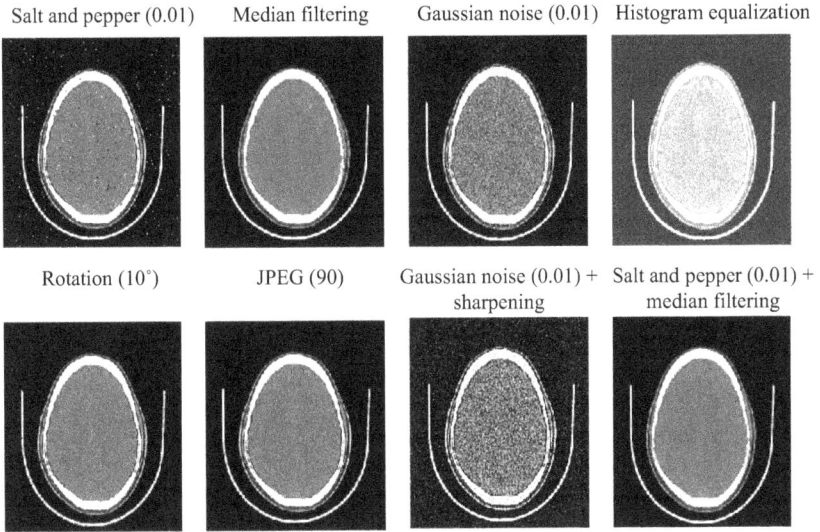

(c)

Salt and pepper (0.01)   Median filtering   Gaussian noise (0.01)   Histogram equalization

Rotation (10°)   JPEG (90)   Gaussian noise (0.01) + Salt and pepper (0.01) +
sharpening   median filtering

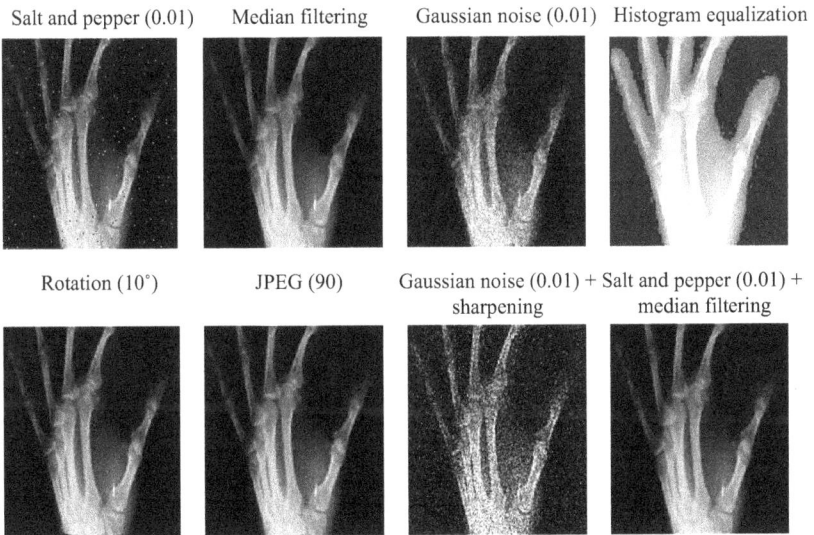

(d)

Figure 3.5   (Continued)

Salt and pepper (0.01)   Median filtering   Gaussian noise (0.01)   Histogram equalization

Rotation (10°)   JPEG (90)   Gaussian noise (0.01) +   Salt and pepper (0.01) +
sharpening   median filtering

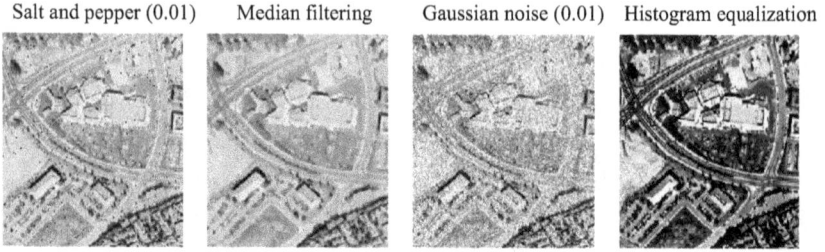
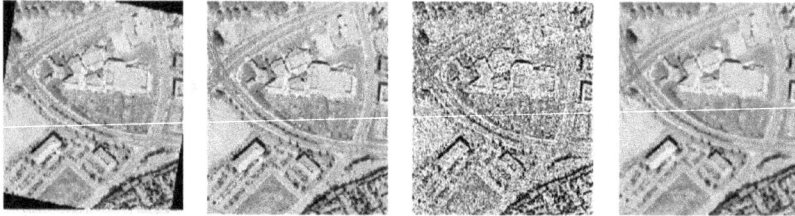

(e)

Salt and pepper (0.01)   Median filtering   Gaussian noise (0.01)   Histogram equalization

Rotation (10°)   JPEG (90)   Gaussian noise (0.01) +   Salt and pepper (0.01) +
sharpening   median filtering

(f)

Figure 3.5   (Continued)

Salt and pepper (0.01)   Median filtering   Gaussian noise (0.01)   Histogram equalization

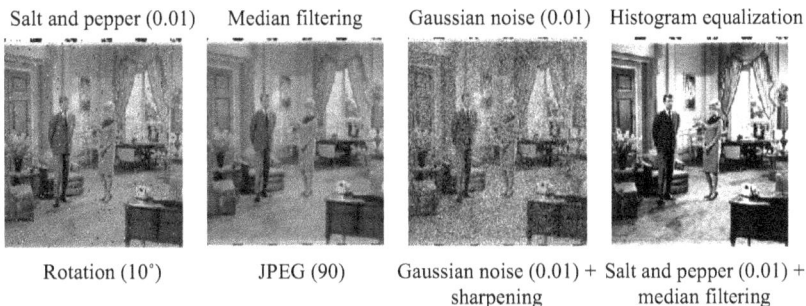

Rotation (10°)   JPEG (90)   Gaussian noise (0.01) +   Salt and pepper (0.01) +
                                          sharpening   median filtering

(g)

Salt and pepper (0.01)   Median filtering   Gaussian noise (0.01)   Histogram equalization

Rotation (10°)   JPEG (90)   Gaussian noise (0.01) +   Salt and pepper (0.01) +
                                          sharpening   median filtering

(h)

Figure 3.5   (Continued)

measure its strength under various attack modes. The obtained average PSNR, BER, and NCC values are presented in Table 3.2. Besides the proposed work has been compared with state-of-the-art schemes, the results of which are presented in Table 3.3.

The subjective results from Figures 3.5(a) to 3.5(h) prove that the watermark embedding does not create visible visual artifacts in the watermarked images. Thus, it can be stated that the watermarking procedure maintains the structural integrity and quality of the host images. Even subjected to different transformations and attacks, the visual difference between the original and watermarked images is negligible, which ensures that the watermark is imperceptible to human eyes. This verifies the success of the proposed scheme in achieving imperceptibility while inserting crucial watermark information, making it very appropriate for real-world applications involving transparent watermarking.

The outcomes of Table 3.2 indicate the vulnerability of the proposed scheme to different attacks. If the NCC values are not equal to 1, validate that distortions caused by attacks have changed the extracted watermark, proving the fragility of the embedding method. In the same manner, BER values not equal to zero confirm bit errors in the extracted watermark, further ensuring the capability of the method to detect changes in the image. Importantly, some attacks, such as Gaussian noise, histogram equalization, Gaussian noise + sharpening, salt and pepper + median filtering, created very high BER values, reflecting extreme distortion of the watermark, which ensures the detection of any unauthorized alteration. This

Table 3.2   Objective assessment of test image "Image1" after subjugating under various attacks

Attacks	PSNR	SSIM	NCC	BER
Salt and Pepper (0.01)	23.7755	0.7477	0.9961	3.4668
Median Filtering	26.5676	0.7771	0.9996	1.3672
Gaussian Noise (0.01)	21.6655	0.2646	0.9381	26.1597
Histogram Equalization	10.5876	0.3033	0.6795	54.9805
Rotation (10°)	20.6587	0.9510	0.9989	6.9214
JPEG (90)	39.9689	0.9047	0.9999	2.2705
Gaussian Noise + Sharpening	15.4307	0.1586	0.8797	29.2603
Salt and Pepper + Cropping (25%)	1.4960	0.0880	0.6795	54.9805
Salt and Pepper + Median Filtering	26.4184	0.7761	0.9996	1.3794

Table 3.3   Comparative analysis of the proposed work

Schemes	Mean PSNR	Mean SSIM	Payload
[31]	51.06	0.9968	65,536
[32]	51.41	0.9952	65,536
Proposed	51.07	0.9959	65,536

Table 3.4   Timing analysis of the proposed work in seconds

Test images	Embedding time	Extraction time
Image1	3.7969	1.3438
Image2	4.7031	1.3906
Image3	4.5000	1.2656
Image4	3.7500	1.1719
Aerial	4.3438	1.4531
Boat	4.0625	1.2344
Couple	4.7188	1.1719
Tank	4.0469	1.3750
Average	4.2027	1.3007

guarantees that even slight or regional variations in the image can be picked up and correctly identified, validating the method's strength in image integrity protection. The high sensitivity of the proposed scheme makes it especially well-suited for applications demanding strict integrity verification. Any attempt at tampering, whether inadvertent or malicious, results in a detectable alteration of the water-mark, such that modifications cannot escape detection.

In addition to this, we have compared our scheme with the other state-of-the-art schemes, the results of which are presented in Table 3.3.

From the results in Table 3.3, it is evident that, for the same EC, the mean PSNR value of our scheme is comparable to that of [31] but slightly lower than [32]. However, our scheme outperforms [32] in terms of mean SSIM, indicating better structural similarity preservation. The overall results of Tables 3.2 and 3.3 demonstrate that our scheme not only efficiently detects temper but also effectively balances imperceptibility and visual quality, making it a viable choice for fragile watermarking applications.

3.4.3   Timing analysis

We conducted experiments using MATLAB 2017a on the Windows 10 operating system with an Intel Core i7-6006U CPU @ 3.6 GHz processor. We recorded the embedding and extraction execution time for processing 256×256. The results presented in Table 3.4 demonstrate that the proposed scheme achieves exceptionally high speeds, making it suitable for real-time applications as well.

3.5   Conclusion

In this work, a fragile image watermarking method has been introduced to ensure the content authentication and tamper detection of digital images. The proposed work successfully detects tampered digital images and hence ensures authentication. This is done by embedding the binary value of DC coefficients into the LSBs of the selected block in a cover image, after applying DCT transformation. The experimental outcome verifies that the designed method provides high imperceptibility, in terms of PSNR and SSIM values, with the added guarantee that even slight changes to the watermarked image can be identified using BER and NCC analysis. It has been shown that the proposed system is highly vulnerable to attacks like compression, filtering, and addition of noise and hence very suitable for applications involving authentication and content verification. This research contributes to the area of fragile watermarking, providing a feasible solution for content verification and authentication in sensitive applications like medical imaging and secure communications. Future developments may investigate hybrid fragile-robust watermarking methods to enhance resistance to unintended alterations at the expense of maintaining fragility for tamper localization.

References

[1]   Sun T, Wang X, Zhang K, Jiang D, Lin D, Jv X, Ding B, and Zhu W (2022) Medical image authentication method based on the wavelet packet and energy entropy. *Entropy (Basel)* 24(6): 798. doi:10.3390/e24060798

[2]   Alam I and Kumar M (2023) A novel authentication protocol to ensure confidentiality among the Internet of Medical Things in COVID-19 and future pandemic scenario. *Internet of Things* 22: 1–16, doi:/10.1016/j.iot.2023.100797

[3]   Hosny KM, Magdi A, Elkomy O, and Hamza HM (2024) Digital image watermarking using deep learning: A survey. *Computer Science Review* 53: 1–12, doi:10.1016/j.cosrev.2024.100662

[4]   Pekerti AA, Sasongko A, and Indrayanto A (2024) Secure end-to-end voice communication: A comprehensive review of steganography, modem-based cryptography, and chaotic cryptography techniques. *IEEE Access* 12: 75146–75168, doi:10.1109/ACCESS.2024.3405317

[5]   Duong P and Lee H (2022) Configurable mixed-radix number theoretic transform architecture for lattice-based cryptography. *IEEE Access* 10: 12732–12741, doi:10.1109/ACCESS.2022.3145988

[6]   Ricci S, Dobias P, Malina L, Hajny J, and Jedlicka P (2024) Hybrid keys in practice: Combining classical, quantum and post-quantum cryptography. *IEEE Access* 12: 23206–23219, doi:10.1109/ACCESS.2024.3364520

[7]   Sajedi H and Yaghobi SR (2020) Information hiding methods for E-Healthcare. *Smart Health* 15: 1–11, doi:10.1016/j.smhl.2019.100104

[8]   Jiang X, Xie Y, Zhang Y *et al.* (2025) Reversible data hiding in encrypted images using reservoir computing-based data fusion strategy. *IEEE Transactions on Circuits and Systems for Video Technology* 35(1): 684–697, doi:10.1109/TCSVT.2024.3459024

[9]   Zhang Y, Ni J, and Su W (2025) HiFiMSFA: Robust and high-fidelity image watermarking using attention augmented deep network. *IEEE Signal Processing Letters* 32: 781–785, doi:10.1109/LSP.2025.3535216

[10] Izumi S, Azuma SI, and Sugie T (2020) Analysis and design of multi-agent systems in spatial frequency domain: Application to distributed spatial filtering in sensor networks. *IEEE Access* 8: 34909–34918, doi:10.1109/ACCESS.2020.2974243

[11] Abdulwahab NM and Basheer NM (2021) Spatial domain block based blind image watermarking for hardware applications. *International Conference on Advanced Computer Applications (ACA)*, Maysan, Iraq, 90–95, doi:10.1109/ACA52198.2021.9626800

[12] Ernawan F, Ariatmanto D, and Firdaus A (2021) An improved image watermarking by modifying selected DWT-DCT coefficients. *IEEE Access* 9: 45474–45485, doi:10.1109/ACCESS.2021.3067245

[13] Ferreira FABS and Lima JB (2021) Watermarking and coefficient scanning for light field images in 4D-DCT domain. *IEEE Access* 9: 32467–32484, doi:10.1109/ACCESS.2021.3060735

[14] Wu D, Zhang X, Wang J, Li L, and Feng G (2024) Novel robust video watermarking scheme based on concentric ring subband and visual cryptography with piecewise linear chaotic mapping. *IEEE Transactions on Circuits and Systems for Video Technology* 34(10): 10281—10298, doi:10.1109/TCSVT.2024.3405558

[15] Wen S, Zhang Q, Hu T, and Li J (2025) Robust audio watermarking against manipulation attacks based on deep learning. *IEEE Signal Processing Letters* 32: 126–130, 2025, doi:10.1109/LSP.2024.3501285

[16] Cao F, An B, Wang J, Ye D, and Wang H (2017) Hierarchical recovery for tampered images based on watermark self-embedding. *Displays* 46: 52–60.

[17] Qin C, Wang H, Zhang X, and Sun X (2016) Self-embedding fragile watermarking based on reference-data interleaving and adaptive selection of embedding mode. *Information Sciences* 373: 233–250.

[18] Swaraja K, Meenakshi K, and Kora P (2020) An optimized blind dual medical image watermarking framework for tamper localization and content authentication in secured telemedicine. *Biomed Signal Process Control* 55: 101665. doi:10.1016/j.bspc.2019.101665

[19] Lin CC, Huang Y, and Tai WL (2017) A novel hybrid image authentication scheme based on absolute moment block truncation coding. *Multimedia Tools and Applications* 76(1): 463–488.

[20] Sarreshtedari S and Akhaee MA (2015) A source-channel coding approach to digital image protection and self recovery. *IEEE Trans Image Process* 24(7): 2266–2277.

[21] Gul E and Ozturk S (2019) A novel hash function based fragile watermarking method for image integrity. *Multimedia Tools and Applications* 78: 17701–17718. doi:10.1007/s11042-018-7084-0

[22] Feng B, Li X, Jie Y, Guo C, and Fu H *et al.* (2020) A novel semi-fragile digital watermarking scheme for scrambled image authentication and restoration. *Mobile Networks and Applications* 25: 82–94. doi:10.1007/s11036-018-1186-9

[23] Tiwari A, Sharma M, and Tamrakar RK (2017) Watermarking based image authentication and tamper detection algorithm using vector quantization approach. *AEU International Journal of Electronics and Communications* 78: 114–123.

[24] Rajput V and Ansari IA (2019) Image tamper detection and self-recovery using multiple median watermarking. *Multimedia Tools and Applications* 79: 35519–35535. doi:10.1007/s11042-019-07971-w

[25] Selvam P, Balachandran S, Iyer SP, and Jayabal R (2017) Hybrid transform based reversible watermarking technique for medical images in telemedicine applications. *Optik* 145: 655–671.

[26] Prasad S and Pal AK (2020) A tamper detection suitable fragile watermarking scheme based on novel payload embedding strategy. *Multimedia Tools and Applications* 79: 1673–1705. doi:10.1007/s11042-019-08144-5

[27] Gull S, Loan NA, and Parah SA (2018) An efficient watermarking technique for tamper detection and localization of medical images. *Journal of Ambient Intelligence and Humanized Computing* 2018: 1799–1808. doi:10.1007/s12652-018-1158-8

[28] Sahu AK (2021) A logistic map based blind and fragile watermarking for tamper detection and localization in images. *Journal of Ambient Intelligence and Humanized Computing* 13(2021): 3869–3881.

[29] Azeroual A and Karim A (2017) Real-time image tamper localization based on fragile watermarking and FaberSchauder wavelet. *AEU – International Journal of Electronics and Communication* 79: 207–218.

[30] Gull S, Mansour RF, Aljehane NO, and Parah SA (2021) A self-embedding technique for tamper detection and localization of medical images for smart-health. *Multimedia Tools and Applications* 80(19): 29939–29964.

[31] Trivedy S and Pal AK (2017). A logistic map-based fragile watermarking scheme of digital images with tamper detection. *Iranian Journal of Science and Technology, Transactions of Electrical Engineering* 41(2): 103–113.

[32] Hussan M, Gull S, Parah SA, and Qureshi GJ (2023) An efficient encoding based watermarking technique for tamper detection and localization. *Multimedia Tools and Applications* 82: 37249–37271.

Chapter 4

Exploring technological trend and collaboration analysis in reversible data hiding in encrypted images

Ankur[1], Sonal Gandhi[2,3], Rajeev Kumar[4] and Ki-Hyun Jung[5]

Abstract

Reversible data hiding in encrypted images (RDHEI) offers a unique dual advantage, benefiting both image owners and data hiders, making it a crucial research area within the domain of privacy and security. This exceptional characteristic has captured the attention of scholars and researchers globally, resulting in an extensive and diverse collection of literature encompassing a wide range of methods. So, this chapter is designed to offer a complete and clear picture of the RDHEI domain by quantitative exploration. For this, the Web of Science database, which boasts an extensive collection of peer-reviewed high-quality research articles, has been explored through a keyword-based search. The collected data have been scrutinized and analyzed along several dimensions to not only detail the theoretical foundations, practical implications, methodological work, and prospects of RDHEI research but also to elucidate the evolutionary trajectory of the domain, identifying key contributors, influential works, and emerging patterns. Therefore, this study serves as an exhaustive and definitive guide by bringing together these complementary viewpoints.

Keywords: Reversible Data Hiding; Encrypted Image; RDHEI; Quantitative; Trend Analysis

4.1 Introduction

In the current digital era, the widespread adoption of cloud services marks a significant trend, offering scalable, flexible, and efficient computing resources

[1]Department of Computer Science and Engineering, NIT Delhi, India
[2]Department of Computer Science and Engineering, Delhi Technological University, India
[3]Department of Computer Science and Engineering, GL Bajaj Institute of Technology and Management, India
[4]Department of Computer Science and Engineering, Delhi Technological University, India
[5]Department of Software Convergence, Andong National University, Republic of Korea

on-demand basis [1]. Therefore, the cloud has emerged as a vast repository for digital data, encompassing secret information in the form of images, videos, and textual content [2–4]. This storage of secret information on the cloud has necessitated the need for advanced data security measures to protect the secret information/data of cloud users [5–7]. To safeguard the data of cloud users, data hiding (DH) has emerged as a prominent methodology as it enables impercep-tible embedding of confidential information [8–10]. For embedding, natural images have proven to be a preferred medium due to their extensive availability and the substantial volume [11,12].

Existing DH methods are usually categorized into two categories, namely: reversible and nonreversible methods, depending on whether the original cover image can be retrieved or not. Reversible data hiding (RDH) is of particular interest because it allows for the recovery of the original cover image after the hidden data has been extracted, making it a compelling area for research [13–18]. Traditionally, RDH has focused on securing the embedded data, often compromising the privacy of the cover image in the process. To address this issue, reversible data hiding in encrypted images (RDHEI) has been developed [19–22]. This advanced metho-dology ensures the secret data are imperceptibly embedded into digital images while also preserving the integrity and privacy of the image during cloud storage and transmission. Thus, RDHEI establishes a balance between data privacy and utility, finding applications in critical areas such as military and medical fields [23–25], where both the secrecy of the embedded data and the accurate recovery along with privacy of the original image are crucial. RDHEI involves three key entities: the content owner, who encrypts the image; the data hider, who embeds data into the encrypted image; and the recipient, who retrieves both the image content and the embedded data using keys as shown in Figure 4.1. In the initial part of Figure 4.1, the procedure starts with the content owner encrypting the image through a specific encryption process. Subsequently, this encrypted image is transmitted to the data hider. Here, the data hider employs a data-specific key to encrypt the data, which is then embedded within the image at a predetermined location that allows for future recovery. The final part of the figure shows the role

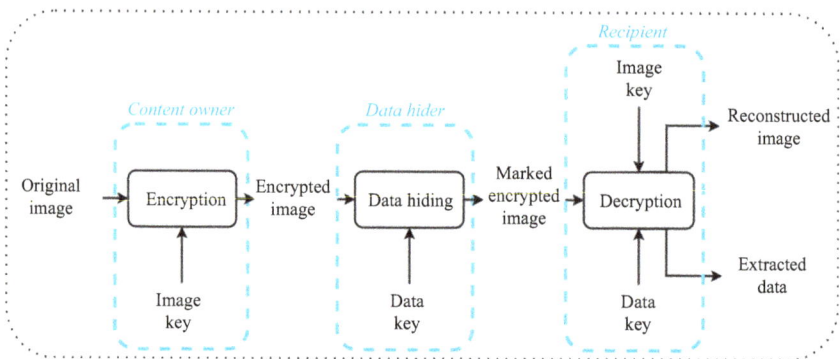

Figure 4.1   Block diagram of RDHEI

of the recipient, who receives the image, now marked with embedded data. Depending on the possession of the necessary keys, the recipient is able to extract the embedded data and successfully reconstruct the original image [7].

The field of RDHEI is evolving, with methodologies like reserving room before encryption (RRBE) [26–29] and vacating room after encryption (VRAE) [30–33] offering distinct approaches to optimize embedding capacity, data reversibility, privacy, and security. This dynamic area of study calls for a detailed review and analysis to map out the intellectual development of the field, identify emerging trends, and spotlight promising research directions. Therefore, this paper presents a detailed investigation into the domain of RDHEI, an area that has experienced remarkable growth in the last five years. This paper not only provides a comprehensive review that displays the domain advancements with a critical assessment of the methodologies employed, but also explores quantitative perspectives by giving the details to uncover the trends, collaborations, and impactful contributions. Thus, this review paper is designed to offer a complete overview of the RDHEI domain by undertaking a two-directional analysis of the current state of RDHEI and illuminating promising avenues for future research. The key contributions of this work can be summarized as follows:

- *Quantitative analysis*: This analysis illuminates the trends, collaborative efforts, and significant contributions within the RDHEI area over the past five years. Employing bibliometric methods, it shows the evolutionary path of RDHEI, highlights the key researchers, institutions, and nations of this domain. Additionally, it investigates co-authorship, institutional, and national collaborations to uncover patterns and networks, while also utilizing keyword co-occurrence and citation analysis to identify central themes, seminal works, and the intellectual structure of the RDHEI field.

These contributions aim to provide a comprehensive resource for researchers, practitioners, and stakeholders to navigate and advance the RDHEI domain.

4.2   Publication trend and productivity analysis

This section offers a comprehensive exploration of the publication trends and productivity analysis applied to RDHEI methodologies through bibliometric studies conducted over the past five years. The quantitative analysis plays a crucial role in this context, providing the scholarly contributions, trends, and advancements within the RDHEI field. The publication trend analysis aims to illuminate the growth and evolution of RDHEI research over time, revealing the increasing significance of this domain in the digital age. This investigation facilitates a deeper understanding of the academic engagement and the spread of knowledge pertaining to the RDHEI domain. Simultaneously, the productivity analysis assesses the impact and scope of RDHEI research through a detailed evaluation of publication outputs from researchers, institutions, and countries. This provides an objective measure of the key contributors and their influence in the domain.

4.2.1   Data collection source and strategy

To investigate the academic trend of RDHEI, this study utilized the Web of Science (WoS) database as its primary source of scholarly articles. WoS is celebrated for its extensive collection of peer-reviewed research, encompassing various indices such as the science citation index expanded, the social sciences citation index, and the book citation index – science. This diversity makes WoS an ideal platform for accessing high-quality studies.

The search for relevant RDHEI literature was conducted with a carefully designed query on January 31, 2024, spanning a period from January 1, 2019 to January 31, 2024. The search string formulated for this purpose was: (RDHEI OR ("reversible data hiding in encrypted images") OR ((hiding OR embedding OR RDH) NEAR/3 encrypted)). This approach resulted in the identification of 455 research articles, each contributing to the discourse on RDHEI. These articles were subsequently analyzed bibliometrically, as discussed subsequently in this paper.

4.2.2   Publication structure analysis

The initial step involved a comprehensive analysis of 455 research articles, categorizing them into various verticals like annual publications, publication type, publishing sources, productive researchers and organization, country trends, and WoS indices.

4.2.2.1   Annual publications and type

Figure 4.2 provides an overview of the annual publication trends and the types of documents that constitute the body of research in the RDHEI domain. Specifically, Figure 4.2 is depicted, which illustrates the upward trajectory of research publications related to RDHEI from 2019 to 2024, as cataloged by the WoS. The initial count of 68 publications in 2019 evidences a steady climb in academic contributions, culminating in a zenith of 113 publications in 2023. This reflects the growing interest in the RDHEI domain. The early count for 2024 is double publication, suggesting continued engagement with RDHEI research, although it is premature to

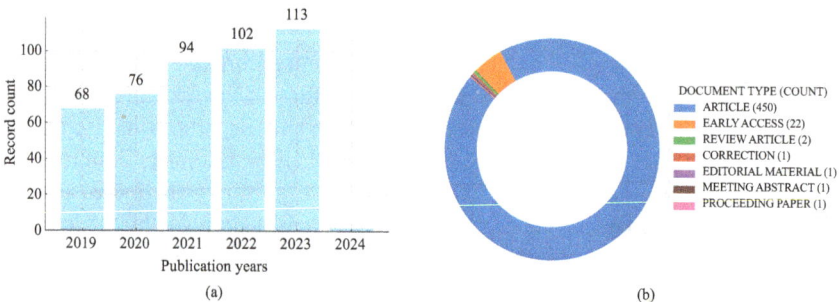

Figure 4.2   *Annual publications and their types from 2019 to 2024 within the domain of RDHEI. (a) Yearly RDHEI publications and (b) document types of publications.*

forecast the total contributions for the year. Such a pattern accentuates the evolving and pertinent nature of RDHEI studies within the scholarly community.

Furthermore, the publication type is illustrated in Figure 4.2(b), which includes a diverse range of document types contributing to the RDHEI research. The majority of this collection is composed of research articles, numbering 450, which constitute 98.901% of the total. Following this, early access papers make up 22 documents, representing 4.835% of the dataset. Review articles, although fewer in number, contribute 2 documents or 0.44%. Other categories of documents, including editorial materials, meeting abstracts, correction notices, and conference proceedings, each add a single document to the total, accounting for a nominal 0.88%. These statistics highlight the extensive variety of publication types that form the RDHEI research compilation, emphasizing that research articles are the predominant medium for the dissemination of new knowledge in the field.

4.2.2.2 Publishing sources

In the section publication source, Figure 4.3 derived from WoS data presents a visual bibliometric analysis of the research landscape in the area of RDHEI. The chart showcases the distribution of publications across various sources, with "Multimedia Tools and Applications" leading at 77 publications, accounting for 16.923% of the total 455 papers analyzed. This is followed by "IEEE Access" and "IEEE Transactions on Circuits and Systems for Video Technology," indicating these journals' significant roles in advancing RDHEI research. Notably, the category "Other 114 Publication Titles" comprises 188 publications, reflecting a diverse range of contributing sources. This graphical representation highlights the breadth of research activity and the key journals contributing to the field of RDHEI.

4.2.2.3 Productive researchers and organizations

In this subsection, the quantifiable overview of the most prolific contributors to the field is shown with the help of Figure 4.4 and Table 4.1. The first chart (i.e.,

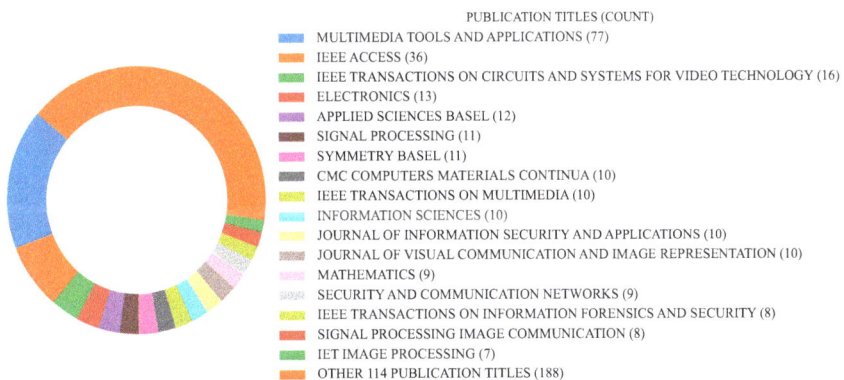

PUBLICATION TITLES (COUNT)
- MULTIMEDIA TOOLS AND APPLICATIONS (77)
- IEEE ACCESS (36)
- IEEE TRANSACTIONS ON CIRCUITS AND SYSTEMS FOR VIDEO TECHNOLOGY (16)
- ELECTRONICS (13)
- APPLIED SCIENCES BASEL (12)
- SIGNAL PROCESSING (11)
- SYMMETRY BASEL (11)
- CMC COMPUTERS MATERIALS CONTINUA (10)
- IEEE TRANSACTIONS ON MULTIMEDIA (10)
- INFORMATION SCIENCES (10)
- JOURNAL OF INFORMATION SECURITY AND APPLICATIONS (10)
- JOURNAL OF VISUAL COMMUNICATION AND IMAGE REPRESENTATION (10)
- MATHEMATICS (9)
- SECURITY AND COMMUNICATION NETWORKS (9)
- IEEE TRANSACTIONS ON INFORMATION FORENSICS AND SECURITY (8)
- SIGNAL PROCESSING IMAGE COMMUNICATION (8)
- IET IMAGE PROCESSING (7)
- OTHER 114 PUBLICATION TITLES (188)

Figure 4.3 Top publishing WoS sources

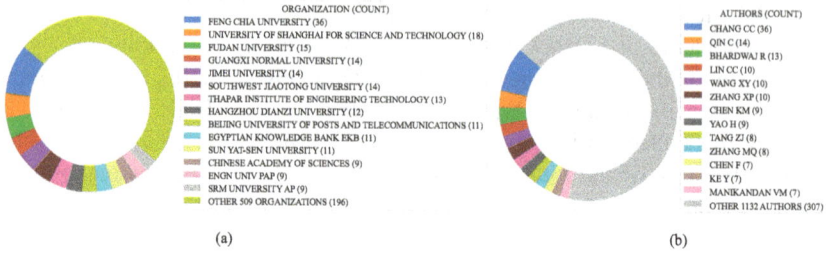

ORGANIZATION (COUNT)
- FENG CHIA UNIVERSITY (36)
- UNIVERSITY OF SHANGHAI FOR SCIENCE AND TECHNOLOGY (18)
- FUDAN UNIVERSITY (15)
- GUANGXI NORMAL UNIVERSITY (14)
- JIMEI UNIVERSITY (14)
- SOUTHWEST JIAOTONG UNIVERSITY (14)
- THAPAR INSTITUTE OF ENGINEERING TECHNOLOGY (13)
- HANGZHOU DIANZI UNIVERSITY (12)
- BEIJING UNIVERSITY OF POSTS AND TELECOMMUNICATIONS (11)
- EGYPTIAN KNOWLEDGE BANK EKB (11)
- SUN YAT-SEN UNIVERSITY (11)
- CHINESE ACADEMY OF SCIENCES (9)
- ENGN UNIV PAP (9)
- SRM UNIVERSITY AP (9)
- OTHER 509 ORGANIZATIONS (196)

AUTHORS (COUNT)
- CHANG CC (36)
- QIN C (14)
- BHARDWAJ R (13)
- LIN CC (10)
- WANG XY (10)
- ZHANG XP (10)
- CHEN KM (9)
- YAO H (9)
- TANG ZJ (8)
- ZHANG MQ (8)
- CHEN F (7)
- KE Y (7)
- MANIKANDAN VM (7)
- OTHER 1132 AUTHORS (307)

(a)   (b)

Figure 4.4   Productive organizations and researchers. (a) Top contributing organizations and (b) top contributing authors.

Table 4.1   Top ten authors based on record count

Author name	Country	Organization	Record count
Chang, Ching-Chun	Taiwan	Feng Chia Univ	36
Qin, Chuan	Peoples R China	Univ Shanghai Sci & Technol	14
Bhardwaj, Rupali	India	Thapar Inst Engn & Technol	13
Lin, Chia-Chen	Taiwan	Natl Chin Yi Univ Technol	10
Zhang, Xinpeng	PR China	Fudan Univ	10
Yao, Heng	PR China	Univ Shanghai Sci & Technol	9
Chen, Kaimeng	PR China	Jimei Univ	9
Tang, Zhenjun	PR China	Guangxi Normal Univ	8
Zhang, Minqing	PR China	Engn Univ PAP	8
Zhang, Xianquan	PR China	Guangxi Normal Univ	7

Figure 4.4(a)) highlights the organizations' contribution, where "Feng Chia University" leading position with 36 publications, evidencing its significant research output. Table 4.1 reinforces this by listing Chang, "Ching-Chun of Feng Chia University" as the top author, further accentuating the university's central role in RDHEI research.

Concurrently, Figure 4.4(b) identifies Chang, Ching-Chun as the most productive author with 36 papers, complementing the organizational data and underscoring the synergy between individual and institutional productivity. Notable contributions from authors affiliated with mainland China and Taiwan signify a regional concentration of expertise in RDHEI, while Thapar Institute of Engineering & Technology's representation points to a growing research interest in India.

4.2.2.4   Country trends

In examining global research patterns within the specialized domain of RDHEI, the provided pie chart depicted in Figure 4.5 offers a concise geographical narrative. Derived from the WoS repository, the chart underscores China's predominant contribution, with a notable 268 publications, indicative of a robust research

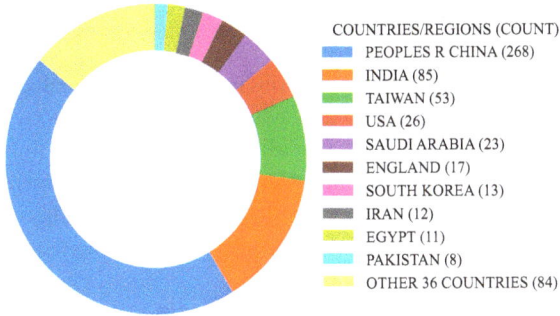

Figure 4.5   Countries' trends in publications

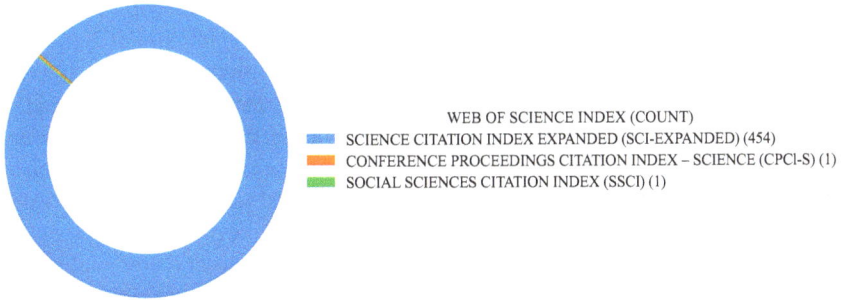

Figure 4.6   Publication counts of Web of Science indices

emphasis in the region. India and Taiwan also emerge as significant contributors, emphasizing the Asian continent's active engagement in the domain. The visualization captures the essence of international collaboration and scholarly interest, with the United States and several European and Middle Eastern countries marking their presence. This bibliometric snapshot, while capturing the 455 papers due to some cross-country collaborations, describes the worldwide momentum toward the RDHEI domain.

4.2.2.5   WoS indices

In the section, the two pie charts depicted in Figure 4.6 illustrate the distribution of research within the indices and categories of the WoS. The first chart (6) shows an overwhelming majority of papers indexed under the Science Citation Index Expanded (SCI-EXPANDED), with 454 out of 455 papers, highlighting the index's prominence in scientific research dissemination. The other indices, conference proceedings citation index – science and social sciences citation index, have a minimal representation, which underscores the predominance of SCI-EXPANDED in the academic citation landscape.

4.3   Co-authorship analysis

This subsection delivers an in-depth examination of co-authorship collaborations across various authors, institutions, and countries within the RDHEI domain. It details the collaboration networks, including the strength of linkages between participants, and visualizes these connections through network maps created with the VoSviewer tool. Such analyzes are instrumental in elucidating the collaborative framework and dynamics, offering a clear view of the interconnectedness that facilitates progress in this area of RDHEI.

4.3.1   Author coauthor linkages

The VoSviewer tool was employed to systematically analyze a dataset containing 455 documents, resulting in the identification of 1,216 authors involved in dataset. This analysis adopted a detailed counting method to evaluate co-authorship links among researchers, applying a publication threshold ranging from a minimum of 2 and 5 for inclusion criteria while excluding publications authored by more than 25 contributors. Following these guidelines, 245 authors qualified under the threshold of 2, and 42 authors met the criteria for the threshold of 5. The co-authorship network maps, depicted in Figures 4.7(a) and (b), visually represent these collaborations for each set criterion. Furthermore, an in-depth analysis of the most comprehensive networks within each group, illustrated in Figures 4.7(c) and (d), revealed that 87 and 19 authors, respectively, formed co-authorship linkages. These visual representations are characterized by distinct colors to differentiate between diverse co-authorship networks. The node sizes within these maps indicate the quantity of an author's publications, with larger nodes signifying a more substantial body of work. Additionally, the thickness of the lines connecting the nodes reflects the co-authorship bond strength, with denser lines indicating more robust collaborations. Prominent authors, such as "Chang, Chin-Chen," "Qin, Chuan," and "Zhang, Xinpeng," are marked by larger, colored circles to denote their significant contribution to the work, while authors with fewer citations are represented by smaller circles, demonstrating a proportional relationship between a node's size and an author's citation volume.

The data presented in Table 4.2 offers a detailed explanation of the top ten authors with the highest co-authorship linkages, where the strength of each linkage is assessed based on the number of jointly authored papers. At the top of this list is "Chang, Chin-Chen," who leads with a remarkable count of 31 documents and 279 citations, showing a linkage strength of 51. Following him, "Qin, Chuan" ranks second with a linkage strength of 37, having contributed to 15 documents and earning 212 citations. "Zhang, Xinpeng" claims the third spot, with a linkage strength of 29. Additionally, a notable observation from the table is the identical linkage strength of 27 shared among "Bu, Zhen Qi," "Huang, Wei Tao," "Liu, Qing Yu," "Lu, Jiao Yang," and "Quan, Min Xia." Each of these authors has contributed to 5 documents and received 36 citations. Furthermore, "Yao, Heng" is listed with a linkage strength of 25, associated with 9 documents and 124 citations, while "Tang, Zhenjun" concludes the list in the tenth position, with contributions to 8 documents, 156 citations, and a linkage strength of 23.

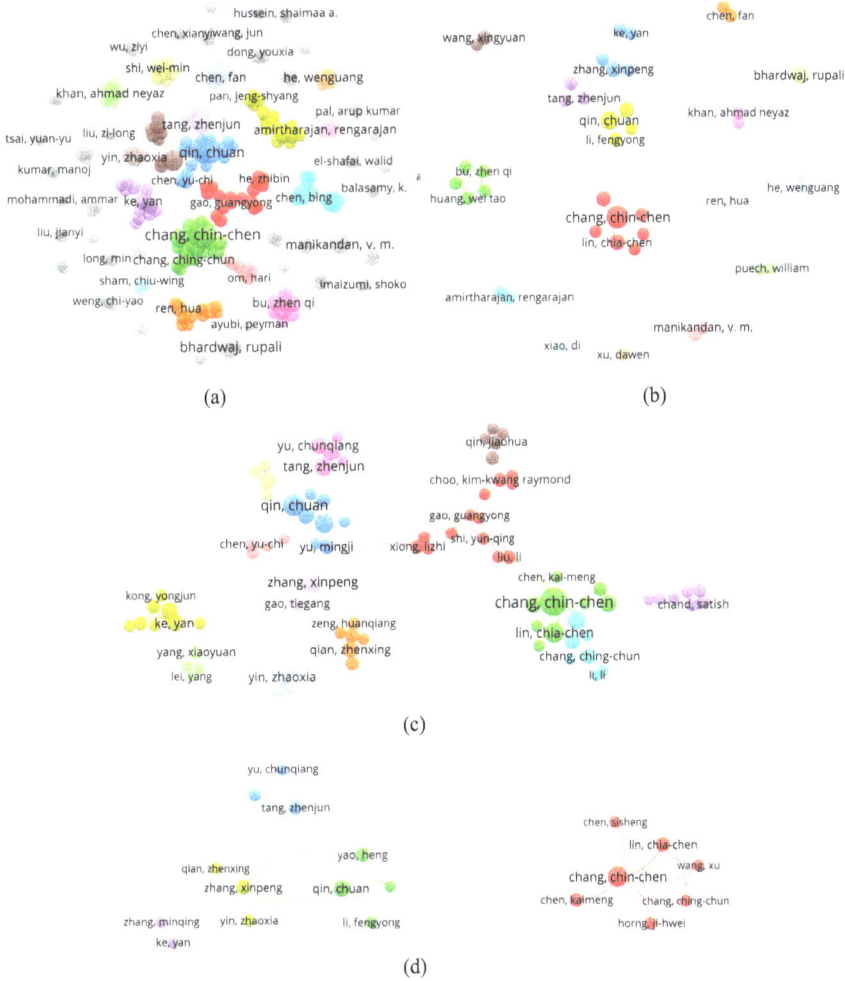

Figure 4.7　Author coauthor networks. (a) Network with minimum publication 2, (b) network with minimum publication 5, (c) most extensive network of (a), and (d) most extensive network of (b).

Table 4.2　Top ten author coauthor based on linkages and citations

	Based on linkages					Based on citations			
Id	Author name	D	C	TLS	Id	Author name	D	C	TLS
112	Chang, Chin-Chen	31	279	51	1,151	Zhang, Xinpeng	10	326	29
704	Qin, Chuan	15	212	37	112	Chang, Chin-Chen	31	279	51

(Continues)

Table 4.2   (Continued)

Based on linkages					Based on citations				
Id	Author name	D	C	TLS	Id	Author name	D	C	TLS
1151	Zhang, Xinpeng	10	326	29	948	Wang, Xingyuan	8	263	21
90	Bu, Zhen Qi	5	36	27	1,086	Yin, Zhaoxia	6	222	9
326	Huang, Wei Tao	5	36	27	704	Qin, Chuan	15	212	37
537	Liu, Qing Yu	5	36	27	1,011	Xiang, Youzhi	3	188	5
559	Lu, Jiao Yang	5	36	27	701	Qian, Zhenxing	5	168	12
715	Quan, Min Xia	5	36	27	371	Jiang, Donghua	5	159	10
1070	Yao, Heng	9	124	25	871	Tang, Zhenjun	8	156	23
871	Tang, Zhenjun	8	156	23	1,204	Zhou, Yicong	3	143	6

Furthermore, Table 4.2 delineates the top ten authors ranked based on citations, offering a distinct perspective on their academic impact and collaboration dynamics. Unlike the previous table, which focused on co-authorship linkages, this table emphasizes citation counts, providing insights into the scholarly influence and reach of these authors' works. "Zhang, Xinpeng" leads this list with 10 documents, amassing an impressive 326 citations and a total linkage strength of 29, showcasing not only prolific collaboration but also significant academic impact. Following him, "Chang, Chin-Chen" reflecting a substantial contribution to the field both in terms of productivity and influence, with a linkage strength of 51. This highlights Chang's central role in collaborative networks and his research's wide acknowledgment. "Wang, Xingyuan" occupies the third position with 8 documents and 263 citations, indicating a notable influence with a relatively lower linkage strength of 21, suggesting impactful yet selective collaborations. Further down the list, authors such as "Yin, Zhaoxia," "Qin, Chuan," and "Xiang, Youzhi" are ranked based on their citation counts, showing the diversity of contributions and the extent of their scholarly impact. Comparing this table to the earlier one, it is evident that while some authors like "Chang, Chin-Chen" and "Qin, Chuan" maintain top positions in both rankings, the criteria shift from co-authorship linkages to citations introduces new names such as "Wang, Xingyuan" and "Yin, Zhaoxia." This indicates that while some authors are central to collaborative networks, others may have a significant impact through highly cited works, despite fewer collaborative links. Moreover, the presence of authors like "Wang, Xingyuan" and "Chang, Chin-Chen" in the top ranks of both tables record their dual role as key collaborators and highly cited scholars, highlighting their comprehensive contribution to the RDHEI field. In contrast, the appearance of authors with fewer documents but high citation counts, such as "Xiang, Youzhi" and "Zhou, Yicong," emphasizes the quality and impact of their research over the number of collaborations. This analysis shows the multifaceted nature of academic contributions, where both collaboration and citation metrics are crucial for understanding an author's influence and role within the scholarly community. Table 4.3 lists the notable

Table 4.3   Top ten documents based on citations

Id	Document title	Authors	Year	Citations	Links
55	Separable and reversible data hiding in encrypted images using parametric binary tree labeling	Yi, Shuang; Zhou, Yicong	2019	114	54
221	A novel triple-image encryption and hiding algorithm based on chaos, compressive sensing and 3D DCT	Wang, Xingyuan; Liu, Cheng; Jiang, Donghua	2021	110	3
39	Reversible data hiding in encrypted images based on multi-MSB prediction and Huffman coding	Yin, Zhaoxia; Xiang, Youzhi; Zhang, Xinpeng	2020	104	42
438	FPGA realization of a RDH scheme for 5G MIMO-OFDM system by chaotic key generation-based Paillier cryptography along with LDPC and its side channel estimation using machine learning technique	Shajin, Francis H.; Rajesh, P.	2022	93	4
115	High-capacity reversible data hiding in encrypted images based on extended run-length coding and block-based MSB plane rearrangement	Chen, Kaimeng; Chang, Chin-Chen	2019	76	40
145	An efficient coding scheme for reversible data hiding in encrypted image with redundancy transfer	Qin, Chuan; Qian, Xiaokang; Hong, Wien; Zhang, Xinpeng	2019	70	32
78	New framework of reversible data hiding in encrypted JPEG bitstreams	Qian, Zhenxing; Xu, Haisheng; Luo, Xiangyang; Zhang, Xinpeng	2019	68	23
149	Effective reversible data hiding in encrypted image with adaptive encoding strategy	Fu, Yujie; Kong, Ping; Yao, Heng; Tang, Zhenjun; Qin, Chuan	2019	60	22
286	Securing data in Internet of Things (IoT) using cryptography and steganography techniques	Khari, Manju; Garg, Aditya Kumar; Gandomi, Amir H.; Gupta, Rashmi; Patan, Rizwan; Balusamy, Balamurugan	2020	60	4
14	An improved reversible data hiding in encrypted images using parametric binary tree labeling	Wu, Youqing; Xiang, Youzhi; Guo, Yutang; Tang, Jin; Yin, Zhaoxia	2020	57	28

contributions to reversible data hiding in encrypted images, with the highest citation impact.

4.3.2   *Organizational coauthor linkages*

The VoSviewer analyzed the dataset of 455 documents, and a total of 505 organizations were identified, as reflected in the provided Figure 4.8. From this dataset, only 160 organizations met the minimum threshold of having published at least 2 documents. This significant filtration shows the specialized nature of the field and the focused contributions of the organizations involved. Figure 4.8(a) presents a more inclusive overview where the minimum document count per organization is 2. This visualization reveals a network of collaboration, with nodes and linkages representing the organizations and their co-authorship ties, respectively. The color coding in the network map facilitates differentiation between various clusters of collaborative groups, indicating the diversity and breadth of the RDHEI domain. Figure 4.8(b), on the other hand, represents the largest set of 8b networks with a total of 83 organizations, highlighting the most prominent organizations within the network. These core entities demonstrate a higher degree of collaboration and influence within the RDHEI field, as evidenced by the thicker and more numerous linkages. Notably, organizations such as "Feng Chia University," "Sun Yat-Sen University," and "Shanghai Sci and Technol" emerge as significant nodes, indicating their central role in RDHEI research.

Table 4.4 complements these findings by listing the top authors within the same domain based on their co-authorship linkages and document counts, respectively. In these tables, organizations such as "Feng Chia University" and "Sun Yat-Sen University" are prominent due to their high document counts and linkage strengths. Their positions in both tables reflect their dual roles as central figures in

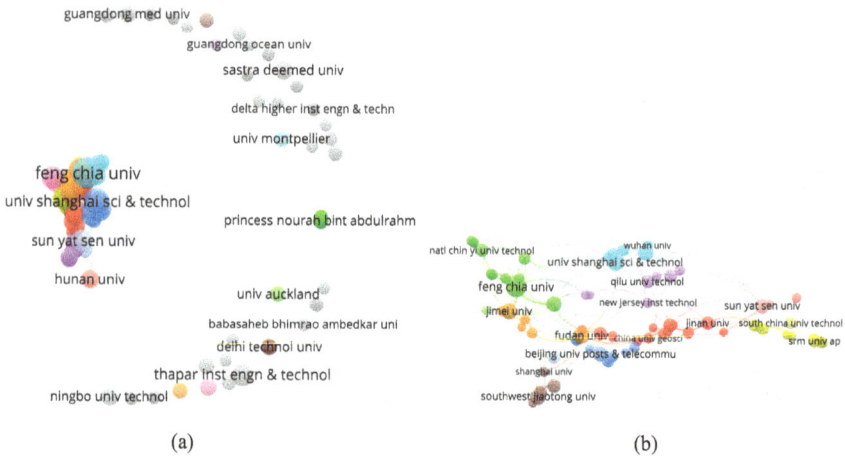

(a)   (b)

Figure 4.8   Organizational coauthor networks. (a) Network with minimum publication 2 and (b) most extensive network of (a).

Table 4.4   Top ten organization and coauthor linkage based on total link strength and documents

Based on total link strength				Based on the number of documents			
ID	Org. name	D	C	TLS	ID	Org. name	D	C	TLS
102	Feng Chia Univ	36	293	58	102	Feng Chia Univ	36	293	58
456	Univ Shanghai Sci & Technol	18	238	25	456	Univ Shanghai Sci & Technol	18	238	25
140	Hangzhou Dianzi Univ	12	89	23	105	Fudan Univ	15	434	17
132	Guangxi Normal Univ	14	291	20	132	Guangxi Normal Univ	14	291	20
384	Sun Yat-Sen Univ	11	166	19	195	Jimei Univ	13	157	16
105	Fudan Univ	15	434	17	370	Southwest Jiaotong Univ	13	124	15
195	Jimei Univ	13	157	16	140	Hangzhou Dianzi Univ	12	89	23
80	Dalian Maritime Univ	8	263	15	398	Thapar Inst Engn & Technol	12	48	0
370	Southwest Jiaotong Univ	13	124	15	384	Sun Yat-Sen Univ	11	166	19
317	Providence Univ	6	80	14	38	Beijing Univ Posts & Telecomm	11	77	6

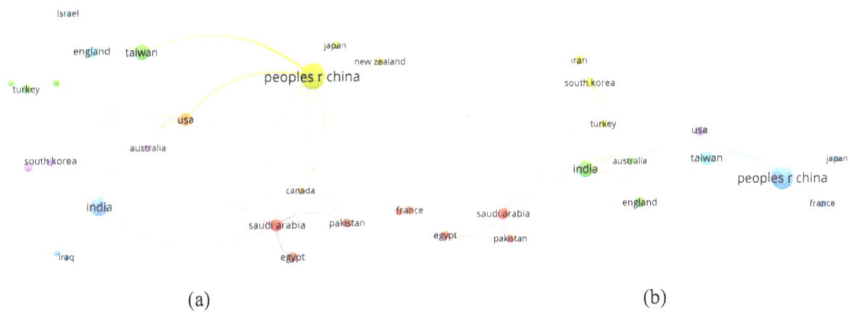

(a)   (b)

Figure 4.9   Country coauthor networks. (a) Network with minimum publication 3 and (b) network with minimum publication 5.

collaboration networks and highly documented publications, signifying their comprehensive contributions to the RDHEI domain. Moreover, the comparative analysis of the table ranking criteria illustrates the varying dimensions of institutional impact. Some organizations stand out for their quantity of research, others for their quality of collaboration, and a select few excel in both.

4.3.3   Country coauthor linkages

The VoSviewer-generated network maps visualized in Figures 4.9(a) and (b), along with the tabulated data in Table 4.5, offer a comprehensive view of country coauthor linkages within a dataset of 455 documents. Figure 4.9(a) displays a threshold of 3 documents per country and encompasses a wider network of 23 countries, with

Table 4.5   Comparison of top ten countries based on total link strength and
number of documents

	Based on total link strength					Based on number of documents			
ID	Country name	D	C	TLS	ID	Country name	D	C	TLS
30	PR China	268	2931	81	30	PR China	268	2931	81
40	Taiwan	53	390	38	14	India	85	633	33
14	India	85	633	33	40	Taiwan	53	390	38
44	USA	26	332	30	44	USA	26	332	30
34	Saudi Arabia	23	157	30	16	Iran	12	228	1
9	England	17	123	23	8	Egypt	11	176	5
37	South Korea	13	166	13	37	South Korea	13	166	13
29	Pakistan	8	40	13	34	Saudi Arabia	23	157	30
2	Australia	6	76	10	9	England	17	123	23
41	Turkey	8	67	8	12	France	6	95	2

the People's Republic of China (PRC), the India and USA, appearing as pivotal
nodes, indicating their extensive collaborative reach and high publication output.
As the threshold increases to 5 documents in 4.9b, the network becomes more
exclusive, focusing on 14 countries. This refined network continues to highlight the
PRC's central role, alongside the USA and India, suggesting a higher degree of
collaboration and academic exchange among these nations.

Table 4.5 further quantifies these relationships by listing the top ten countries
based on TLS and the number of documents, which allows for a comparative
analysis of the countries' collaborative impact. For example, the PRC's central
position in the network maps is quantitatively supported by its high TLS, reflecting
not just the quantity of its research output, but also the strength and frequency of its
international collaborations. Similarly, the USA and India's nodes are corroborated
by their strong linkages and high document counts, showing their substantial roles
in global research networks.

4.4   Citation analysis

Citation analysis is a powerful tool that uncovers the interconnections among various participants in a particular research field, based on citations. The following
subsections of this study showcase and elucidate such linkages among authors,
nations, documents, organizations, and sources. These analysis offer valuable
insights into the collaborative and influential relationships within the research
domain being explored.

4.4.1   Author citation linkages

In the domain of academic research, analyzing citation and author linkages provides
a multifaceted view of influence and collaboration. Figures 4.10(a) and 4.10(b)

generated from a set of 455 documents, which include contributions from a total of 1216 authors, illustrate these linkages under 2 thresholds: a minimum of 2 documents per author and a minimum of 5 documents per author.

With the lower threshold of 2 documents, Figure 4.10(a) reveals a broad network consisting of 245 authors, indicating a wide range of research interests and contributions. Out of these, 210 authors belong to the large set showcased in the figure. This extensive network suggests a vibrant academic community, with numerous authors contributing to a variety of topics. Raising the bar to a minimum of 5 documents per author, refines the network to include 48 authors, with 35 in the large set as depicted in Figure 4.10(b). This trimmed network highlights the most prolific authors and their associated documents, pinpointing the central figures within the community whose work is frequently referenced and who have a substantial impact on the direction of research within the field.

The accompanying Table 4.6 presents a clear view of author citation linkages of the top ten authors. Similar to Table 4.2, "Zhang Xinpeng" and "Chang Chin-Chen" are prominent in the categories, indicating their extensive influence and productivity. Zhang's work, while comprising fewer documents, has garnered a high number of citations, suggesting a deep impact on the academic domain.

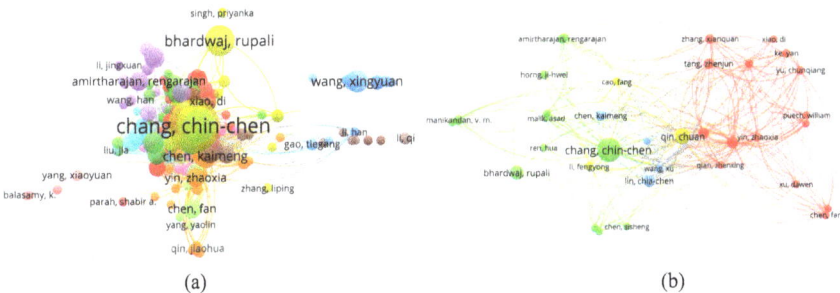

(a) (b)

Figure 4.10   Author citation networks. (a) Extensive network minimum of two documents and (b) extensive network minimum of five documents.

Table 4.6   Top ten author citation linkages

ID	Author name	D	C	TLS
1151	Xinpeng Zhang	10	326	597
112	Chin-Chen Chang	31	279	623
948	Xingyuan Wang	8	263	21
1086	Zhaoxia Yin	6	222	373
704	Chuan Qin	15	212	548
1011	Youzhi Xiang	3	188	327
701	Zhenxing Qian	5	168	255
371	Donghua Jiang	5	159	18
871	Zhenjun Tang	8	156	323
1204	Yicong Zhou	3	143	289

Conversely, Chang's substantial document count paired with a significant number of citations highlights a productive author whose work is both widespread and influential. The table also reveals that some authors, despite having fewer documents, have a high citation count (e.g., "Xiang Youzhi" and "Zhou Yicong"), indicating that their research, while less voluminous, is highly regarded within the community.

This examination highlights the complex network of academic contributions, illustrating how citation counts and document volume provide distinct perspectives on influence. The provided figures and tables show both emerging and established scholars within the field, underscoring the variety of contributions that drive the advancement of the research community.

4.4.2   Country citation linkages

The investigation of citation linkages among countries, based on a dataset of 455 documents, offers a detailed map of global research interconnected and scholarly influence. Figure 4.11(b) focused on the collaborative dynamics with a threshold of at least 3 documents per country, showing a network of 23 countries. This network reveals a rich collection of international research efforts, with a total of 21 countries emerging in the large set. With a higher threshold of ten documents, Figure 4.11(b) presents a more concentrated network of nine countries, all of which are part of the large set. This representation spotlights the most progressive nations whose research not only has significant volume but also high citation impact, marking them as influential players in this field.

Table 4.7 provides a complementary quantitative perspective, listing the top ten countries based on citations. PR China leads with the highest number of documents and citations, reflecting its dominant role in producing and disseminating influential research. India and Taiwan also feature prominently, indicating their substantial contributions to the volume of work and the scholarly

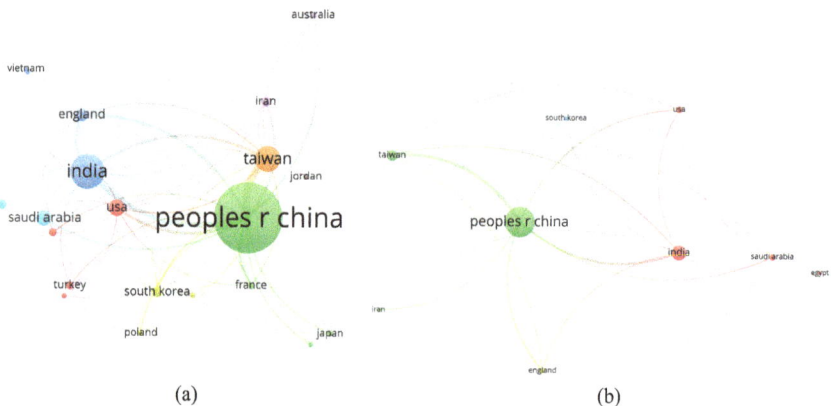

(a)   (b)

Figure 4.11   Country citation networks. (a) Network minimum with three documents and (b) network minimum with ten documents.

Table 4.7   Top ten country citation linkages

ID	Country name	D	C	TLS
30	PR China	268	2,931	839
14	India	85	633	296
40	Taiwan	53	390	438
44	USA	26	332	109
16	Iran	12	228	64
8	Egypt	11	176	5
37	South Korea	13	166	43
34	Saudi Arabia	23	157	62
9	England	17	123	114
12	France	6	95	91

discourse. Conversely, the United States, despite a lower volume of publications, demonstrates a significant citation count and link strength, emphasizing the impactful nature of its scholarly work.

The figures and tables present a comprehensive view of the RDHEI country citation, where countries like the People's Republic of China, India, and the United States emerge as central nodes of research activity and citation. Additionally, they reveal a core consortium of nations that, although producing a smaller volume of publications, wield significant influence through their scholarly contributions.

4.4.3   Document citation linkages

The VoSviewer visualizations provided in Figure 4.12, alongside the table, offer a multifaceted look at citation and document linkages within a body of 455 documents. This figure depicts the citation landscape based on varying citation thresholds, highlighting the spread and concentration of research impact. Figure 4.12(a) includes documents with a minimum citation threshold of 0, effectively including almost the entire dataset of 455 documents. This visualization likely presents a dense network of document connections, signifying the foundational and comprehensive scope of the research corpus. Raising the citation threshold to five for Figure 4.12(b) narrows the focus to documents that have garnered more attention in the academic community. With 224 documents selected, this figure presents a more targeted view of the research that has achieved a noticeable level of recognition and citation, indicating influential works within the dataset. Figure 4.12(c), which likely corresponds to Figure 4.12(a), large set with a zero-citation threshold, includes a subset of 312 documents. These represent the core publications that form the bulk of the research connections, providing a baseline for the network of research activity. In Figure 4.12(d), the large set is further concentrated into 136 documents, as seen from the 5 citation threshold. This subset represents the most cited and, presumably, the most impactful research within the dataset, highlighting key papers that have significantly contributed to the field's development.

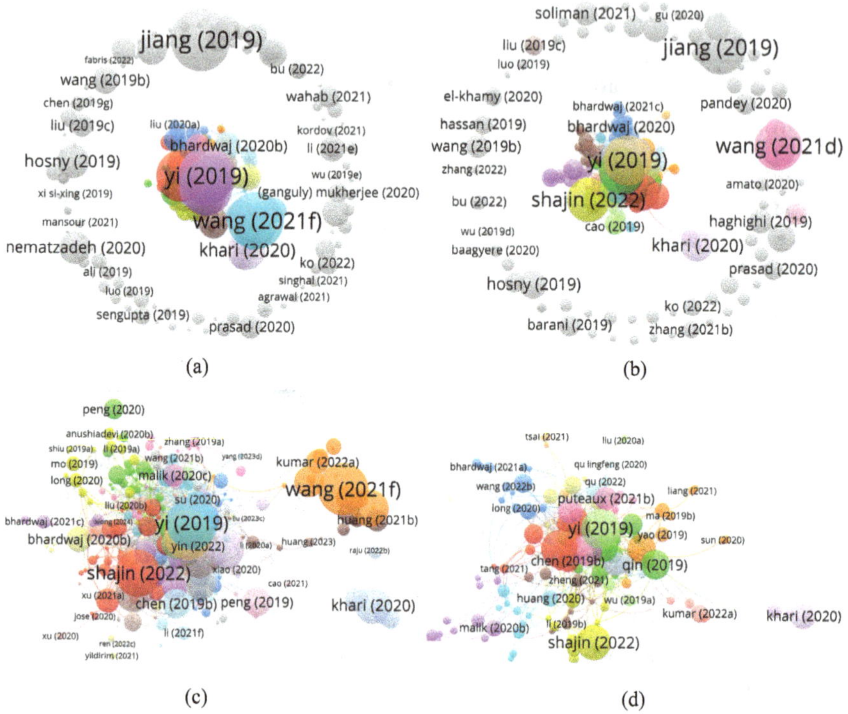

(a)

(b)

(c)

(d)

Figure 4.12   Document citation linkage networks. (a) Network with minimum citation 0, (b) network with minimum citation 5, (c) extensive network of (a), and (d) extensive network of (b).

Table 4.8   Document citations and links

Id	D	C	TLS
55	[34]	114	54
221	[35]	110	3
39	[36]	104	42
438	[37]	93	4
115	[38]	76	40
145	[39]	70	32
78	[40]	68	23
149	[41]	60	22
286	[42]	60	4
14	[43]	57	28

Table 4.8 complements the visual data by providing specific details on the top ten documents based on citations and the number of links to other documents. It gives a clearer picture of individual document impact, with papers like Jiang (2019) having a high number of citations but no links, possibly indicating a seminal work

that is frequently cited but not necessarily part of a larger discussion network. In contrast, documents like Yi [34], which have a high number of both citations and links, suggest papers that are central to the discourse, influencing and interacting with a wider body of research.

The integration of visual and tabular data showcases the dynamic character of research impact. The documents represented in the figures and table span a range of influence, encompassing both seminal works with broad recognition and those with specific significance within a network of interconnected research. The figures elucidate the interwoven nature of scholarly contributions, whereas the table quantifies the extent and engagement of individual articles.

4.4.4   Organizational citation linkages

The analysis of organizational citation linkages through VoSviewer offers a visual representation of how institutions interconnect within the academic domain of RDHEI. With a minimum threshold of 2 documents per organization, encompasses a network of 160 organizations, with 141 in the large set as depicted in Figure 4.13(a). This broad visualization likely includes a diverse range of institutions, indicative of a widespread and collaborative research environment. The large set included within the visualization suggests a robust network where multiple organizations are actively contributing and citing each other's work, laying out a map of significant academic exchange. Figure 4.13(b) with a higher threshold of 5 documents per organization, reveals a more focused network of 40 organizations. This concentration of institutions highlights those with a stronger presence in the dataset, suggesting that these organizations are not only prolific in their output but also central in academic citations. The density of the network in this visualization would be indicative of core research hubs with extensive citation linkages, reflecting their leading roles in the field.

The accompanying Table 4.9 provides quantitative data that complements the visual analysis, listing the top organizations. Institutions like "Fudan University" and "Feng Chia University" have a strong presence in both the number of citations and documents (visualized in 4.4), pointing to their significant influence and active participation in research dissemination. The table also reveals the TLS, which is a

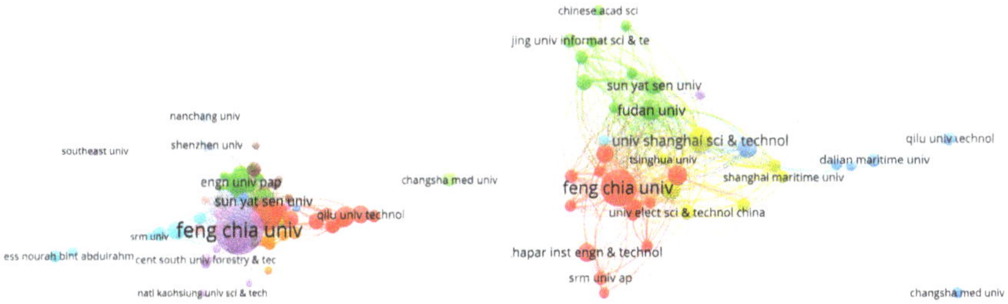

Figure 4.13   Organization citation networks. (a) Network with minimum two documents and (b) network with minimum five documents.

Table 4.9   Organization citation linkages

Id	Org. name	D	C	TLS
105	Fudan Univ	15	434	576
102	Feng Chia Univ	36	293	557
132	Guangxi Normal Univ	14	291	285
80	Dalian Maritime Univ	8	263	22
456	Univ Shanghai Sci & Technol	18	238	414
17	Anhui Univ	7	231	295
423	Univ Elect Sci & Technol China	8	225	110
439	Univ Macau	5	172	247
384	Sun Yat-Sen Univ	11	166	198
195	Jimei Univ	13	157	245

measure of the overall connectivity of an organization within the network. For instance, "Fudan University," despite having fewer documents than "Feng Chia University," shows a higher TLS, suggesting that its contributions are highly central within the citation network.

The contribution of visual and tabular data reveals the complex network of research connections among institutions. Organizations like the University of Shanghai for Science and Technology, characterized by a high TLS and numerous citations, highlight their central importance in the academic community. In contrast, institutions producing a large volume of documents but exhibiting lower link strength may be introducing new insights that have not yet been thoroughly assimilated into broader scholarly discussions. This analysis also illustrates the dual role of certain organizations: they not only contribute a significant body of work but also serve as pivotal agents in the dissemination and citation of research, establishing themselves as foundational pillars within their respective disciplines.

4.5   Conclusion

This detailed review of the RDHEI domain has provided a meaningful perspective through quantitative analysis. The key observations of this review are as outlined as follows:

1. There has been a continuous growth in terms of the year-wise number of publications. More specifically, approximately 94 RDHEI-related articles are published annually in the WoS database.
2. The journal "Multimedia Tools and Applications" has published the highest number of RDHEI articles (77).
3. Feng Chia University, Taiwan, has published the maximum number of articles (36) among all institutions.
4. Chang, Ching-Chun (China) has authored the highest number of articles (36).
5. Researchers from China have been publishing the most RDHEI articles annually since inception.

6. The top-cited document in the last five years is "Separable and Reversible Data Hiding in Encrypted Images Using Parametric Binary Tree Labeling" by Zhou *et al.* [33].
7. Chang, Chin-Chen; Qin, Chuan; and Zhang, Xinpeng; exhibit the most prominent co-authorship linkages.
8. Zhang, Xinpeng has been recognized as a notably influential author, distinguished by a considerable volume of co-citations alongside the highest TLS. Furthermore, the document titled "Reversible Data Hiding in Encrypted Image" by Yi and Zhou [34] has obtained 326 citations within this period.

References

[1] Buyya R, Yeo CS, Venugopal S, *et al.* Cloud computing and emerging IT platforms: Vision, hype, and reality for delivering computing as the 5th utility. *Future Generation Computer Systems*. 2009;25(6):599–616.

[2] Sharma D, Kumar R, and Jung KH. A bibliometric analysis of convergence of artificial intelligence and blockchain for edge of things. *Journal of Grid Computing*. 2023;21(4):79.

[3] Kaushal A, Lin CC, Chauhan R, *et al.* Charting the growth of text summarisation: A data-driven exploration of research trends and technological advancements. *Applied Sciences.* 2024;14(23):11462.

[4] Kaushal A, Kumar S, and Kumar R. A review on deepfake generation and detection: Bibliometric analysis. *Multimedia Tools and Applications*. 2024; 83:1–41.

[5] Branco Jr T, de Sá-Soares F, and Rivero AL. Key issues for the successful adoption of cloud computing. *Procedia Computer Science*. 2017;121:115–122.

[6] Chang V, and Ramachandran M. Towards achieving data security with the cloud computing adoption framework. *IEEE Transactions on Services Computing*. 2015;9(1):138–151.

[7] Kumar R, Caldelli R, KokSheik W, *et al. Advancements in multimedia security in the context of artificial intelligence and cloud computing*. Elsevier; 2025.

[8] Wu M, and Liu B. Watermarking for image authentication. In: *Proceedings 1998 International Conference on Image Processing. ICIP98 (Cat. No. 98CB36269)*. vol. 2. IEEE; 1998. pp. 437–441.

[9] Langelaar GC, Setyawan I, and Lagendijk RL. Watermarking digital image and video data. A state-of-the-art overview. *IEEE Signal Processing Magazine*. 2000;17(5):20–46.

[10] Hu P, Peng D, Yi Z, *et al.* Robust time-spread echo watermarking using characteristics of host signals. *Electronics Letters*. 2016;52(1):5–6.

[11] Kersten D. Predictability and redundancy of natural images. *Journal of the Optical Society of America A*. 1987;4(12):2395–2400.

[12]   Kumar R, Ranjan P, Jung KH, *et al.* Leveraging rANS for synchronized high capacity reversible data hiding in encrypted image. *Expert Systems with Applications.* 2025;267:126181.

[13]   Kumar R, Chand S, and Singh S. An optimal high capacity reversible data hiding scheme using move to front coding for LZW codes. *Multimedia Tools and Applications.* 2019;78(16):22977–23001.

[14]   Sahu M, Padhy N, Gantayat SS, *et al.* Local binary pattern-based reversible data hiding. *CAAI Transactions on Intelligence Technology.* 2022;7(4):695–709.

[15]   Ni Z, Shi YQ, Ansari N, *et al.* Reversible data hiding. *IEEE Transactions on Circuits and Systems for Video Technology.* 2006;16(3):354–362.

[16]   Kumar N, Kumar R, and Caldelli R. Local moment driven PVO based reversible data hiding. *IEEE Signal Processing Letters.* 2021;28:1335–1339.

[17]   Gandhi S, and Kumar R. A high-capacity reversible data hiding with contrast enhancement and brightness preservation for medical images. *Multimedia Tools and Applications.* 2024;84:1–26.

[18]   Gandhi S, and Kumar R. Survey of reversible data hiding: Statistics, current trends, and future outlook. *Computer Standards & Interfaces.* 2025;94:104003.

[19]   Liao X, and Shu C. Reversible data hiding in encrypted images based on absolute mean difference of multiple neighboring pixels. *Journal of Visual Communication and Image Representation.* 2015;28:21–27.

[20]   Puteaux P, and Puech W. A recursive reversible data hiding in encrypted images method with a very high payload. *IEEE Transactions on Multimedia.* 2020;23:636–650.

[21]   Ankur, Kumar R, and Sharma AK. High capacity reversible data hiding with contiguous space in encrypted images. *Computers and Electrical Engineering.* 2023;112:109017.

[22]   Ren F, Wu Z, Xue Y, *et al.* Reversible data hiding in encrypted image based on bit-plane redundancy of prediction error. *Mathematics.* 2023;11(11):2537.

[23]   Bender W, Butera W, Gruhl D, *et al.* Applications for data hiding. *IBM Systems Journal.* 2000;39(3.4):547–568.

[24]   Fridrich J. Applications of data hiding in digital images. In: *ISSPA'99. Proceedings of the Fifth International Symposium on Signal Processing and its Applications (IEEE Cat. No. 99EX359).* vol. 1. IEEE; 1999. p. 9.

[25]   Ankur, Kumar R, and Sharma AK. Link chain driven reversible data hiding in encrypted images for high payload. Signal, *Image and Video Processing.* 2024;18:1–16. https://doiorg/101007/s11760-024-03275-1.

[26]   Ma K, Zhang W, Zhao X, *et al.* Reversible data hiding in encrypted images by reserving room before encryption. *IEEE Transactions on Information Forensics and Security.* 2013;8(3):553–562.

[27]   Malik A, He P, Wang H, *et al.* High-capacity reversible data hiding in encrypted images using multi-layer embedding. *IEEE Access.* 2020; 8:148997–149010.

[28]   Gao H, Zhang X, and Gao T. Hierarchical reversible data hiding in encrypted images based on multiple linear regressions and multiple bits prediction. *Multimedia Tools and Applications.* 2023;83:1–27.

[29] Ankur, Kumar R, and Sharma AK. Bit-plane based reversible data hiding in encrypted images using multi-level blocking with quad-tree. *IEEE Transactions on Multimedia*. 2023;26.

[30] Puech W, Chaumont M, and Strauss O. A reversible data hiding method for encrypted images. In: *Security, forensics, steganography, and watermarking of multimedia contents X*. vol. 6819. SPIE; 2008. pp. 534–542.

[31] Zhang X. Reversible data hiding in encrypted image. *IEEE Signal Processing Letters*. 2011;18(4):255–258.

[32] Ankur, Kumar R, and Sharma AK. Adaptive two-stage reversible data hiding in encrypted images using prediction error expansion. In: *2023 Third international conference on secure cyber computing and communication (ICSCCC)*. IEEE; 2023. pp. 427–432.

[33] Zhou J, Sun W, Dong L, *et al.* Secure reversible image data hiding over encrypted domain via key modulation. *IEEE Transactions on Circuits and Systems for Video Technology*. 2015;26(3):441–452.

[34] Yi S, and Zhou Y. Separable and reversible data hiding in encrypted images using parametric binary tree labeling. *IEEE Transactions on Multimedia*. 2019;21(1):51–64.

[35] Wang X, Liu C, and Jiang D.A novel triple-image encryption and hiding algorithm based on chaos, compressive sensing and 3D DCT. *Information Sciences*. 2021;574:505–27.

[36] Yin Z, Xiang Y, and Zhang X. Reversible data hiding in encrypted images based on multi-MSB prediction and Huffman coding. *IEEE Transactions on Multimedia*. 2019;22(4):874–84.

[37] Shajin FH, and Rajesh P, FPGA realization of a reversible data hiding scheme for 5G MIMO-OFDM system by chaotic key generation-based paillier cryptography along with LDPC and its side channel estimation using machine learning technique, *Journal of Circuits, Systems and Computers*. 2022;31:2250093.

[38] Chen K, Chang C C. High-capacity reversible data hiding in encrypted images based on extended run-length coding and block-based MSB plane rearrangement. *Journal of Visual Communication and Image Representation*. 2019;58:334–44.

[39] Qin C, Qian X, Hong W, and Zhang X. An efficient coding scheme for reversible data hiding in encrypted image with redundancy transfer. *Information Sciences*. 2019;487:176–92.

[40] Qian Z, Xu H, Luo X, and Zhang X. New framework of reversible data hiding in encrypted JPEG bitstreams. *IEEE Transactions on Circuits and Systems for Video Technology*. 2018;29(2):351–62.

[41] Fu Y, Kong P, Yao H, Tang Z, and Qin C. Effective reversible data hiding in encrypted image with adaptive encoding strategy. *Information Sciences*. 2019;494:21–36.

[42] Khari M, Garg A K, Gandomi A H, Gupta R, Patan R, and Balusamy B. Securing data in Internet of Things (IoT) using cryptography and

steganography techniques. *IEEE Transactions on Systems, Man, and Cybernetics: Systems*. 2019;50(1):73–80.

[43] Wu Y, Xiang Y, Guo Y, Tang J, and Yin Z. An improved reversible data hiding in encrypted images using parametric binary tree labeling. *IEEE Transactions on Multimedia*. 2019;22(8):1929–38.

Chapter 5

Blockchain architecture, consensus mechanisms, and their applications in watermarking

Gandharba Swain[1], Anita Pradhan[1], Satish Muppidi[2], Pramoda Patro[3] and Monalisa Sahu[4]

Abstract

In 2008, the idea of Bitcoin, a peer-to-peer electronic cash system, was proposed by Satoshi Nakamoto. It describes a distributed system for managing digital transactions. Based on this idea, the blockchain concept has evolved. Blockchain is a distributed ledger. The ledger is immutable and shareable among all the user nodes. The ledger/ blockchain contains several blocks chained by hash values. If we try to modify a block, its hash value will change; the hash value is already stored in the neighbor node, so the neighbor node will not allow it to change. Thus, immutability is achieved. The blockchain is worthy because of its good characteristics, such as data decentralization and a high level of trust. This chapter represents a detailed study on blockchain architecture, consensus mechanisms, and their application in digital image watermarking. Digital image watermarking is used for copyright protection, ownership claim, and image tamper detection. If we use blockchain with watermarking, the technique becomes more secure and robust. Though the applications of blockchain technology in image watermarking are in a nascent stage at present, the disruptive and revolutionary nature of the blockchain will make it a significant force shortly.

Keywords: blockchain architecture; blockchain consensus; blockchain for watermarking; types of blockchain; blockchain generations

5.1 Introduction

There are three types of systems available for information sharing: (i) centralized system, (ii) decentralized system, and (iii) distributed system [1]. Centralized systems use client and server architecture that connect one or more client nodes to a

[1]Department of CSE, Koneru Lakshmaiah Education Foundation, India
[2]Department of Computer Science and Engineering, GMR Institute of Technology, India
[3]School of Computer Science and Artificial Intelligence, SR University, India
[4]School of Computer Science and Engineering (SCOPE), VIT-AP University, India

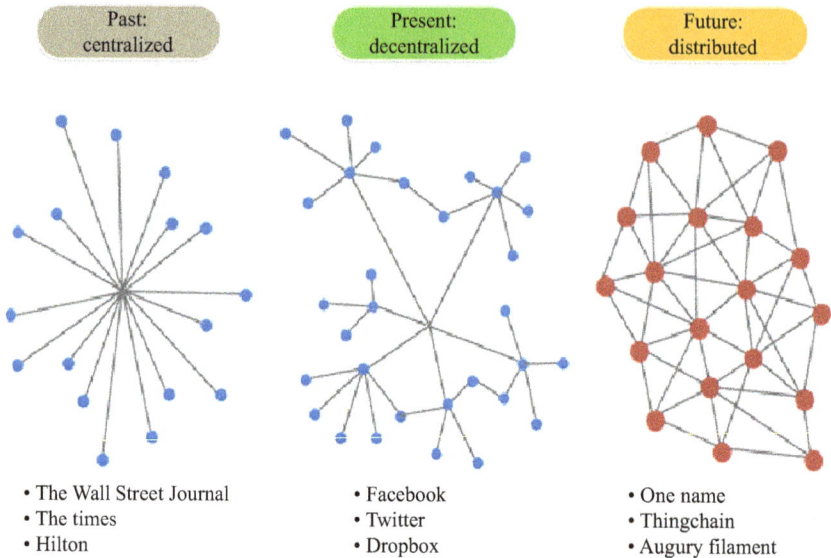

Figure 5.1   (a) Centralized system, (b) decentralized system, and (c) distributed system

central server, as shown in Figure 5.1(a) [2]. The node at the center is the server, and the nodes surrounding the server are clients. The client sends a request to a server, and the server responds. Some drawbacks of this system are (i) if the server fails, the total system fails, and (ii) if the server gets many requests, then the system becomes slow. Due to these reasons, a decentralized system came into existence. Figure 5.1(b) represents a decentralized system [2], wherein there are a few server nodes rather than only one. Clients connected to different servers can communicate through the server nodes. Under each server, a group of clients communicates. The drawbacks of this model are (i) the coordination problem, as it is hard to accomplish tasks that are collectively done by the system, and (ii) it is not an ideal solution for small systems, as it is not helpful for small decentralized systems due to high cost and low benefit. In a distributed system, as shown in Figure 5.1(c), there is no central server [2]. A distributed system consists of independent computers linked to each other. No node is a server; all are of equal status. They are termed as peers. The characteristics of this model are (i) all nodes apply a consensus protocol to agree on some value of the transactions, and (ii) failure of one node does not cause the entire system to fail.

Bitcoin is a distributed network of cryptocurrency [3]. A cryptocurrency is a digital property designed to function as an exchange medium for secure financial asset transfers. Blockchain technology has evolved due to the success of Bitcoin. Electronic trading depends on finance-related organizations for electronic payments as trusted parties. The salient features of Bitcoin are (i) distributed, (ii) peer-to-peer, and (iii) permissionless. The distributed network completely avoids centralization. The main idea for the distributed network is connectivity for everyone and equal access for everyone. All the stakeholders form a system called a "peer-to-peer" network. Permissionless means that anyone can join in the network without any authorization. Bitcoins are based on cryptography to transact directly with no requirement of a third-party. The cryptographic hash algorithms like MD5, and SHA256 has been used for security purposes [4]. They are used to map a data of variable size to a fixed size which are one-way function. For a given x, we can compute $H(x)$, but no deterministic algorithm is available to compute x when $H(x)$ is given, where H is a hash function like MD5 or SHA256.

5.2 Blockchain architecture

The architecture of blockchain is depicted in Figure 5.2. A blockchain is a distributed ledger shared among a group of nodes. These nodes are peers. All the peers have equal rights on the ledger.

The ledger is a collection of blocks that are linked together to form a chain, as shown in Figure 5.3. In the blocks, the information is permanently stored. Each block typically consists of two parts: (i) the block header and (ii) the data or transactions. The block header contains (i) nonce, (ii) timestamp, (iii) hash of the previous block, and (iv) hash of the present block [2]. The data in the block contains records of exchanges, medical records, certificates, property rights, etc. The hash of the present block is used as the previous hash of the next block. The hash of a block is computed over the time stamp, previous hash, nonce, and data. So, if one tries to tamper with the data in a block, then its hash value will change, and the hash values of all the further blocks in the chain will change. This is a very difficult task to do. The chain of blocks is stored in the nodes. A node can be any device like laptop, or computer. The nodes are connected in a distributed system and the nodes constantly exchange the information, to update the blockchain from time to time.

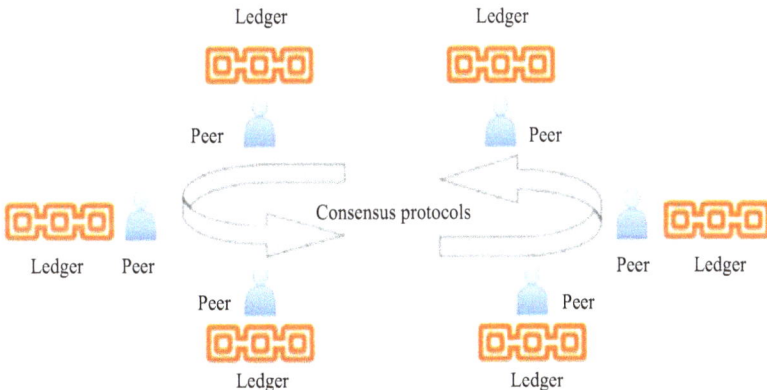

Figure 5.2 Architecture of blockchain

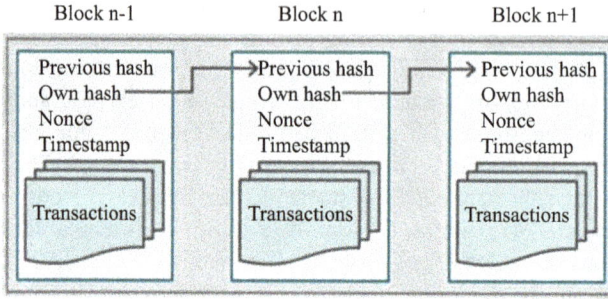

Figure 5.3   Linking of blocks by the previous hash

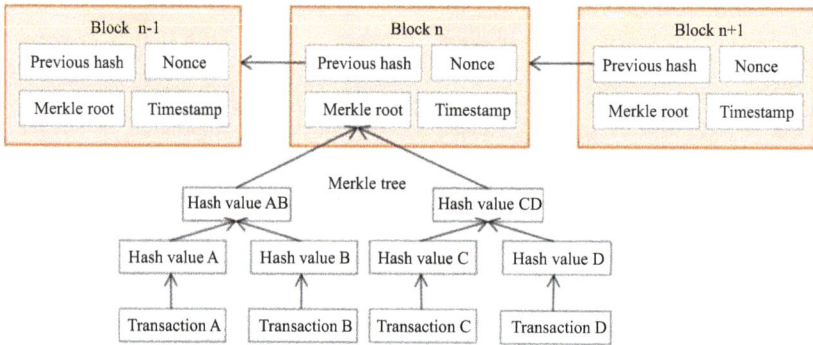

Figure 5.4   Transactions hashed in Merkle tree

This is called distributed ledger technology [5]. The need of a blockchain is due to its advantageous characteristics like, security, collaboration, decentralized, reliability, unchangeable transactions, and time reduction.

A node in a Bitcoin network contains the complete history of the blockchain transactions [3]. It is a permissionless model. Each node in it has equal privileges as it is "peer-to-peer." The transactions occur between the nodes in the network, and when they are verified by the other nodes and approved, they are successfully inserted into the blockchain in a new block. A set of transactions is formed as a single block and added to the blockchain. When a transaction occurs, it is validated by the remaining nodes in the network. The hash value of the block is calculated by using a hash algorithm on the four components: (i) previous hash, (ii) nonce, (iii) Merkle root, and (iv) timestamp. Now this hash is set as the previous hash value for the next block. Figure 5.4 depicts a chain of three blocks. The block $n - 1$ is the previous block of block n, and the block n is the previous block of block $n + 1$. The transactions in each block are not shown; only the transactions in block n are shown. A, B, C, and D are the four transactions in block n. The transactions are organized in a Merkle tree. The leaf nodes of the Merkle tree contain the transactions. The hash value of these transactions is appended and hashed again repeatedly up to the root. The hash of the root is known as the Merkle root. The Merkle root becomes a part of the block header.

Now, let us see how a block is generated from transactions and added to the blockchain (refer to Figure 5.5 [6]). A transaction generator shall be a node in the network that has some data to store in the blockchain. Suppose DN is a transaction generator, BN is a blockchain node, T is a transaction, and B is a block. The DN need not be a BN.

Step 1: Suppose DN creates a new transaction, where the data are D, and the reward for the miner is R. *Step 2*: As soon as this transaction is created, node DN broadcasts it to all nodes in the system. So all the nodes will update their transaction pool named memPool. It is a queue where the transactions wait before being packaged into a block. *Step 3*: The nodes in the blockchain now conduct a consensus mechanism to choose a miner. Suppose, BN_3 wins the race for generating the next block, that is, the first node that gets the desired Nonce. *Step 4*: Now BN_3 selects some transactions from memPool and forms a new block B_t. B_t is stored locally at BN_3 and BN_3 gets the reward point R. *Step 5*: Furthermore, BN_3 also broadcasts the new block B_t to all other nodes in the blockchain network. Other nodes will verify B_t and update it in their local blockchain. Thus, the transaction is finally stored in the ledger.

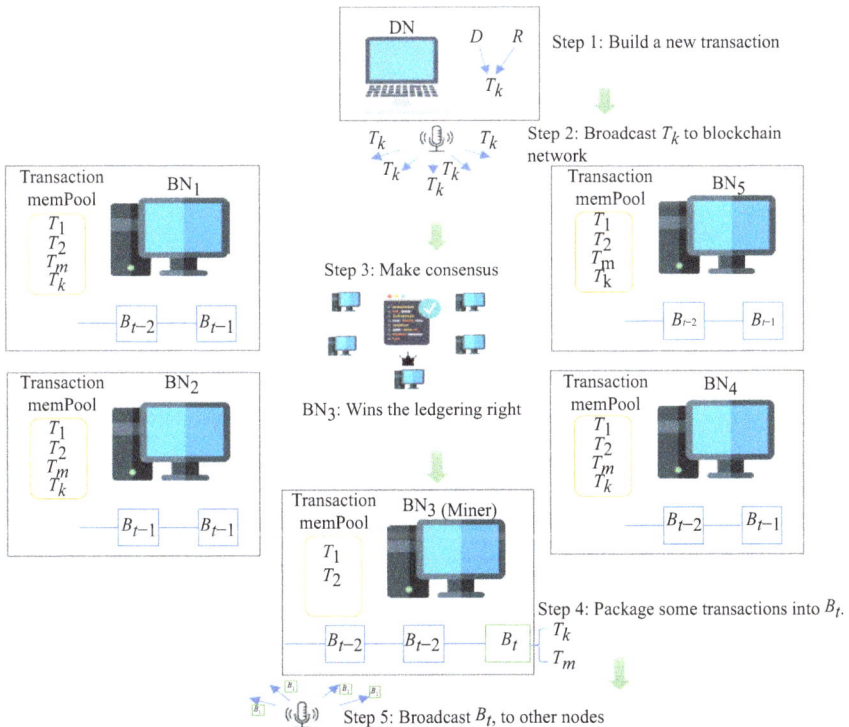

Figure 5.5   A typical data processing flow in the blockchain system

5.3   Consensus mechanisms

The addition of the blocks into the blockchain is done only upon a common agreement among the peers in the network. So, the consensus is used to achieve reliability and establish trust among unknown peers in the distributed network. The consensus is achieved by using a consensus algorithm. As shown in Figure 5.6, the consensus algorithms are classified into two broad categories: (1) permissioned blockchain and (2) permissionless blockchain. Most of the digital currencies available in the market work under the category of permissionless blockchain, where anyone can become part of the chain without requiring any authentication or other barriers. Users can simply create their personal addresses and start interacting with the blockchain network without any censorship. Decentralization, anonymity, and transparency are the key issues in permissionless blockchain. Bitcoin is an example of a permissionless blockchain. Permissioned blockchain, on the other hand, is a closed or private network where users are not allowed to join without permission/authorization/censorship. Users are expected to know each other in this category of blockchain. Permissioned blockchain is used for any organization, such as a private corporation or a consortium group, where some authorized entities are permitted to participate. Unlike permissionless algorithms, where the miners need to use power, time, and/or cryptocurrency, permissioned blockchain avoids the mining (computational) overhead. However, the consensus among the users is a primitive challenge that could be handled through the concept of state machine replication. The major challenge in achieving the distributed consensus in permissioned blockchain is a fault such as a crash fault, network fault, and byzantine fault.

The various consensus algorithms used for permissionless blockchains are: (i) proof of work (PoW), (ii) proof of stake (PoS), (iii) proof of burn (PoB), and (iv) proof of elapsed time (PoET) [7]. Recent literature [8,9] describes some more

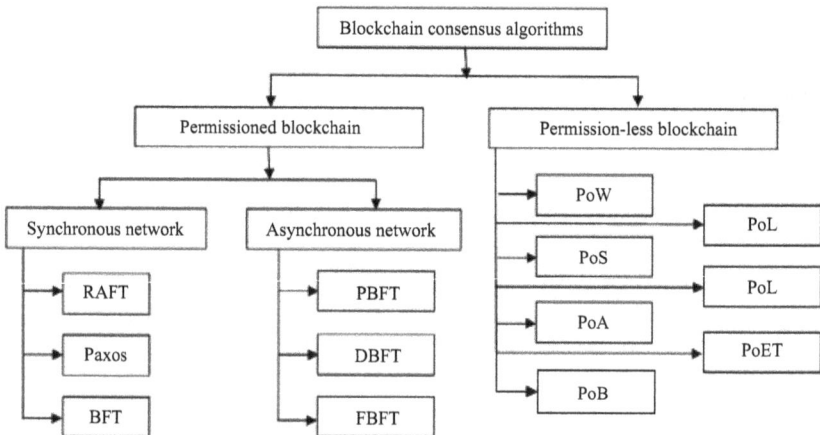

Figure 5.6   Taxonomy of consensus algorithms

consensus mechanisms: (v) delegated proof of stake (DPoS), (vi) proof of authority (PoA), (vii) proof of location (PoL), and (viii) proof of importance (PoI).

In the PoW consensus algorithm, the network asks the miners to solve a cryptographically challenging response-type of puzzle [2]. A miner is a node with higher computational resources to solve the assigned problem. A miner who solves the problem first shall add the block to the blockchain and will get some Bitcoins as a reward from the network. In the PoS consensus algorithm, a miner who has a higher stake of Bitcoins can get more chance to add a block to the blockchain. For example, if a node holds 2% of the stake value of the network, then only 2% of the blocks can be mined by that node. In the PoB consensus algorithm, the miners shall burn some Bitcoins and then be able to add a block to the blockchain. PoB works by burning PoW-mined currencies. In the PoET consensus algorithm, the idea is that each participant in the network waits a random amount of time; the first participant to finish becomes the leader for the new block. These consensus algorithms are complex in nature. In the DPoS consensus algorithm, the system presents some block producers called witnesses. Witnesses shall produce new blocks and validate transactions. The witnesses are chosen by the system users, that is, the token holders. In PoA consensus mechanism, small designated blockchain actors are empowered to update the ledger and validate the transactions. These actors are registered and trusted stakeholders possessing a node. These trusted entities or validating nodes secure the blockchain in PoS-based model from 51% attack.

For permissioned models, we have other algorithms, which are used traditionally in distributed systems. There are two scenarios, synchronous and asynchronous [9]. In a synchronous communication system, run under a common time clock with a finite delay. In an asynchronous environment like the internet, there is no bound like delay, so time constraints should not be there. In synchronous scenarios, the consensus algorithms used are (i) PAXOS, (ii) RAFT, and (iii) byzantine fault tolerance (BFT). In an asynchronous environment, the consensus algorithms are (iv) practical BFT (PBFT), (v) delegate BFT, and (vi) federated BFT. The PAXOS and RAFT tolerate crash and network faults, but not the Byzantine faults. The BFT and PBFT algorithms tolerate crash faults, network faults, and byzantine faults. Essentially, all nodes in the BFT system are grouped in a series where one node is the primary node (leader) and the others are the backup nodes. All network nodes communicate with one another and must not only prove that messages originated from a peer node, but also check that the message was not changed during transit. What distinguishes the BFT model from other forms of consensus algorithms is its ability to confirm final transactions without the need for extensive checks, such as those found in the PoW models. If the nodes in a BFT process agree on a proposed block, the block is final. This is possible because, as a result of their interaction with each other, all the honest nodes agree on the state of the network at that time.

The four main features of blockchain are (i) trust, (ii) reliability, (iii) security, and (iv) efficiency [10]. The blockchain network ensures trust. Unlike the centralized trust, such as central currency issuing governments and commercial banks, the blockchain network functions as new trust holder with decentralized ledgers.

These ledgers are shared in a tamper-proof network. The decentralized nature of the blockchain network shifts the entire transaction record database from closed and centralized ledgers maintained by just a few certified organizations to open, distributed ledgers maintained by tens of thousands of nodes. A single node failure does not influence the entire network's operation. So, the reliability is provided by avoiding a single point of failure.

The blockchain network utilizes a one-way hash function, which is a mathematical function that takes a variable-length input string and transforms it to a fixed-length binary sequence to ensure security. The output does not have any input correlation. The process is difficult to reverse because the input cannot be determined with the output alone. The blockchain is efficient because all information will pass through predetermined processes automatically. Therefore, blockchain technology can not only decrease labor costs considerably but also enhance effectiveness. For example, in Bitcoin, the decrease in intermediaries made the transaction process quicker and more effective; blockchain technology could speed up the clearing and settlement of certain economic transactions.

5.4   Types of blockchains

Basically, blockchains are of three types: (i) public blockchain, (ii) private blockchain, and (iii) consortium-based blockchain [11]. Figure 5.7 depicts these three kinds of blockchains. Public blockchains are open source. They are also called a permissionless model. They allow anyone to be involved as users, miners, or members of the network. All transactions that occur on them are completely transparent, which means that the transaction details can be examined and accessed

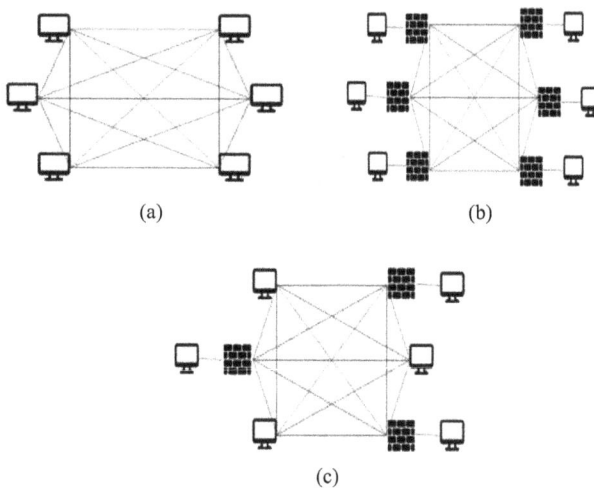

(a)   (b)

(c)

Figure 5.7   (a) Public blockchain model, (b) private blockchain model, and (c) consortium blockchain

by anyone in the network. Public blockchains are all connected with a token, usually designed to encourage and reward network users. Bitcoin is a public blockchain. Private blockchains are also known as permissioned blockchains, which have several significant variations from public blockchains [2]. Participants require permission to join the network. They are more centralized than the public blockchain, and are useful to business houses that want to cooperate and share data. The identities of the nodes are known, so there is no need for any token. The private blockchain is governed by a single organization. The consortium-based blockchains are treated as a distinct classification from private blockchains [10]. The main difference between them is that consortium blockchains are governed not by a single entity, but by a group of entities. This cooperative model is useful for the frenemies – companies that work together but also compete against each other. By partnering on certain aspects of their business, they can be more effective, both individually and collectively.

The evolution of blockchain can be classified into three phases, namely Blockchain 1.0, Blockchain 2.0, and Blockchain 3.0, as shown in Figure 5.8. The Blockchain 1.0 is confined mainly to cryptocurrencies like Bitcoin, Zcash, and Ripple [3]. It focuses mainly on cryptocurrency transactions. The application areas are related to digital payment systems, currency transfer, and remittance, which are based on transactions. Blockchain 2.0 was for smart contracts and decentralized applications [5]. The smart contracts are used to resolve the transactions as soon as the specified criteria and conditions are met without the involvement of trusted third parties.

Blockchain 3.0 is mainly for the development of applications in education, administration, and health care [12], as well as enterprise applications. The versatile nature of the blockchain makes it the most sought-after technology in the future. The next section describes the various applications of blockchain in detail. The Blockchain 4.0 is an industry infrastructure-based echo system [9]. Blockchain 4.0 is expected to involve artificial intelligence for autonomous decision-making.

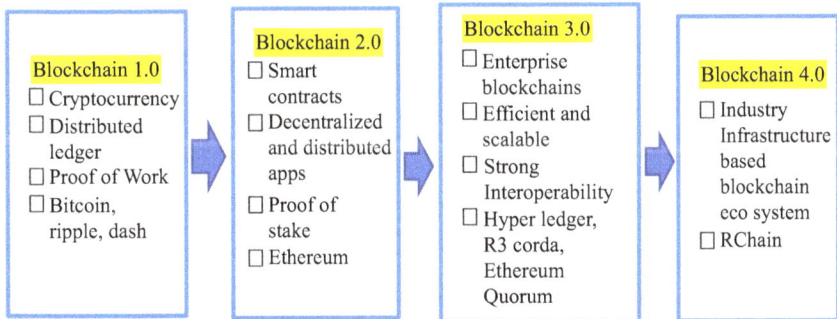

Blockchain 1.0
☐ Cryptocurrency
☐ Distributed ledger
☐ Proof of Work
☐ Bitcoin, ripple, dash

Blockchain 2.0
☐ Smart contracts
☐ Decentralized and distributed apps
☐ Proof of stake
☐ Ethereum

Blockchain 3.0
☐ Enterprise blockchains
☐ Efficient and scalable
☐ Strong Interoperability
☐ Hyper ledger, R3 corda, Ethereum Quorum

Blockchain 4.0
☐ Industry Infrastructure based blockchain eco system
☐ RChain

Figure 5.8   Generations of blockchains

5.5   Applications of blockchain in image watermarking

The disruptive and revolutionary nature of the blockchain makes it useful in various sectors. The predominant use of blockchain has been seen in, (1) finance, (2) health care, (3) education, (4) identity management, (5) smart contract (6) e-commerce, (7) e-voting, (8) KYC, (9) crowd funding, (10) supply chain, and (11) image watermarking. Recently, researchers have viewed the applications of blockchain in image watermarking, too (Figure 5.9). Image watermarking is applied for various security aspects like copyright protection, image tamper detection and correction, ownership claim, etc. Numerous research articles are available wherein the authors used blockchain in association with watermarking to strengthen the purpose of watermarking or to provide additional security to it. These researches are discussed below.

While doing watermarking, we make changes in the pixels, so distortion happens. Furthermore, some trusted third parties may be necessary to act as an arbitrator. To avoid these two issues, Wang *et al.* [13] suggested using blockchain with watermarking. As per blockchain technology, the third party can be de-trusted, and by using zero-watermarking, the distortion can be minimized. In zero-watermarking, the watermark is embedded not by changing the pixels, but by the pixel features. So, distortion is negligible. Lu *et al.* [14] also used a blockchain-based system and zero-watermarking for copyright registration, copyright query, and copyright trading. The trading comprises two phases: (i) ensuring the trading details, and (ii) approving it by the buyer and the seller. This strategy also permits the seller to ensure whether the buyer has the right to resell the copyright. As zero-watermarking is used, it is not prone to many image processing attacks.

Many media owners earn by sharing their media through an online trusted platform or by sharing individual copies. The online platform will take a good

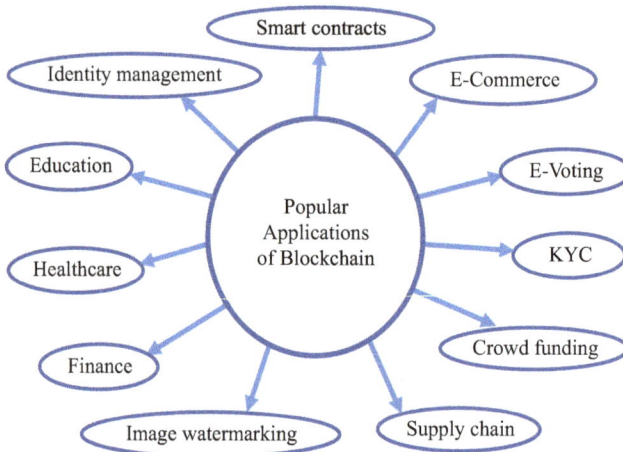

Figure 5.9   Applications of blockchain

percentage of the earnings as commission, and also lacks of transparency. Sharing individually to the buyer is not always possible. So, Xiao *et al.* [15] proposed a blockchain-based watermarking system to solve this problem. Blockchain can gather the media from multiple owners and can make a decentralized management that prevents illegal distribution and malicious users. Some earlier researchers have used symmetric digital watermarking in their blockchain, which cannot protect the malicious users. So, Xiao *et al.* recommended a design for media sharing by a blockchain network in which the asymmetric fingerprinting is with user-side embedding. There is a huge demand for buying and selling digital media due to the widespread use of social media. The traditional approaches used for buying and selling digital images may disclose some classified information about the users. Yu *et al.* [16] proposed a blockchain-based copyright protection scheme for secure trading of digital media, wherein they combined locality-sensitive hashing with symmetric encryption to retrieve the image from the blockchain securely. Also used digital fingerprinting to achieve copyright protection of images. The smart contract is used to achieve fairness in transaction processes.

The security and privacy of the photos in social media are protected by various cryptographic means. But when the image is spreading into other platforms, the copy control is lost. So, Zhang *et al.* [17] used a blockchain-based approach for photo sharing in social media. The centralized servers do not trust each other, so a consensus mechanism shall be consistently operated on photo dissemination through a smart contract. Their approach created a platform-independent dissemination tree for each photo, which provides dissemination control. This will avoid the conflict between the original owners and re-posters of the photos. There is no standard copyright protection mechanism for the cross-border distribution of images. So, Islam and In [18], the concept of global ledger (blockchain) for storing copyright data. using a consortium blockchain. This system defines a synchronized platform to register and trade copyright information without using a cloud platform. Different countries can be members and participate in block creation by executing the PoA consensus algorithm. The countries are nodes in the blockchain platform to validate transactions conducted by users. Only registered users can perform transactions, but anyone can investigate copyrighted information. A payment system is also attached for paying the copyright fee. This is a good idea for international copyright management, but the consensus protocol should be energy efficient, very important.

In the traditional design of copyright protection, we appoint an authority to impose a copyright penalty. But there is a chance of a single point of failure. The earlier blockchain-based approaches decide the copyright penalty after infringement is discovered. So, a blockchain-based approach is very useful, as proposed by Chen *et al.* [19], wherein they propose a proactive defense strategy that can stop infringement before its occurrence. In this strategy, copyright registration and copyright transfer are considered as transactions in blockchain, and are subject to prior infringement penalty. The infringement penalty is enforced by a smart contract. Li *et al.* [20] proposed a blockchain-based copyright protection mechanism with multiple parties on the blockchain. It involves the seller, the buyer, a cloud platform, and a blockchain. The man-in-the-middle attack is avoided between the buyer and seller by introducing certificateless secret key negotiation.

Liu *et al.* [21] proposed a new digital watermarking mechanism using inter-planetary file system (IPFS), and fast Walsh Hadamard transform (FWHT) for watermark embedding and extraction. The dependence on a third-party platform is avoided. Transactions among the parties are managed by a smart contract. IPFS stores the watermark data, and FWHT is used to embed the watermark into the host image. Results are good. Mannepalli *et al.* [22] also proposed a blockchain-based watermarking using discrete wavelet transform (DWT). Here, edge detection of DWT coefficients is used to achieve watermarking. Xu *et al.* [23] raised that the zero-watermarking schemes for remote sensing images rely on features, which are prone to attacks. So, they proposed a blockchain-based watermarking for copyright protection using a stacked denoising autoencoder (SDAE). SDAE is used to exploit deep and robust features, and Hyperledger fabric and IPFS is used for watermark registrations. Ferik *et al.* [24] proposed a blockchain-based watermarking system using DWT for protecting medical images. This approach used DWT to embed the compressed watermark into the image. The encrypted watermark is stored in the blockchain for an integrity check. This approach also achieves secure sharing among multiple parties. It achieves a peak signal-to-noise ratio of 63.24 dB and a structural similarity of 1.

While an image is in transit from sender to receiver, it is prone to tampering. Aberna and Agilandeeswari [25] proposed a blockchain-based watermarking for image tamper detection and correction by combining a PoW consensus blockchain scheme with a convolutional attention model system. This system uses a convolution attention model to create watermarks. A quaternion graph-based transform for embedding, ensuring imperceptibility and robustness. A fuzzy inference system optimizes embedding regions and factors based on human visual system characteristics. The security is ensured only when the embedded hash key is authentic with its previous block to proceed with the further extraction process. The PSNR value is 63.84 dB, and the SSIM value is 1.0.

These advantages of the blockchain system are: (i) real-time tracking, (ii) security, (iii) no single point of failure, (iv) transparency, (v) reduced cost, (vi) trusted transactions, (vii) no third-party involvement, and (viii) unalterable copies only [7,12]. The data or records in blockchain can be tracked in real-time, and the records cannot be altered, so they are secure. If a node fails, no problem, the remaining nodes can continue. The addition of new blocks is done by mutual consent, so transparency is improved. The cost of managing the data is also less, and trust is established among the users. In smart contract-like applications, there is no need for any third party in the business process.

5.6 Conclusion

The blockchain is a distributed ledger. Due to its immutable property, it solves various security issues in different applications. Only new blocks can be added to the existing blockchain if consensus is achieved among all the peer nodes. The existing blocks cannot be altered, but new blocks can be added to them. Unlike a centralized system, if a node fails, the blockchain will not fail. There are many applications of blockchain available today. The applications of blockchain in the field of image watermarking are in their infancy. It is still growing day by day. The very important applications of blockchain in image watermarking are copyright protection, ownership claim, and tamper detection. Unlike traditional watermarking systems, a blockchain-based watermarking system does not require a trusted third party. As the trust is established, the cost of data management is less.

References

[1] F. Chahlaoui and H. Dahmouni, A taxonomy of load balancing mechanisms in centralized and distributed SDN architectures, *SN Computer Science*, 1 (268), pp. 1–16, 2020. https://doi.org/10.1007/s42979-020-00288-8

[2] H.T.M. Gamage, H.D. Weerasinghe, and N.G.J. Dias, A survey on blockchain technology concepts, applications, and issues, *SN Computer Science*, 1 (114), pp. 1–15, 2020. https://doi.org/10.1007/s42979-020-00123-0

[3] P. Ciaian, M. Rajcaniova, and d. Kancs, The digital agenda of virtual currencies: can Bitcoin become a global currency? *Information Systems and e-Business Management*, 14, pp. 883–919, 2016. https://doi.org/10.1007/s10257-016-0304-0

[4] S. Haber and W.S. Stornetta, How to timestamp a digital document, *Journal of Cryptology*, 3(2), pp. 99–111, 1991. https://doi.org/10.1007/BF00196791

[5] V. Gatteschi, F. Lamberti, C. Demartini, C. Pranteda, and C. Santamaría, Blockchain and smart contracts for insurance: is the technology mature enough, *Future Internet*, 10(2), 20, pp. 1–16, 2018. https://doi.org/10.3390/fi10020020

[6] X. Li and W. Wu, Recent advances of blockchain and its applications, *Journal of Social Computing*, 3(4), pp. 363–394, 2022. https://doi.org/10.23919/JSC.2022.0016

[7] A.R. Khettry, K.R. Patil, and A.C. Basavaraju, A detailed review on blockchain and its applications, *SN Computer Science*, 2(30), pp. 1–9, 2021. https://doi.org/10.1007/s42979-020-00366-x

[8] I. Abrar and J.A. Sheikh, Current trends of blockchain technology: architecture, applications, challenges, and opportunities, *Discover Internet of Things*, 4(7), pp. 1–17, 2024. https://doi.org/10.1007/s43926-024-00058-5

[9] B. Shrimali and H.B. Patel, Blockchain state-of-the-art: architecture, use cases, consensus, challenges and opportunities, *Journal of King Saud University-Computer and Information Sciences*, 34, pp. 6793–6807, 2022. https://doi.org/10.1016/j.jksuci.2021.08.005

[10] J. Zarrin, H.W. Phang, L.B. Saheer, and B. Zarrin, Blockchain for decentralization of internet: prospects, trends, and challenges, *Cluster Computing*, 24, pp. 2841–2866, 2021, https://doi.org/10.1007/s10586-021-03301-8

[11]   S.K. Pani, R. Chatterjee, and N.R. Mahapatra, Towards trusted, transparent and motivational professional education system through blockchain, *International Journal of Information Systems and Social Change*, 10(2), pp. 62–73, 2019. https://doi.org/10.4018/IJISSC.2019040105

[12]   F. Casino, T.K. Dasaklis, and C. Patsakis, A systematic literature review of blockchain-based applications: current status, classification and open issues, *Telematics and Informatics*, 36, pp. 55–81, 2019. https://doi.org/10.1016/j.tele.2018.11.006

[13]   B. Wang, S. Jiawei, W. Wang, and P. Zhao, Image copyright protection based on blockchain and zero-watermark, *IEEE Transactions on Network Science and Engineering*, 9(4), pp. 2188–2199, 2022, https://doi.org/10.1109/TNSE.2022.3157867

[14]   Z. Lu, J. Wei, C. Li, J. Zhai, and D. Tong, Robust copyright tracing and trusted transactions using zero-watermarking and blockchain, *Multimedia Tools and Applications*, 84, pp. 10687–10724, 2025, https://doi.org/10.1007/s11042-024-19325-2

[15]   X. Xiao, Y. Zhang, Y. Zhu, P. Hu, and X. Cao, FingerChain: copyrighted multi-owner media sharing by introducing asymmetric fingerprinting into blockchain, *IEEE Transactions on Network and Service Management*, 20(3), pp. 2869–2885, 2023, https://doi.org/10.1109/TNSM.2023.3237685

[16]   F. Yu, J. Peng, X. Li, C. Li, and B. Qu, A copyright-preserving and fair image trading scheme based on blockchain, *Tsinghua Science and Technology*, 28(5), pp. 849–861, 2023, https://doi.org/10.26599/TST.2022.9010066

[17]   M. Zhang, Z. Sun, H. Li, B. Niu, F. Li, Z. Zhang, Y. Xie, and C. Zheng, Go-sharing: a blockchain-based privacy-preserving framework for cross-social network photo sharing, *IEEE Transactions on Dependable and Secure Computing*, 20(5), pp. 3572–3587, 2023, https://doi.org/10.1109/TDSC.2022.3208934

[18]   M.M. Islam and H.P. In, Decentralized global copyright system based on consortium blockchain with proof of authority, *IEEE Access*, 11, pp. 43101–43115, 2023, https://doi.org/10.1109/ACCESS.2023.3270627

[19]   X. Chen, A. Yang, J. Weng, Y. Tong, C. Huang, and T. Li, A blockchain-based copyright protection scheme with proactive defense, *IEEE Transactions on Services Computing*, 16(4), pp. 2316–2329, 2023, https://doi.org/10.1109/TSC.2023.3246476

[20]   M. Li, L. Zeng, L. Zhao, Y. Wang, and G. Liu, Multiparty watermarking protocol based on blockchain, *Multimedia Tools and Applications*, 83, pp. 367–379, 2024, https://doi.org/10.1007/s11042-023-15691-5

[21]   T. Liu, S.-N. Lai, X. Yuan, Y. Liu, and C.-T. Lam, A novel blockchain-watermarking mechanism utilizing interplanetary file system and fast Walsh Hadamard transform, *iScience*, 27(110821), pp. 1–25, 2024, http://creative-commons.org/licenses/by/4.0/

[22]   P.K. Mannepalli, V. Richhariya, S.K. Gupta, P.K. Shukla, P.K. Dutta, S. Chowdhury, and Y.-C. Hu, A robust blockchain-based watermarking using edge detection and wavelet transform, *Multimedia Tools and Applications*, 84, pp. 12739–12763, 2025, https://doi.org/10.1007/s11042-024-18907-4

[23]   D. Xu, N. Ren, and C. Zhu, High-resolution remote sensing image zero-watermarking algorithm based on blockchain and SDAE, *IEEE Journal of Selected Topics in Applied Earth Observations and Remote Sensing*, 17, pp. 323–339, 2024, https://doi.org/10.1109/JSTARS.2023.3329022

[24]   B. Ferik, L. Laimeche, A. Meraoumia, O. Aldabbas, M. Alshaikh, A. Laouid, and M. Hammoudeh, A multi-layered security framework for medical imaging: integrating compressed digital watermarking and blockchain, *IEEE Access*, 12, pp. 187604–187622, 2024, https://doi.org/10.1109/ACCESS.2024.3514668

[25]   P. Aberna and L. Agilandeeswari, PoWBWM: proof of work consensus cryptographic blockchain-based adaptive watermarking system for tamper detection applications, *Alexandria Engineering Journal*, 112, pp. 510–537, 2025, https://doi.org/10.1016/j.aej.2024.10.016

Chapter 6

Cloud-based analysis with quantum cryptography-based cloud security model (QC-CSM) for enhanced data security in storage and access

Amit Kumar Chandanan[1], Vivek Kumar Sarathe[2], Akhilesh Dwivedi[3], Raja Chandrasekaran[4], Vandana Roy[5] and Aditya Kumar Sahu[6]

Abstract

Data security challenges in cloud computing have grown as a fundamental issue because of rising cyber threats during this period. The proposed model QC-CSM relies on the quantum key distribution process (QKDP) and ABE to use quantum cryptography in developing a cloud security framework that enhances security measures. The model provides key distribution together with authentication and encryption security through the application of the quantum no-cloning theorem alongside quantum mechanics principles. The experimental results conducted through CloudSim and iQuantum generate data that confirms the model exhibits efficient performance across storage, encryption, and processing speed. The key generation process takes 35% less time than conventional multi-authority attribute-based encryption (MAABE) and pairing-based provable multi-copy data possession (PB-PMDP) encryption methods. Additionally, the encryption and decryption operations operate at optimized speeds. The reduction of storage overhead (SO) amounts to 20% which leads to improved memory performance. The model demonstrates effective detection abilities to identify quantum key distribution (QKD) network eavesdropping which prevents secure network breaches. The

[1]Department of Computer Science and Engineering, Guru Ghasidas Vishwavidyalaya (A Central University), India
[2]Department of Computer Science and Information Technology, Guru Ghasidas Vishwavidyalaya (A Central University), India
[3]Department of Electrical and Electronics, IES College of Technology, India
[4]Department of Electronics and Communication Engineering, Vel Tech Rangarajan Dr. Sagunthala R&D Institute of Science and Technology, India
[5]Department of Electronics Communication, Gyan Ganga Institute of Technology and Sciences, India
[6]Department of Computer Science and Engineering, SRM University-AP, Andhra Pradesh, India

experimental data demonstrate that QC-CSM delivers both a secure platform and a scalable and efficient cloud security infrastructure. The chapter shows that quantum cryptography enables cloud system protection when it incorporates QKD components with contemporary encryption technologies.

Keywords: Quantum Key Distribution Protocol; Quantum Cryptography; Cloud Computing; Quantum Key Distribution; Attribute-Based Encryption; Encryption; Decryption; CloudSim; iQuantum

6.1   Introduction

The storage management features of cloud computing provide businesses and individuals with flexible and economical solutions which changed how data maintained and accessed remotely. Remote data storage allows users to keep large amounts of information secure as they can access the computing resources through internet connections. The numerous benefits of cloud computing face major security problems that focus on protecting data confidentiality with added risks for maintaining data integrity and data availability [1]. The crucial security challenge arises from storing data in third-party cloud service provider (CSP) networks because companies need to protect their information from unauthorized access and breaches. Historically traditional cryptographic methods consisting of symmetric and asymmetric approaches provide widespread security to cloud data storage [2]. The efficient AES encryption system depends on secure key transfer techniques but these arrangements attract intercepting attacks during key exchange distribution. The cryptographic security provided by Rivest–Shamir–Adleman (RSA) together with elliptic curve cryptography (ECC) allows better protection yet generates substantial processing strains. Progress in cryptographic science does not safeguard encryption methods from cyber attackers who free themselves annually as quantum computing becomes a future threat as shown in Figure 6.1[3].

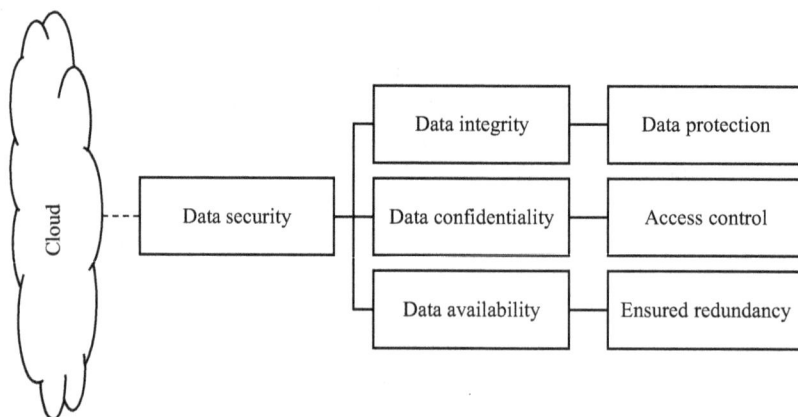

Figure 6.1   Important security considerations in the cloud architecture

The fast progression of quantum computing technology threatens modern encryption standards to become useless soon. The operation of quantum computers by utilizing superposition and entanglement provides efficient ways to break encryption schemes that are currently in wide use [4]. The behavior of Shor's algorithm proves that quantum computers achieve prime number factorization of large numbers at a speed that surpasses classical computers, making RSA encryption vulnerable. Grover's algorithm decreases brute-force attack complexity, which reduces the security strength of symmetric encryption [5]. The threat from imminent quantum computing technology requires organizations to create advanced security structures that defeat quantum-based assaults to preserve the enduring protection of cloud-stored information.

Quantum cryptography (Figure 6.2) established itself as a viable method that resolves the weaknesses in current encryption systems. The underlying principles of quantum mechanics define quantum cryptography instead of classical cryptography, which depends on mathematical complexity. The quantum key distribution (QKD) technology stands as a core advantage of quantum cryptography because it creates protected key exchanges for two communicating parties [6]. Any attempted eavesdropping of communication channels through QKD leaves detectable disturbances that stop unauthorized access. Quantum cryptography stands out because of its exclusive characteristics, which recommend it as the best option for cloud communication security [7]. The security of key distribution systems relies on two essential components, which together with encryption for data protection, maintain cloud security. In cloud environments, attribute-based encryption (ABE) serves as a popular encryption approach that develops sophisticated access control methods. Traditional encryption methods differ from ABE, which enables users to set encryption access rules by assigning user attributes that serve as decryption

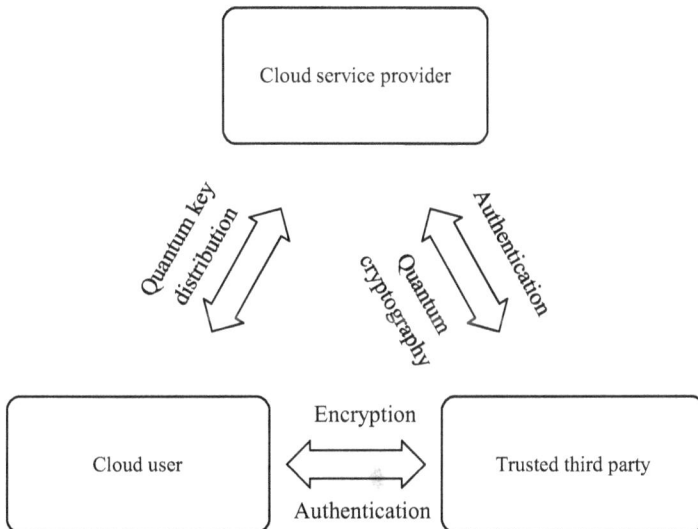

Figure 6.2   Quantum cryptography for securing information in the cloud

authorizations [8]. Cloud security receives additional strength from the ABE-quantum integration because this combination establishes resistant keys and encryption that withstand advancing cyber threats.

Cloud services expansion has made secure cloud security models with efficiency and scalability the main priorities for businesses worldwide. Security professionals have introduced multi-authority attribute-based encryption (MAABE) alongside pairing-based provable multi-copy data possession (PB-PMDP) to address particular cloud computing security issues [9]. The existing security models encounter obstacles while dealing with computational resource requirements, as well as storage requirements, and their ability to withstand quantum attacks. A reliable security system has to protect data storage and transmission, and must also enhance performance measurements involving encryption speed, decryption speed, key creation speed, and efficiency. Cloud security faces a critical challenge because attackers perform eavesdropping along with man-in-the-middle (MITM) attacks. Unauthorized attackers try to intercept channels for communication as a method to acquire sensitive data without proper authorization. Public key infrastructure (PKI) with digital certificates serves as a traditional security practice, yet advanced persistent threats (APTs) demonstrate their potential for failure during these instances [10]. The quantum no-cloning theorem in quantum cryptographic systems ensures perfect security against all forms of MITM attacks since it blocks attackers from conducting state duplications. Key exchange protection remains tamper-proof because of these methods, which lead to improved security for cloud-based systems.

More organizations need secure cloud environments because they adopt Internet of Things (IoT), 5G networks, and edge computing systems in their operations. Secure transmission and storage methods become necessary because IoT devices produce tremendous volumes of data. The computational constraints of IoT devices make traditional encryption methods hardly suitable for their operations [11]. Quantum cryptography functions as a security solution that creates lightweight yet powerful encrypted communication networks for IoT platforms operating within cloud environments. The present chapter investigates cloud computing security issues while establishing the need to develop quantum encryption systems because cloud security problems continue to escalate. This chapter completes the connection between quantum cryptography and cloud security by presenting the quantum cryptography-based cloud security model (QC-CSM). The proposed model leverages quantum key distribution protocol (QKDP) for secure key exchange and ABE for access control and data encryption [12]. The model combines integrated techniques for delivering optimum security that safeguards quantum data stored across clouds against attacker exploits from conventional and quantum cyber threats. Cloud security models need performance evaluation as an important aspect, which also serves security purposes. The proposed model undergoes evaluation through assessment of important performance criteria that include storage effectiveness and computational overhead, together with encryption and decryption speed and key generation time (KGT) [13]. The security and efficiency performance of different cloud computing scenarios becomes possible through analysis through simulation tools CloudSim and iQuantum. This chapter delivers findings about actual implementation possibilities and practicality of quantum cryptography methods in modern cloud platform deployments.

This work strives to promote cloud security development through the introduction of a new quantum-secure encryption framework model. The proposed model establishes cloud security for the future because it resolves traditional encryption problems through quantum-safe encryption measures. The chapter outcomes should support CSPs alongside enterprises and researchers in their development of protected and efficient cloud systems prepared to fight present-day and impending cyber threats.

6.2 Related work

Data security has experienced revolutionary changes because of cloud and quantum computing systems for efficient, secure data sharing. Data security depends on maintaining complete integrity and protecting both privacy and confidentiality, despite being a substantial challenge. Cloud security systems use encryption as their main element, together with precise access regulation and advanced key governance strategies. Advanced encryption standard (AES) and identity-based encryption (IBE) function as main encryption approaches to protect access to data, although their implementation suffers from performance issues, security concerns, and potential key management vulnerabilities [14]. The recent developments in searchable encryption have resulted in secure keyword search solutions that demand reduced storage space. The security issues, which include unauthorized access, data tampering, and key management problems in cloud computing, require further extensive research efforts.

The security approach for cloud-stored data gets transformed with quantum computing technology. QKD establishes extremely secure data protection through the application of entanglement and superposition principles. The enhancement of data security through quantum cryptography has included the development of various experiments, including measurement-device independent QKD (MDI–QKD) and quantum bit commitment protocols [15]. Quantum cryptography presents obstacles for industrial deployment since it requires solutions to problems associated with scalability difficulties, resource limitations, and technical implementation challenges. The literature review, as shown in Table 6.1, examines current models of cloud and quantum computing with secure data sharing capability, while it recognizes their weaknesses and points out research areas requiring improvement for future progress.

The evaluation of relevant studies demonstrates how researchers have achieved progress in cloud data protection using innovative cryptocurrency methods with quantum computing technology. Cloud security models CP-ABE and IBE ensure fine access control, but users encounter problems with excessive computation and storage issues, and the requirement for external authorities. Hybrid cryptographic models try to achieve both efficiency and security protection in the system, yet they create time-based and complexity-related challenges. The implementation of QKD

Table 6.1   The existing work done in the same field

Model	Methodology	Advantages	Research gaps identified
Ciphertext-policy attribute-based encryption (CP-ABE) [16]	Uses attribute-based policies for fine-grained access control	Provides flexible access control for multiple users	High computational cost, limited scalability, and complex key management
Identity-based encryption (IBE) [17]	Encrypts data based on user identity attributes such as email or date of birth	Eliminates the need for traditional public-key infrastructure	Trusted third-party security concerns, key escrow problem, and lack of user privacy
Hierarchical attribute-based encryption (HABE) [18]	Multi-level key generation for secure structured data access	Reduces storage complexity and improves hierarchical access control	Computational complexity increases with more users, and real-time processing difficulties
Certificate-less proxy re-encryption (CL-PRE)	Uses symmetric and asymmetric encryption for secure group data sharing [19]	Eliminates the need for a trusted certificate authority	Computational overhead due to bilinear pairings, risk of cloud service provider compromise
Hybrid cryptographic models (AES + RSA)	Combines symmetric (AES) and asymmetric (RSA) encryption for enhanced security [20]	Ensures confidentiality and authentication with strong encryption standards	Increased encryption/decryption time, storage overhead (SO), not suitable for large-scale data processing
Secure multi-owner data sharing model	Uses RSA–CRT (Chinese remainder theorem) and access control policies for data sharing [21]	Enables controlled multi-user access with revocation support	Inefficient for multimedia file security, complex key revocation process
Quantum key distribution (QKD)	Uses quantum principles such as entanglement and superposition for secure key exchange [22]	Provides theoretically unbreakable encryption due to quantum properties	High resource requirements, limited real-world deployment, and hardware limitations
Measurement-device independent QKD (MDI–QKD)	Secure QKD using decoy state methodology [23]	Eliminates measurement-based attacks, enhancing security	Implementation complexity in large-scale networks makes it difficult to integrate with existing infrastructures
Quantum secure direct communication (QSDC)	Uses hyper-entangled states for secure direct communication [24]	Reduces the risk of key interception	Susceptible to noise and decoherence in quantum channels, practical implementation challenges

(Continues)

Table 6.1 (*Continued*)

Model	Methodology	Advantages	Research gaps identified
Cloud security with blockchain	Decentralized ledger technology for data integrity and authentication [25]	Prevents unauthorized modifications and ensures transparency	High storage costs, latency issues in processing transactions, and scalability challenges
Provable data possession (PDP)	Ensures data integrity in cloud storage without retrieving full data	Reduces SO and enables third-party verification	Limited real-time auditing and key management complexities
Multi-authority attribute-based encryption (MA-ABE) [26]	Uses multiple trusted authorities to issue attribute-based encryption keys	Enhances trust and reduces single-point-of-failure issues	Not suitable for resource-constrained devices, increased computation due to bilinear pairing
Searchable encryption for secure data retrieval [27]	Enables encrypted search using keyword-based retrieval	Improves efficiency in retrieving encrypted data	Vulnerable to statistical attacks, high computational costs for frequent updates
Authenticated key exchange (AKE) [28]	Establishes secure session keys for communication between parties	Prevents replay and man-in-the-middle attacks	Session key compromise risks, high memory utilization
Quantum cryptography in cloud security [29]	Uses quantum-resistant algorithms for securing cloud-stored data	Provides future-proof security against quantum threats	Scalability issues, dependency on quantum hardware, compatibility with classical systems

and MDI–QKD quantum cryptographic methods faces difficulties for practical use because they need large resource availability, which leads to deployment barriers. Multiple studies prove that encryption models need development to establish effective security defense strategies. The development of better secure data-sharing systems for cloud and quantum environments requires attention to challenges that will lead to practical security solutions.

6.3 Objective of the chapter

The fundamental goal of this investigation involves the creation of the QC-CSM model that leverages QKDP and ABE to boost encryption security throughout cloud environments. The chapter aims to:

- Providing safe key distribution methods ensures complete protection against unauthorized access.

- The model should reduce both computational overhead and achieve maximum security performance.
- The storage efficiency increases through QKDP and ABE methods, which minimize memory consumption during encryption-related processes.
- The system detects attempts at eavesdropping through quantum principles for its eavesdropping mitigation process.
- The model performance requires assessment for its encryption abilities, decryption processes, and processing duration.

The chapter shows that quantum cryptography can build a secure cloud security framework that is both scalable and operationally efficient.

6.4   Motivation

The fast growth of cloud computing creates a fundamental difficulty in protecting data securely. Encryption methods become less secure because of rising cybercrimes and accelerating computing power developments. Quantum cryptography introduces QKDP as a protocol to achieve uncrackable encryption by using the QKD Protocol. The chapter pursues a development goal for a cloud security system that uses secure, powerful, and expandable security measures preventing unauthorized access and breaches alongside eavesdropping attempts. The chapter develops a secure cloud environment framework through quantum cryptography integration with ABE to improve security mechanisms for key distribution, along with authentication and encryption capabilities.

6.5   Proposed work

The QC-CSM represents a modern security architecture that defends cloud-stored data through integration between QKD and ABE approaches. The encryption methods currently in use depend on computational difficulty, but they become susceptible to quantum computing developments. The fundamental principles of quantum mechanics enable QC-CSM to distribute secure keys for cloud environments, which ensures confidentiality and integrity throughout the system. QKDP enables cloud users and CSPs to conduct secure encryption key exchanges through QKDP, which combines with ABE access controls to establish fine-grained protection over encrypted data, as shown in Figure 6.3. At the subatomic scale, quantum mechanics provides the rules of behavior and basic principles that quantum cryptography uses as its foundation. Key principles include:

- One main quantum principle states that quantum particles, including qubits, maintain simultaneous multiple states until detection occurs.
- Through entanglement, two quantum particles link their states together, which creates an immediate effect on one another even when they exist at vast distances from each other.
- A measurement of a quantum state necessarily changes it, thus making it possible to detect unauthorized information interceptors.

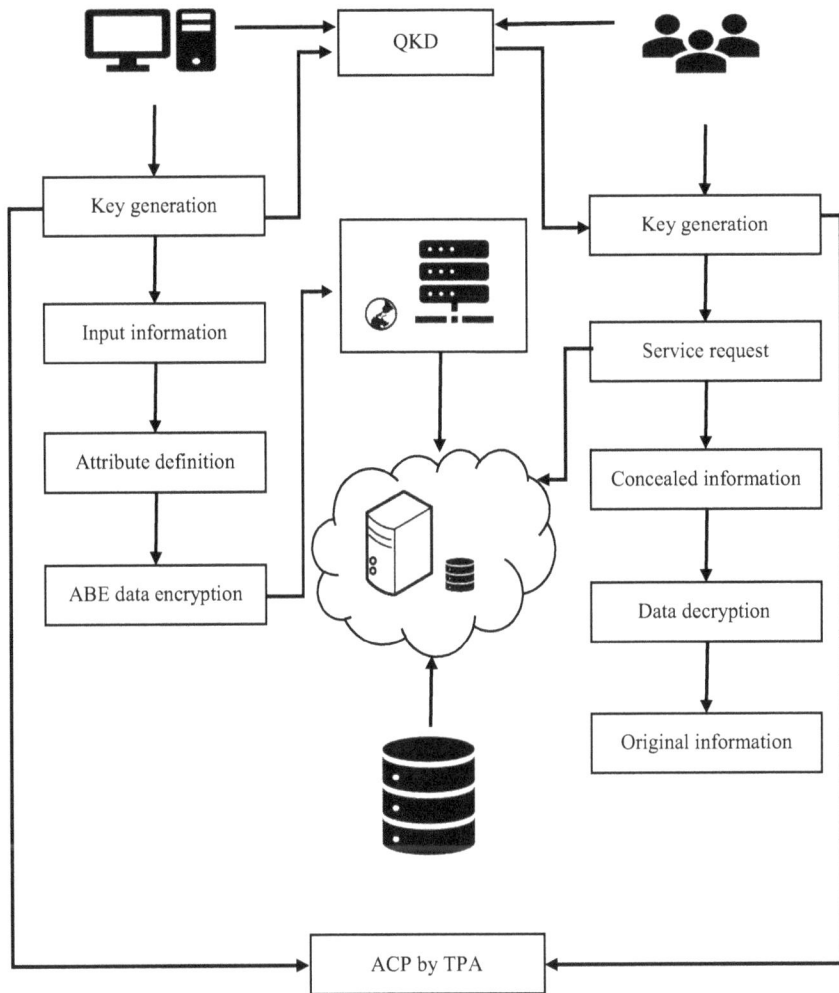

Figure 6.3   Operation process of the suggested model

6.5.1   Quantum key distribution (QKD)

Through QKD, service providers can securely establish encryption keys by using quantum mechanics principles to connect with their users. QKD establishes information-theoretic security because it depends on fundamental laws of physics rather than classical encryption, which depends on computational hardness assumptions for security. The primary mechanism of QKD depends on applying quantum superposition alongside quantum entanglement for cryptographic key distribution. BB84 represents the most recognizable QKD protocol that Bennett and Brassard created in 1984. The BB84 protocol uses Alice to transmit randomly polarized photons, which she encrypts by means of two polarization bases.

Rectilinear basis $(\oplus)$ | : 0⟩(horizontal)and | 1⟩(vertical)   (6.1)

Diagonal basis $(\otimes)$ | : +⟩(45°)and | −⟩(−45°)   (6.2)

After transmission, the receiver Bob carries out photon measurements by selecting bases at random. Alice and Bob reveal their basis selection publicly following transmission, but they abandon the results obtained through incompatible basis choice. Shared secret keys are formed by the remaining bits after the security check. The No-Cloning Theorem protects quantum states so that when an eavesdropper (Eve) listens in on the photons, she disrupts the states, which informs both Alice and Bob.

$$K = \{k_1, \ k_2, \ ..., \ k_n$$ (6.3)

The basis of QKD as a fundamental technology structure for future quantum communication derives from its ability to provide absolute forward secrecy while protecting against quantum computing breaches.

6.5.2   Quantum no-cloning theorem

According to the quantum no-cloning theorem, it is not feasible to make an exact clone of a quantum state that is unknown and arbitrary. The theorem may be represented mathematically in the following way: For each given quantum state $\psi\rangle$, there is no unitary function U that satisfies the following condition:

$$U(|\psi\rangle \otimes |0\rangle) = |\psi\rangle \otimes |\psi\rangle$$ (6.4)

The auxiliary blank quantum state 0⟩ serves as the representation in this context. The impossibility of cloning occurs because quantum states remain in superposition before measurement until the Heisenberg Uncertainty Principle requires an observation to collapse them. Quantum information security for QKD becomes vulnerable when cloning is feasible because eavesdroppers could intercept and duplicate quantum signals without detection. Quantum cryptography especially relies on this theorem to protect QKD protocols because it ensures eavesdroppers cannot duplicate quantum keys without creating detectable disturbances. Quantum information cloning attempts made by unauthorized parties always cause disturbances that identify the attack to genuine users. The no-cloning theorem maintains secure quantum communication by protecting it from adversaries, which makes it essential for present-day quantum security approaches.

6.5.3   Secure key distribution using QKD

The secure digital transfer through cloud platforms keeps all confidential information protected by ensuring its integrity and authentication during transfer between users and their cloud provider. The standard encryption approaches AES and RSA provide secure data transmission through encryption but quantum computing threats make their security vulnerable. QC-CSM uses QKD along with ABE through an integrated approach to provide enhanced cloud security. The first stage of QC-CSM involves ABE-based encryption that specifies access policies for user authorization. The security exchange of the encryption key happens through QKD protocols and detects all eavesdropping activities by using the No-Cloning Theorem. A cloud-based transmission process protects the data from MITM attacks while the information remains encrypted.

6.5.4 Attribute-based encryption (ABE) for secure data storage

The established encryption key through QKD enables ABE to perform data encryption within the cloud environment. Data access through ABE depends on preassigned user attributes for giving access. The encryption process adopts bilinear pairing methods as part of elliptic curve cryptosystems (Figure 6.4).

The system works with G_0, which represents a bilinear group that contains prime-number elements and employs $e: G_0 \times G_0 \rightarrow G_T$ as its bilinear map function. Security setup commences when security parameters are selected first:

$$PK = g^a, MK = (g, a) \tag{6.5}$$

Here, g = generator, a = random secret exponent.

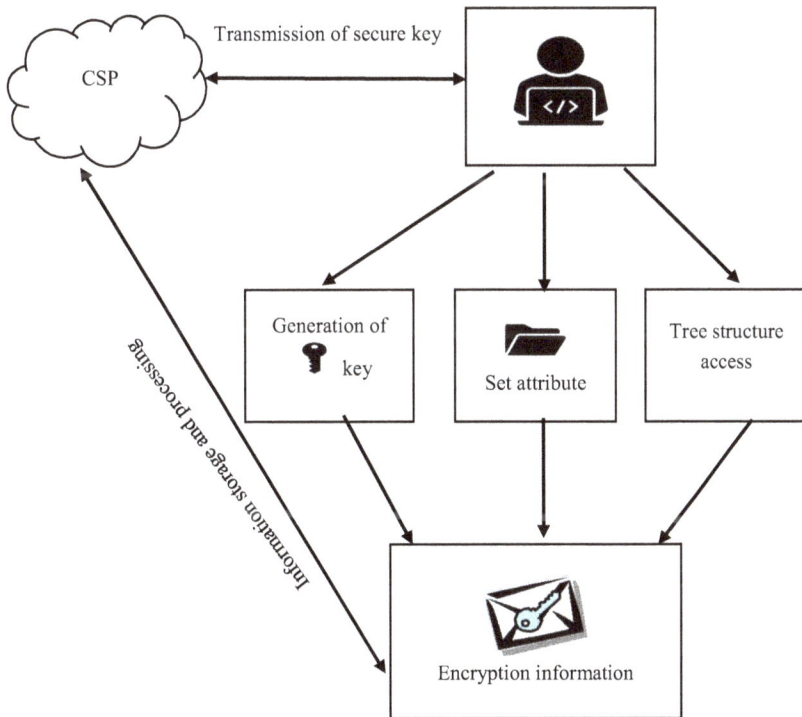

Figure 6.4   Generation of keys using the ABE procedure

The system assigns attributes $A = \{A1, A2, \ldots, A_n\}$ to each user for enforcing access policies through access trees (AT).

Encryption is performed as:

$$C = (D \cdot e(g,g)^{xyz}, T) \tag{6.6}$$

Here, D = information, xyz = random secrets, and T = encodes access policy.

If a user's qualities meet the requirements of the policy, they may decode the information. The key that unlocks the code is created in the following way:

$$SK = g^{(x+H(A))} \tag{6.7}$$

Here, $H(A)$ = hash of user attributes.

The decoding function recreates the following if the characteristics are in accordance with the accessibility regulations:

$$D = C/e(g^x, g^y) \tag{6.8}$$

This guarantees that only users with permission may view the information at hand.

6.5.5   Quantum authentication mechanism

A cryptographic system that uses quantum mechanics enables user identity verification while blocking unauthorized access in secure communication networks, operates under the name quantum authentication mechanism. User authentication methods based on passwords, together with digital signatures, remain exposed to quantum computer attacks because quantum technology enables the decryption of RSA and ECC encryption schemes. The security model of quantum authentication relies on quantum states for establishing unbreakable user identification. The implementation of quantum authentication can be established through quantum challenge-response methods. A quantum state denoted as ψ) originates from the verifier to the user with the role of prover during this process. The user carries out a predetermined quantum operation that allows the transformed state to be transferred to the verifier. According to the quantum no-cloning theorem and its impossibility to achieve perfect cloning, any adversary will make detectable errors during quantum state impersonation attempts.

A method of authentication known as entanglement-based uses pre-shared entangled pair (EPR) connections between the user and the verifier for identity confirmation. The authentication request's interception causes entanglement disruption, thus exposing successful intrusion attempts. QKD authentication through mathematical methods establishes secure handshakes to stop MITM attacks. The BB84 and E91 protocols act as secure mechanisms to exchange authentication keys, which protect encrypted cloud data by allowing only authorized users to access it. Security for the cloud faces significant enhancement through quantum authentication techniques because they stop identity theft while providing resistant authentication methods for secure access, which will secure next-generation quantum-secured networks.

6.5.6 Quantum certificates and digital signatures for integrity and non-repudiation

Quantum certificates, together with digital signatures, establish next-level integrity protection through quantum cryptography principles for authentication and integrity protection and non-repudiation in secure cloud exchanges. RSA, together with ECDSA, represent key asymmetric encryption methods used for digital signatures within classical cryptography, but they face vulnerability to quantum attack methods. Signature verification and certification obtain higher security through quantum cryptographic methods, which establish their certification using quantum states. The quantum certificate functions as a digital credential that connects cryptographic keys to identity using quantum elements. The Quantum certificate contains two parts: metadata and public key in classical format, while using quantum states for signature authentication. Quantum certificates implement QKD technology for protecting their signatures against tampering. The no-cloning theorem protects the certificate through its security mechanism, which prohibits illegal duplications.

The encoding process for quantum digital signatures (QDS) produces signature information that spreads throughout multiple verifiers by using GHZ entangled states. Signature verification requires state measurements that lead to result comparisons between multiple verifiers. QDS security stems from quantum entanglement combined with Bell's theorem, which detects any attempt at modification through alterations to the quantum state that reveal forgery. Mathematically, the process for quantum signatures operates through the following steps:

$$S(Q) = H(Q)^a \tag{6.9}$$

The recipient verifies the integrity using the public key:

$$H(Q) = e(g, g)^{H(m)} \tag{6.10}$$

Here, $H(Q)$ = quantum hash function, a = signer's private key.

Quantum certificates and digital signatures deliver both non-repudiation, which blocks senders from rejecting their message ownership, and integrity, which maintains data integrity. The mechanisms provide indispensable security measures that enable secure transactions during cloud operations and quantum-secured network development.

6.5.7 Secure key distribution using QKD

The encryption process in secure key distribution using QKD protects the safe transfer of keys between parties through quantum mechanics protocols. QKD implements security through both the Heisenberg Uncertainty principle and the no-cloning theorem, thereby making it possible to detect any attempted interceptions by eavesdroppers. QKD protocols use BB84 and E91 protocols for transmitting qubits through polarized photon transmissions. During QKD exchange, Alice selects her polarization basis randomly or to encode digital communication, which she sends to Bob. During the operation, Bob picks mea-surement bases at random. Alice and Bob follow basis comparison through public channels before keeping the bits successfully measured from each other to create a shared key. The quantum disturbances that occur during an eavesdropping attempt by Eve will reveal that someone has intercepted the transmission. The third key plays a role in protecting cloud data by providing unconditional security that stands against quantum attacks.

6.5.8   Quantum key distribution protocol process flow

The QKDP is an essential part of QC-CSM. The procedure is as follows:

- Alice, the information owner, generates qubits by employing random polarization.
- The CSP gets the qubits and then measures them using quantum detection.
- If the measures are the same, an identical key is created.
- The password for encryption is created and used to ensure that all commu-nication is safe.

Quantum encryption is used to assure the security of QKDP. The validity of the qubits that are transferred is confirmed by a breach of Bell's inequalities.

The likelihood that a listener has been detected in the QKD transmission is calculated as follows:

$$P(eavesdropper) = 1 - (1 - \epsilon)^n \tag{6.11}$$

Here ϵ = error rate, n = no of transmitted qubits.

QC-CSM offers a strong and safe architecture for cloud security that utilizes quantum cryptography. QC-CSM guarantees secure exchange of keys, role-based access control, and information integrity by combining QKD and ABE. The system of protection is made even more secure by the incorporation of Quantum Authentication and electronic Signatures, which make it immune to both conven-tional and quantum assaults.

6.6   Results and analysis

We conduct thorough computations and evaluations to assess the efficacy of the suggested QC-CSM. Effectiveness in generating quantum keys, effectiveness when encrypting and decrypting, memory expenses, and computational com-plexity are some of the important security elements that are evaluated in this section. To ensure a realistic assessment of the model's performance, it is eval-uated in different cloud computing environments utilizing iQuantum and CloudSim. The model's effectiveness in protecting information in the cloud from conventional and quantum attacks is evaluated by examining critical character-istics, including encryption time (ET), decryption time (DT), and key generation speed. Further, to emphasize the enhancements in efficiency and security, we compare it to current cryptographic simulations, such as MAABE and PB-PMDP. Safe, scalable, and quantum-resistant security measures for the cloud may be assured with the help of QC-CSM.

Accuracy (Acc) is an estimation of how well the security system classifies information exchanges as either safe or unsafe. The calculation is as follows:

$$Acc = \frac{CD + CR}{Total\ prediction} \tag{6.12}$$

Precision (Pre) measures how well the algorithm classifies interactions as secure. It figures out that the reported safe transactions were indeed safe.

$$Pre = \frac{CD}{CD + ID} \tag{6.13}$$

Sensitivity (Sen) quantifies how well the algorithm can identify all real safe transactions, making sure that no encrypted transfer is incorrectly labeled as unsafe.

$$Sen = \frac{CD}{CD + IR} \tag{6.14}$$

The model's ability to accurately identify insecure transactions is determined by specificity (*Spe*). It assesses the capacity to refuse transactions that are not trustworthy.

$$Spe = \frac{CR}{CR + ID} \tag{6.15}$$

The F1-score is the average of precision and sensitivity, which means it balances both erroneous successes and incorrect negatives. It is particularly helpful if there is a discrepancy between safe and insecure interactions.

$$FS = 2 * \frac{Pre * Sen}{Pre + Sen} \tag{6.16}$$

KGT: Calculates the amount of time it takes to create a safe quantum key employing QKD.

$$KGT = \frac{N_q}{R_q} \tag{6.17}$$

ET: The amount of time it takes to encrypt information utilizing ABE.

$$ET = C_E \times S_d \tag{6.18}$$

DT: The amount of time it takes to decrypt information is determined by user characteristics.

$$DT = C_D \times S_d \tag{6.19}$$

SO: Calculates the extra space in storage needed for information that is encrypted.

$$SO = S_E - S_O \tag{6.20}$$

Here, CD, correct detection; CR, correct rejection; ID, incorrect detection; IR, incorrect rejection; N_q, number of qubits; R_q, quantum key rate; C_E, encryption complexity; S_d, data size; C_D, decryption complexity; S_E, encrypted data size; and S_O, original data size.

The QC-CSM is superior to all other methods when it comes to every assessment measure, as shown in Table 6.2 and Figure 6.5. It has an accuracy of 97.2%, which guarantees that safe and insecure transactions are classified correctly. It has a precision of 95.8%, which means that it effectively reduces the number of false positives, and a recall of 96.5%, which means that it accurately identifies almost all

Table 6.2   Comparison of the performance of the existing approach with the suggested approach

Approach	Accuracy (%)	Precision (%)	Recall (%)	Specificity (%)	F1-score (%)
PB-PMDP	88.40	84.30	86.10	88.90	85.20
MAABE	89.90	86.70	88.30	90.50	87.50
PDP	86.60	83.20	84.50	86.80	83.80
AKE	88.30	85.90	87.10	89.40	86.40
QC-CSM (proposed)	97.20	95.80	96.50	98.10	96.10

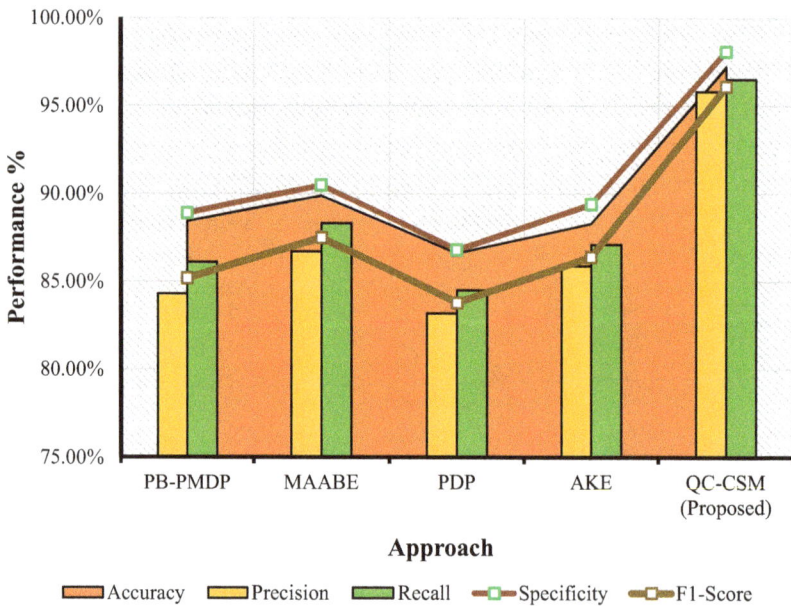

Figure 6.5   Illustration of the compared performance

safe transactions. Its specificity (98.1%) demonstrates that it is better at detecting insecure attempts, which decreases the number of breaches. The F1-score (96.1%) indicates that QC-CSM is its most efficient cloud security strategy since it is a regulated, exceptionally well-secured architecture.

The QC-CSM is more efficient than other methods in terms of processing time, as shown in Table 6.3 and Figure 6.6. QC-CSM consistently produces the lowest time across all file sizes, with 5.6 ms for 100 MB and 20.2 ms for 500MB, greatly surpassing PB-PMDP, MAABE, PDP, and AKE. As the size of the file rises, QC-CSM continues to operate at its best, which reduces the amount of computing overhead. The distribution of quantum keys and optimized encryption algorithms are responsible for this efficiency, which makes QC-CSM an extremely accessible and powerful cloud safety mechanism available.

The QC-CSM has the quickest encryption speeds for files of all sizes when compared to PB-PMDP, MAABE, PDP, and AKE, as shown in Table 6.4 and

Table 6.3 Comparison of key generation time of the existing approach with the suggested approach

File size (MB)	PB-PMDP	MAABE	PDP	AKE	QC-CSM (proposed)
100	10.5	12.8	10.2	11.1	5.6
200	12.3	21.1	10.8	11.5	6.4
300	17.5	26.1	15.2	16.1	12.6
400	22.9	28.3	21.5	22.7	18.8
500	26.1	31.0	23.1	24.5	20.2

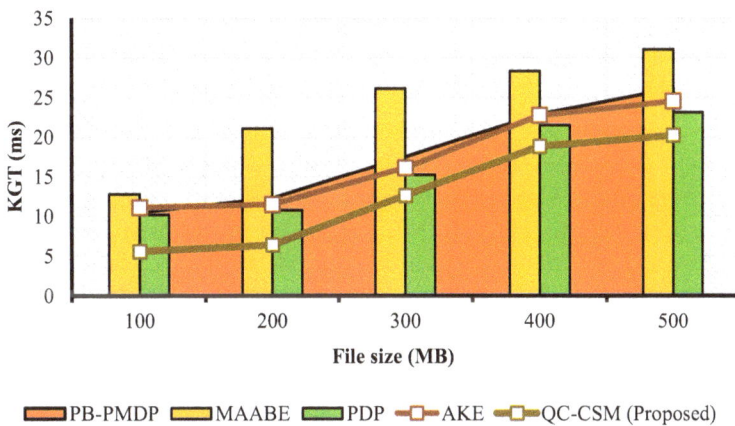

Figure 6.6 Illustration of the compared key generation time

Table 6.4   Comparison of encryption time of the existing approach with the
suggested approach

File size (MB)	PB-PMDP	MAABE	PDP	AKE	QC-CSM (proposed)
100	36.1	38.9	31.2	33	27.8
200	42.3	46.2	39.1	40.5	34.2
300	51	54.9	47.1	49.2	40.8
400	58.2	63.5	52.7	55.8	48
500	65.1	71	60.5	62.3	53.5

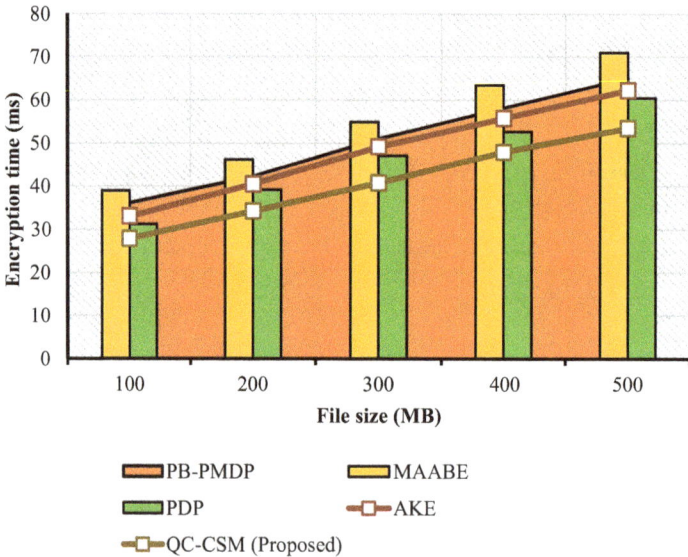

Figure 6.7   Illustration of the compared encryption time

Figure 6.7. QC-CSM encrypts 100 MB in 27.8 ms, whereas other programs take between 31.2 ms and 38.9 ms. QC-CSM is a highly efficient option as the file size rises, taking just 53.5 ms for a 500MB file, which is far faster than the other options. QC-CSM is a cloud solution that is both extremely scalable and safe because of its optimized QKD and ABE, which contribute to its effectiveness.

The QC-CSM has the quickest decoding speeds for files of all sizes, surpassing PB-PMDP, MAABE, PDP, and AKE, as shown in Table 6.5 and Figure 6.8. For a 100 MB, QC-CSM decrypts in 22.1 ms, substantially lower than others. As the file size increases, it remains efficient, taking 46.8 ms for 500 MB, whereas MAABE takes 60 ms. The improved efficiency is a result of QKD and optimized ABE, which provide quick, safe, and accessible decoding for applications in the cloud.

Table 6.5 Comparison of decryption time of the existing approach with the suggested approach

File size (MB)	PB-PMDP	MAABE	PDP	AKE	QC-CSM (proposed)
100	28.7	31.2	25.9	27.8	22.1
200	34.5	37.3	32.3	33.9	27.5
300	41.8	44.7	38.9	40.6	34.2
400	49.2	52.5	45.1	46.7	41.3
500	56.1	60	51.6	53.2	46.8

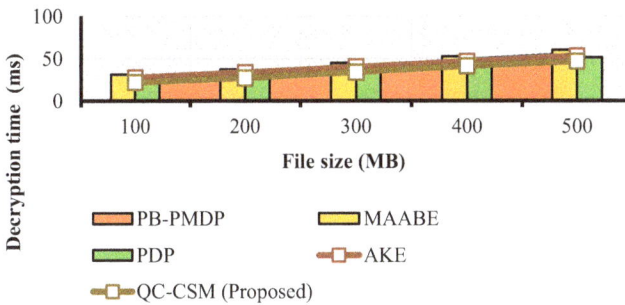

Figure 6.8 Illustration of the compared decryption time

Table 6.6 Comparison of storage overhead of the existing approach with the suggested approach

File size (MB)	PB-PMDP	MAABE	PDP	AKE	QC-CSM (proposed)
100	87.12	81.05	78.45	86.5	66.3
200	190.25	170.1	158.2	168.05	149.75
300	288.4	253.25	240.75	249	229.2
400	389.1	335	320.1	328	310.5
500	491	424	402.5	408.3	390

In comparison with the PB-PMDP, MAABE, PDP, and AKE, the QC-CSM greatly decreases the amount of storage space needed, as shown in Table 6.6 and Figure 6.9. QC-CSM only takes 66.3 MB for 100 MB, whereas other programs demand between 78.45 MB and 87.12 MB. QC-CSM keeps its overhead low as file size rises, utilizing 390 MB for 500 MB files, while competing programs use more than 400 MB. QKD and optimized encryption algorithms are responsible for this efficiency. They provide safe but storage-efficient computing in the cloud, providing QC-CSM the best option for large-scale safeguarding of information.

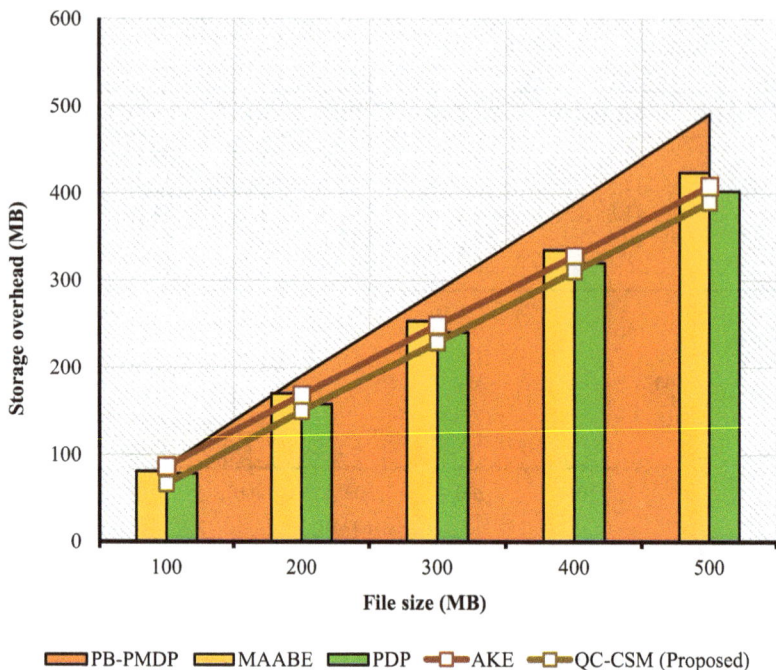

Figure 6.9   Illustration of the compared storage overhead

6.7   Conclusion

The Quantum Cryptography-Based Cloud Security Model (QC-CSM) is introduced in this chapter as a very effective and secure method for protecting data in the cloud. QC-CSM combines QKD with ABE to provide secured exchange of keys, authorization, and extremely fine control of access, which greatly improves the safety of the cloud. Performance assessments show that QC-CSM is superior to current cryptographic models, including PB-PMDP, MAABE, PDP, and AKE, when considering KGT, encryption and decryption speed, computing efficiency, and SO. QC-CSM has an accuracy rate of 97.2%, which shows that it is effective at properly identifying encrypted and unsafe activities. Furthermore, its decreased processing time and lower SO make it extremely scalable and resource-efficient in large-scale cloud systems. QC-CSM avoids surveillance, MITM assaults, and unauthorized access by using the fundamental concepts of quantum physics, such as the no-cloning principle and quantum entanglement. It is a future-proof option for cloud security since it can identify and reduce security risks in real time. In general, this chapter provides a safety net for cloud computing that is quantum-secure, scalable, and powerful. Future chapters may investigate the integration of post-quantum cryptographic algorithms and hybrid quantum-classical encryption to further improve the flexibility of QC-CSM in developing cloud systems.

References

[1] S. Ullah, J. Zheng, N. Din, M. T. Hussain, F. Ullah, and M. Yousaf, "Elliptic curve cryptography; applications, challenges, recent advances, and future trends: A comprehensive survey," *Comput. Sci. Rev.*, vol. 47, 2023, Art. no. 100530.

[2] B. White, D. Andre, G. Arquero, R. Bajaj, J. Cronin, A. Dames, H. Lyksborg, A. Miranda, and M. Weiss. (2022). *Transitioning To Quantum-Safe Cryptography on IBM Z. IBM.*

[3] H. Yi. "A post-quantum secure communication system for cloud manufacturing safety," *J. Intell. Manuf.* vol. 32, no. 3, 2021, pp. 679–688.

[4] I. Negabi, S. Ait El Asri, S. El Adib, and N. Raissouni, "Convolutional neural network based key generation for security of data through encryption with advanced encryption standard," *Int. J. Electr. Comput. Eng. (IJECE)*, vol. 13, no. 3, 2023, p. 2589.

[5] P. Mishra, T. K. Pandey, V. Roy, R. Kashyap, Chhaya, and R. Ahluwalia, "Advancements in language processing algorithms transforming linguistic computing," *2024 International Conference on Advances in Computing Research on Science Engineering and Technology (ACROSET)*, Indore, India, 2024, pp. 1–5.

[6] P. Catania and M. Cartsidimas, "APRA's prudential standard CPS 234 on information security: What does the way forward hold?" *Privacy Law Bull.*, vol. 17, nos. 1–2, 2020, pp. 6–8.

[7] L. Sim, S. Ren, S. Keoh, and K. Aung, A cloud authentication protocol using one-time pad, in *IEEE Region 10 Conference (TENCON)*, Singapore, 2016, pp. 2513–2516.

[8] V. A. Thakor, M. A. Razzaque, A. D. Darji, and A. R. Patel, "A novel 5-bit S-box design for lightweight cryptography algorithms," *J. Inf. Secur. Appl.*, vol. 73, 2023, Art. no. 103444.

[9] F. I. Lessambo, "The Australian prudential regulation authority," in *Fintech Regulation and Supervision Challenges Within the Banking Industry: A Comparative Study Within G-20*. Cham, Switzerland: Palgrave Macmillan, 2023, pp. 187–198.

[10] D. Zhu, X. Li, and J. Wu, "A quantum key-based mobile security payment scheme," *Int. J. Perform. Eng.* vol. 15, no. 8, 2019, pp. 21–65.

[11] V. Roy, S. V. Kumar, V. H. Raj, S. Lakhanpal, D. K. Yadav and R. A. Alzuhairi, "Benign Non-Convex Optimization Techniques for Training Neuro-Inspired Architectures," *OPJU International Technology Conference (OTCON) on Smart Computing for Innovation and Advancement in Industry 4.0*, Raigarh, India, 2024, pp. 1–6, doi: 10.1109/OTCON60325.2024.10687503

[12] N. Lata and R. Kumar, "DSIT: A dynamic lightweight cryptography algorithm for securing image in IoT communication," *Int. J. Image Graph.*, vol. 23, no. 4, 2023, *Art.* no.2350035.

[13] N. Bindel, U. Herath, M. McKague, and D. Stebila, "Transitioning to a quantum-resistant public key infrastructure," in *Post-Quantum*

Cryptography, 8th International Workshop, PQCrypto, Utrecht, The Netherlands: Springer, 2017, pp. 384–405.

[14]   G. Sharma and S. Kalra, "A novel scheme for data security in cloud computing using quantum cryptography," in *Proceedings of the International Conference on Advances in Information Communication Technology & Computing*, Bikaner, India, 2016.

[15]   D. Rosiyadi, A. I. Basuki, T. I. Ramdhani, H. Susanto, and Y. H. Siregar, "Approximation-based homomorphic encryption for secure and efficient blockchain-driven watermarking service," *Int. J. Electr. Comput. Eng. (IJECE)*, vol. 13, no. 4, 2023, p. 4388.

[16]   W. Barker, M. Souppaya, and W. Newhouse, "Migration to postquantum cryptography," *NIST National Institute of Standards and Technology. National Cybersecurity, Center Excellence*, Gaithersburg, MD, USA, Tech. Rep., 2021, pp. 1–15. Accessed: February 12, 2024. https://csrc.nist.gov/pubs/pd/2021/08/04/migration-to-postquantum-cryptography/final

[17]   P. A. Patil, S. Patil, and S. Joshi, "Hidden CP-ABE to enhance patient data privacy in smart healthcare systems," *Int. J. Appl. Eng. Res.*, vol. 12, no.13, 2017, pp. 3950–3960.

[18]   V. Roy. "An improved image encryption consuming fusion transmutation and edge operator." *J. Cybersec. Inform. Manage.*, vol. 8, no. 1, pp. 42–52, 2021.

[19]   K. Rama Devi and E. Bhuvaneswari, "An enhancement in data security using trellis algorithm with DNA sequences in symmetric DNA cryptography," *Wireless Pers. Commun.*, vol. 129, no. 1, 2023, pp. 387–398.

[20]   J. Suo, L. Wang, S. Yang, W. Zheng, and J. Zhang, "Quantum algorithms for typical hard problems: A perspective of cryptanalysis," *Quantum Inf. Process.*, vol. 19, no. 6, pp. 1–26, 2020.

[21]   T.V.X. Phuong, G. Yang, and W. Susilo, "Hidden cipher-text policy attribute-based encryption under standard assumptions," *IEEE Trans. Inf. Forensics Secur.*, vol. 11, no. 1, pp. 35–45, 2016.

[22]   N. N. Anandakumar, M. S. Hashmi, and S. K. Sanadhya, "Field programmable gate array based elliptic curve Menezes-Qu-Vanstone key agreement protocol realization using physical unclonable function and true random number generator primitives," *IET Circuits, Devices Syst.*, vol. 16, no. 5, pp. 382–398, 2022.

[23]   S. S. Gupta, T. K. Pandey, V. P. Raju, R. Shrivastava, R. Pandey, A. Nigam, and V. Roy, "Diabetes estimation through data mining using optimization, clustering, and secure cloud storage strategies," *SN Comput. Sci.*, vol. 5, pp. 781, 2024. doi:10.1007/s42979-024-03158-9

[24]   K. F. Hasan, A. Overall, K. Ansari, G. Ramachandran, and R. Jurdak, "Security, privacy, and trust of emerging intelligent transportation: cognitive internet of vehicles," in *Next-Generation Enterprise Security and Governance*, Boca Raton, FL: CRC Press, 2022, pp. 193–226.

[25]   Z. Sakhi, R. Kabil, A. Tragha, and M. Bennai, "Quantum cryptography based on Grover's algorithm," *IEEE, Second International Conference,*

18–20 September 2012 Mitch Leslio "Quantum Cryptography via satellite", 2017.

[26] W. T. Meshach, S. Hemajothi, and E. A. M. Anita, "Retraction note to: real-time facial expression recognition for affect identification using multi-dimensional SVM," *J. Ambient Intell. Human. Comput.*, vol. 14, 2023, p. 203. doi:10.1007/s12652-022-04015-4

[27] R. Maruthamuthu, D. Dhabliya, G. K. Priyadarshini, A. H. R. Abbas, A. Barno, and V. V. Kumar, "Advancements in compiler design and optimization techniques," *E3S Web of Conferences*, vol. 399, 2023, p. 8. doi:10.1051/e3sconf/202339904047

[28] M. Victor, D. D. W. Praveenraj, R. Sasirekha, A. Alkhayyat, and A. Shakhzoda, "Cryptography: advances in secure communication and data protection," *E3S Web of Conferences*, vol. 399, 2023, p. 9. doi:10.1051/e3sconf/202339907010

[29] M. Tamilselvi, M. K. Parameshwari, X. M. Raajini, G. Chamundeeswari, K. Kalaiselvi, and V. S. Pandi, "IoT based smart robotic design for identifying human presence in disaster environments using intelligent sensors," *2024 International Conference on Automation and Computation, AUTOCOM*, 2024, pp. 399–403. doi: 10.1109/AUTOCOM60220.2024.10486106

Chapter 7

An improved deep reinforcement learning approach for security-aware virtual network embedding algorithm

G. Yogarajan[1] and G. Rajasekaran[1]

Abstract

In network virtualization, which includes mapping virtual network (VN) requests onto real network infrastructure, virtual network embedding (VNE) is a significant issue. VNE plays a vital role in providing on-demand network services, enabling the efficient sharing of network resources among multiple tenants. However, mapping VNs onto physical infrastructure is a complex task that requires solving various optimization problems, such as routing, capacity allocation, and resource allocation. This chapter proposes a novel approach for security-aware VNE using policy network-based deep reinforcement learning (DRL) and graph convolutional neural network (GCNN) techniques. The motivation behind this work is the increasing demand for secure and reliable network services, which require effective mechanisms for detecting and mitigating security threats. The proposed approach employs RL to train the optimal VNE policy that increases the overall network security while satisfying various resource constraints. The performance of the RL algorithm is improved by using GCNN to derive relevant feature matrices from the network topology. Comparing the suggested methodology to existing approaches, the experimental findings show that it is more successful in attaining high levels of network security and resource utilization. Overall, this work highlights the potential of combining DRL and GCNN techniques for addressing security challenges in VNE.

Keywords: Virtual network embedding (VNE); policy network; deep reinforcement learning (DRL); graph convolution neural network (GCNN); feature matrix

7.1 Introduction

In recent years, the virtual network embedding (VNE) issue has been a difficult optimization problem that entails mapping virtual networks (VNs) onto a physical

[1]Department of Information Technology, Mepco Schlenk Engineering College, India

network infrastructure while satisfying the VNs' resource needs and minimizing the actual network infrastructure's resource utilization [1]. There has been a lot of interest recently in employing machine learning (ML) approaches to tackle the VNE problem, such as reinforcement learning (RL) and its upgraded variant, deep reinforcement learning (DRL), which employs neural networks to perform better. DRL is a subset of ML that teaches an agent how to make choices in a world where there are rewards and punishments. The agent attempts to increase the cumulative benefits over time while learning by making mistakes.

RL-based VNE algorithms can adapt to changing network conditions and provide better performance than traditional heuristic-based or optimization-based algorithms [2]. A survey of VNE algorithms and identifies the main challenges facing VNE, such as the scalability of the algorithms and the need for handling dynamic network conditions, has been presented in Xu *et al.* [3]. The report also makes suggestions for future research areas, such as the application of ML strategies to improve the efficiency of VNE algorithms. Virtual connections connect the various virtual nodes (such as a remote server and router) that make up the VN. The difficulty with VNE is mapping the simulated network to the underlying substrate connection and providing adequate bandwidth and processing capacity for dealing with requests [4,5].

It is unavoidable that NV increases network architectural flexibility while also introducing certain new security issues. Some virtual network requests (VNRs) in the virtual environment demand great security. For instance, recent times have seen a rise in the popularity of online shopping and payments, which are both strongly tied to money. Online video chat and other VNRs have relatively lax security requirements [6]. It is simple to ignore security concerns when the network is "busy" since many terminal devices must seek network resources. Currently, the network could come under attack from malicious software or suffer a data leak. Consequently, the security issue in network virtualization mapping must be taken into account [7,8]. To combat these challenges, metaheuristics – in some cases, evolutionary algorithms – have been introduced [9]. Although effective, these methods can be slow because of convergence problems. However, you can use these techniques to solve issues without first learning how to do it, so you cannot gain from prior knowledge. DRL-based methods have recently been applied [10].

Every technique has some effectiveness. The usage of neural networks or manual feature extraction, which are ineffective for dealing with data having a graph structure, is still sporadic in these methods. They also include some presumptions about how the VNs and the substrate are structured. These techniques are also exceedingly sluggish, which would preclude their application in real networks. RL and DRL are able to address the issue of VNE since they are superior examples of ML. In this frame, we present an end-to-end method for training an agent to tackle the VNE issue based on DRL and graph convolutional neural networks (GCNNs) [11]. Sensibly retrieved attributes from the substrate networks are then utilized to train and verify the RL agent. We took the security requirements of VNRs into consideration while setting the level of security attribute for substrate nodes and the safety requirement level parameter for virtual nodes. Last but not least, the experimental results show that our technique has achieved successful results. We are aware of no work that attempts to solve the VNE issue by fusing the security of VNE with an advanced RL system.

The following is a summary of the developments made by this chapter:

1. This research proposes a security-aware RL-based VNE method. The fundamental VN (the node's computational capacity restriction and connection bandwidth limitation) must be incorporated in order for the necessary security level restriction to be bound to each virtual node. The security level imposes restrictions on each substrate node. Virtual nodes can only be included in substrate nodes that do not fall below the security criterion level. This can ensure the safe functioning of the VNE algorithm.

2. We include GCN as part of the VNE issue. Through the convolution kernel, the multiple-layer GCN may obtain large-order spatial arrangement data from substrate nodes for end-to-end performance. Here, we present the policy network gradient approach to develop the policy network's parameters, which significantly enhances training efficiency.

3. We assess our algorithm's average revenue, revenue consumption ratio, acceptance rate, and loss against those of other representative algorithms. The experimental findings demonstrate that the RL-based algorithm outperforms other proposed algorithms. The security criterion level has some application to the VNE issue as well; therefore, it has some physical and logical use.

The rest of the chapter is organized as follows: Section 7.2 examines the work associated with the VNE algorithms. Section 7.3 discusses the network model and evaluation metrics. The advanced RL method based on the policy agent network is thoroughly introduced in Section 7.4. The parameter settings and analysis of the simulation experiments' findings are covered in Section 7.5. Finally, section 7.6 concludes the chapter.

7.2   Backgrounds

The associated research on VNE, security-relevant VNE, and graph convolutional networks (GCNs) will be introduced in this section.

7.2.1   Virtual network embedding (VNE)

There is a substantial and expanding corpus of research on the VNE issue. The embedding of links and nodes are two steps that are often included in the implementation of VNE. In other writings, these two phases are carried out independently [12,13]. A multiple path-based VN approach was proposed by Nguyen and Huang [11]. It made it possible for the supporting network to dynamically split and shift the virtual connection using multiple paths. The results of the trial showed a considerable rise in the bandwidth's resource usage. A nearest node-based VNE technique was further put out by Haeri and Trajkovic [12] to solve the issue of scarce resources in the VNE procedure. This method finds the shortest pathways for virtual linkages that satisfy its specifications and positions virtual vertices as near to the substrate as feasible. Nguyen and Huang [11] employ the Markov Random Walk framework to rank nodes according to their topographical and resource properties. The author then provided two ways to determine the connection priority based on the order of importance of the virtual node. Based on node grade and clustering frequency data, Kipf and Welling used node important metrics in [13] to determine the most important virtual servers in each VN demand by base nodes. The execution of these two steps is combined in several articles [14–17]. Yu *et al.* represented the VNE issue as a mixed-integer algorithm by extending the substrate structure in [14]. The author also suggested stochastic and unpredictable rounding techniques for network dynamic embedding. Razzaq and Rathore in [15] introduced a subgraph isomorphic behavior identification method that can complete nodes and links encapsulation in the same phase. The simulation results show that this strategy is faster than two-stage alternatives.

The heuristic technique, however, has a problem with computing complexity as the network scale grows. Numerous ML-based studies on VNE architecture have recently been conducted as a result of the development of artificial intelligence. Shanbhag *et al.* used the advanced neural network in DRL to tackle the VN problem in [18]. All these experimental results showed how well the neural network-based solution performed. The Monte Carlo tree was employed by Bruna *et al.* in [19] to find the best VNE solutions. Additionally, in [20,21], DRL algorithms were added to VNE.

7.2.2 Security-based virtual network embedding algorithms

The safety issues with VNs have recently been explored in certain research studies. A security-based VNE technique based on multiple attribute assessment and path optimization was suggested by Lu *et al.* [22]. He created a multiple-objective mixed-integer linear programming algorithm to represent this mapping process of the security-based VN, and he finished the embedding of the VN by creating a node map function and link map functions. Whatever, this algorithm only gives the safety level condition to the whole VNR, which is not done very carefully and with great attention to detail. and does not fully take into account the safety performance of every VN and physical network. A method for trust-aware security VNE was proposed by Li *et al.* [23]. They added the concepts of trust connection and trust degree to the allocation of VN resources and conducted a quantitative analysis of the security issues in the NV environment. The VNE regulations are created manually by the algorithm, which is inconsistent with reality.

Along with the previous two algorithms, the security of VNE is also taken into consideration by the RL-based security-aware VNE algorithm. The primary remedy is to give the substrate a net and the virtual net security properties. Instead of setting the security characteristics for the whole VNR, as is the case with the aforementioned algorithms, which specify the safety requirement condition for each virtual node and the safety level criteria for every substrate node. The technique described above also employs a heuristic approach to the VN problem. To address all the practical issues, we employ the RL method.

7.2.3   Machine learning-based virtual network embedding algorithms

The aforementioned heuristic approaches to the VNE problem can't entirely capture the network's actual state. The majority of them rely on fictitious rules and are unable to automatically optimize network parameters, which might result in local optimization of embedding outcomes. Many academics have currently used ML algorithms to address the VN problem and suggested a Monte Carlo search algorithm in the source [24]. The node map function procedure is seen by the algorithm as a Markov decision process (MDP). This node is connected using the Monte Carlo search tree after the VNR has arrived. The link map function is then followed using the multiple commodity flow algorithm or any shortest path technique.

The neural network approach was applied to the VNE issue in reference [25]. To increase the VN's mapping effectiveness, this strategy suggests an autonomous, artificial neural network-based system. The optimum reward mechanism learning mapping method employing the Q-learning algorithm is used to address the VN problem [26]. According to reference [27], the optimum mapping mechanism may be gradually learned by the RL agent by including the policy gradient algorithm in the RL algorithm.

This approach primarily uses the node embedding stage of the VNE domain while using the policy gradient technique. This model investigates how to balance the development of current models with the search for better solutions. Our method also uses a network of policy nodes as a learning agent to determine the likelihood of each substrate node. It should be highlighted that our work differs from the studies mentioned above. The security characteristic and the intelligent learning algorithm are being used for the first time to explore the VN problem. In order to train learning agents, we first extract five crucial node features, involving the essential safety attributes. Most importantly, when all the node attributes are extracted sensibly, the algorithm's overall performance is maintained at a high level. As most feature matrix extraction permits the policy agent to comprehend additional information regarding the substrate network, the VNE approach becomes more and more beneficial. Furthermore, the aforementioned algorithms do not consider how security factors may affect the VNE algorithm. Our approach gives nodes security properties, which is a great way to satisfy VNE's security criteria.

7.2.4   Graph convolutional network (GCN)

Recently, graph neural networks (GNN) have attracted greater attention. GNN adapts the deep neural network to data that has a graph structure in order to extract non-Euclidean knowledge from that data. The two basic types of spatial feature retrieval approaches are spatial domain methodology and spectral domain method. Using the spatial domain technique, topological graphs' spatial characteristics are intuitively extracted. The innovative Deep Walk algorithm is put out by Yao *et al.* in [28]. DeepWalk learned a representation of structural regularities using local information from the random walk method as its inputs. Zhao and Parhami presented a unique graph convolutional technique dubbed line in [29], where the empirical probabilities of nodes and their first-order similarity are determined separately. The results of the trial demonstrated that the LINE worked well in all types of information networks. Perozzi *et al.* suggested node-to-vector for processing graph data and approaching learning the continuous presentation of features matrix extraction of nodes [30]. The Struc2Vec approach was first out by Dolati *et al.* in [25] and it successfully recovered the prospective model of the node graphical identity. The Graph SAGE induction approach was established by William *et al.* and successfully exploits other nodes to create new node vectors that are embedded.

As a result, the spectral domain must be used to accomplish the process of convolution over the graph using the help of spectrum theory. In order to extend the convolutional neural network (CNN) to broader domains, Hammond *et al.* suggested a hierarchical segmentation based on the range of the graph Laplacian techniques [31]. A first-generation GCN formula was provided by the author after parameterizing the graph convolution kernel in light of this. In [32], Cheng *et al.* presented a unique approach for applying wavelet transformations based on Chebyshev polynomial approximation to any finite weighted big network. The convolution kernel was computed using parameter hypotheses and re-regularization approaches, in accordance with Ghazar and Samaan's partially supervised learning methodology on graphically structured data presented in [33].

In GCN–VNE, the embedding of simulated nodes onto actual nodes is the outcome of the VN's embedding challenge, which is phrased as a graph classification issue, which involves the input graph representing the request for the VN to be embedded. The VN request's graph framework is encoded using the GCN, and the embedding procedure makes use of both graph concentration and RL methods.

7.3   Network models and evaluation indicators

In this part below, the system concept and formulation of problems are first introduced. The performance of network embedding is then assessed using three assessment metrics.

7.3.1   *System model*

Modeling the substrate network using an undirected weighted graph, $G^S = \{N^S, L^S\}$, is possible. Here, L^S denotes the collection of all substrate connections, and N^S, the collection of all substrate nodes. Computing power CPU (n^s) and security SEC (n^s) level features of the substrate node $n^s \in N^S$. One significant example of substrate network security is the security constraint level of each substrate node. The evidence maps to the substrate node in a more secure manner and is less susceptible to security problems, the higher the level. Broadband capability BW (l^s), which describes the characteristics of the substrate connection $l^s \in L^S$.

To simulate the VN, the same undirected weighted graph is denoted as $G^V = \{N^V, L^V\}$. And N^V stands for the collection of all virtual nodes, while L^V stands for the collection of all virtual connections. The properties of the virtual node $n^v \in N^V$

are estimated resource restriction CPU (n^v) and security requirement level SEC (n^v). The security need for a virtual node is represented by its degree of security. Virtual nodes can able to map with the substrate node that meets the security criteria in order to maintain the safety of VNRs. The characteristics of the virtual connection $l^v \in L^V$ is represented by the bandwidth resource demand BW (l^v).

Hence, the VN challenges can be calculated by $E: G^V(N^V, L^V) \rightarrow G^S (N, L)$, thus $N \in N^S$ and $L \in L^S$. The node embedding process will be designed by $A = \{a_{ij} \mid n_i \in N^V, n_j \in N^S\}$.

when

$$\sum_{j=1}^{|NS|} a_{ij} = 1. \tag{7.1}$$

It shows that the substrate node n^j and node n^i were correctly integrated. Similar to that, the embedded link may be identified by $B = \{b_{ij} \mid l_i \in L^V, l_j \in L^S\}$. When

$$\sum_{j=1}^{|LS|} b_{ij} \geq 1. \tag{7.2}$$

It shows that this link l_i is correctly integrated into one or more physically mapped links.

In a substrate networks, two VNs are shown in Figure 7.1. The first value in brackets next to the node for two VNs denotes the node's estimated resource needs. The node's level of required security is indicated by the second number. The quantity on the virtually linked link denotes the needed bandwidth for this links. The first parameter in brackets after to a node in this substrate network denotes node's capacity for computation. The security measurement of each node is indicated by the second parameter. The figure on the substrate links denotes the bandwidth allocations that the links are capable of providing. Two VNRs were successfully embedded onto the base networks, as denoted in Figure 7.1. Specific nodes' correspondence connection is $a \rightarrow E$ and $b \rightarrow D$ Table 7.1 provides the summary of various symbols used in the proposed system model.

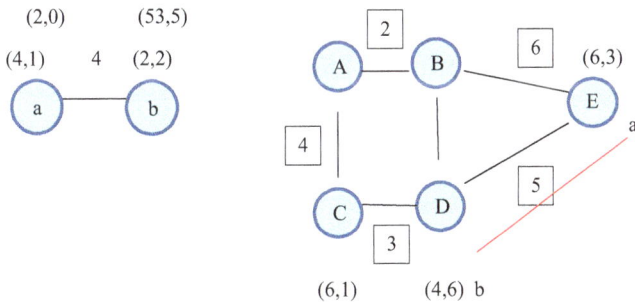

Figure 7.1   An example of VNE

Table 7.1 Symbol summary

Notation	Description
G^S, G^V	Substrate networks and virtual networks
N^S, L^S	Collection of all substrate nodes and links
N^V, L^V	Collection of all virtual nodes and links
CPU (n^v), CPU (n^s)	Computation power of the substrate and the virtual node
BW (l^s), BW (l^v),	Bandwidth power of the substrate and the virtual link
SEC (n^s), SEC (n^v)	Security level of substrate and virtual node
DEG (n^s)	Degree of substrate node

7.3.2 Evaluation metrics

In this research, three assessment criteria are developed in order to assess the usefulness of network embedding. The node allocation expenditure CPU (n^v) and link allocation expenditure BW (l^v), which may be written as follows, are connected to the usefulness of the VNE:

$$\mathrm{RE}\left(G^V, t, t_d\right) = t_d \cdot \left[w_l \sum_{nv \in NV} \mathrm{CPU}(n^v) + w_n \sum_{lv \in LV} \mathrm{BW}(l^v) \right], \qquad (7.3)$$

where w_l and w_n indicate the weights of the link and the node, and t_d represents the available time of the requesting node G^V. Correspondingly, the required consumption of G^V can be denoted by:

$$\mathrm{RC}\left(G^V, t, t_d\right) = t_d \cdot \left[\sum_{nv \in NV} \mathrm{CPU}(nv) + \sum_{lv \in LV} \sum_{ls \in LS} \mathrm{BW}(l^v) \right]. \qquad (7.4)$$

In order to assess how lucrative the algorithm has been, an average revenue metric, which may be determined by:

$$\mathrm{REV} = \frac{\lim_{T \to \infty} \sum_{t=0}^{T} RE(GV, t, tp)}{T}, \qquad (7.5)$$

where the time period T is indicated.

The revenue-to-cost statistic is then created, and it can be calculated by:

$$\mathrm{R2C} = \frac{\lim_{Time \to \infty} \sum_{t=0}^{Time} \mathrm{Re}(GV, t, tp)}{\sum_{t=0}^{Time} Co(GV, t, tp)}. \qquad (7.6)$$

This measure shows how well an algorithm uses resources. Another significant indicator of the algorithm's reliability is the success rate of virtual requests when embedded. though. As a result, we develop the average acceptance ratio measure, which is denoted by:

$$AR = \frac{\lim\limits_{Time\to\infty} \sum_{t=0}^{Time} Acc(GV,t,tp)}{\sum_{t=0}^{Time} Arr(GV,t,tp)}, \tag{7.7}$$

where $Acc(GV,t,tp)$ is the no. of correctly embedded networks and $Arr(GV,t,tp)$ denotes the total number of requests.

7.4 Introduction to a deep reinforcement learning algorithm for the policy-based network

7.4.1 Feature extraction

For the first layer input in this work, we chose five feature vectors to denote the physical node states.

1. *Computation power (CPU)*: CPU (n^S) reflects the node's remaining computing capacity n^S that can be formulated as:

$$CPU(n^s) = CPU(n^s) - \sum_{n^v \to n^s} CPU(n^s), \tag{7.8}$$

where CPU (n^s) denotes the remaining resource allocation of n^s. The $\sum_{n^v \to n^s} CPU(n^s)$ denotes the cumulation of CPU criteria of n^v.

2. *Bandwidth (BW)*: The formula below may be used to calculate the total bandwidth of all connections connecting to node n^s:

$$SUM(n^s) = \sum_{ls \in LSn} BW(l^s), \tag{7.9}$$

where L^S denotes substrate links linked to the node n^s, l^S belongs to $L_N{}^S$. There will be more link alternatives and a greater mapping impact, the higher the nodes with bandwidth and the higher the virtual nodes that are connected to those substrate nodes.

3. *Degree (DE)*: DE represents the number of links to be mapped with each node. The DE will be calculated as follows:

$$DE(ns) = \sum_{nsi \in NS} Link(n^s, nis), \tag{7.10}$$

where Link(n^s, n) denotes that the n^s and n are mapped. The $L(n^s, n)$ is 1 if the connection is made or 0 if it is not. A node is connected to more nodes, the higher its degree.

4. *Average distance (DST)*: We also need to take other virtual nodes' embedding locations into account while network embedding. The closer the nodes are to one another within the same request, the less bandwidth will be used. In this publication, the Floyd–Warshall [34] method, which determines the smallest distance between the different nodes, is introduced. Thus, breaking into different half the entire distance with the number of nodes, the DST may be

determined as follows:

$$DST = \frac{\sum_{NsV \in NS} DST\left(n^s, n_v^s\right)}{count + 1},$$ (7.11)

where $DST(n^s, n_v^s)$ shows the separation between ns and the mapped node. count is the total number of mapped nodes plus one to keep the denominator from becoming zero.

5. *Security level (SEC)*: Thus, mapping to a node is safer than a higher substrate node's security level. Only the substrate node with the greater degree of safety requirements can be translated to virtual nodes.

Then, these four features can form the feature vector F of physical nodes:

$$F = [CPU, \ SUM, \ DE, \ DST, \ SEC].$$ (7.12)

7.4.2 Policy network

As a learning agent, we utilize a policy network, which is just a CNN that is extensively used in ML [35,36]. The learning agent is trained using the input policy matrix to provide the mapping probability for each substrate node. As probability rises, it becomes more likely that the substrate nodes will be mapped. The policy network consists of a node selector, a convolution layer, an input layer, and a softmax layer.

The convolution layer receives the feature extracted matrices, which are then relocated from that input layer and utilized as the policy network's input. The convolution layer largely convolved the feature matrix. The term "convolutional operation" initially referred to the process of creating a third function out of two existing ones. Here, that feature matrix may be convolved to produce the feature vectors of every node. We refer to it as the available resource vector, which is

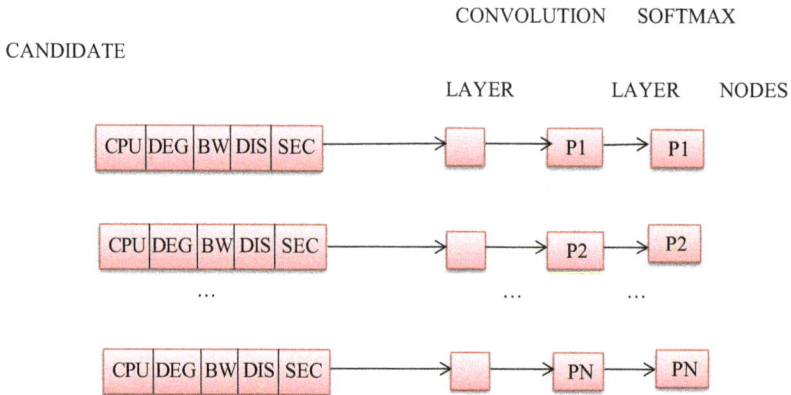

Figure 7.2 Policy network

formally stated as:

$$\mathrm{con}_i = a.W_i + b, \tag{7.13}$$

where con_i is the convolution layers' ith output, W is the convolution kernel's weight vector, and b is the difference between the two values.

The likelihood that virtual nodes will map to that node increases as the probability increases. Because certain substrate nodes might not have the processing power or security to connect all of the virtual nodes, it is impossible to estimate the likelihood that this portion of the substrate nodes will be mapped. In order to choose a set of selector nodes with sufficient processing power and safety level, we add a node selector.

7.4.3 Graph convolutional network

The GCN–VNE algorithm uses multiple layers of GCNs to perform VNE. Each layer of the GCN algorithm involves a set of learnable parameters that are used to transform the hidden feature representation of the node in a VN request.

Here is an overview of the layers in the GCN algorithm:

1. *Input layer*: The input layer of the GCN algorithm takes as input the adjacency matrix and feature matrix of a VN request.
2. *Hidden layers*: The hidden layers of the GCN algorithm perform graph convolution and graph attention operations to transform the hidden features presentation of the nodes in this VN request. Each hidden layer typically consists of multiple GCN blocks, which are composed of a graph convolution operation followed by a graph attention operation.
3. *Output layer*: The output layer of the GCN algorithm produces the embedded of that virtual node into physical nodes in the VN. This is done using a combination of graph attention and RL techniques to learn the optimal mapping between virtual node and physical node.

Each layer on the GCN algorithm is designed to learn a different aspect of VNE challenges. The input layer encodes the initial information about the VN request, while the hidden layers learn to capture the complex relationships among virtual node and the physical node in the network. Finally, output layer produces the optimal embedding of the virtual nodes into physical nodes, which maximizes network utilization and resource allocation efficiency.

7.4.4 Training and testing

The learning agents in our system are the policy-based network. An unlearned state is initially established for the policy network. After the agent is entered, we use the feature matrix as its learning environment. By carefully studying each node attribute in the feature matrix, the agent selects those substrate nodes that satisfy the virtual node's requirements for computing resources as well as the security performance requirements. A list of substrate nodes that are reachable and the possibility that each of them will be mapped to by the virtual nodes is generated by the final policy network.

A substrate node is selected as the node to be mapped from a sample of the substrate network set that was created using the probability distribution model. We start by calculating each substrate node's probability. The initialization of the policy network is random, so just because a node has the greatest probability does not guarantee that it will map to the best result. This process is continued until all virtual nodes are provided or the VNE is terminated due to insufficient resources on the substrate node. If all virtual node mappings are successful, link mapping will start.

Setting reward standards for the learning agent is essential because, in real life, the learning impact is determined by the action the learning agent chooses. If the agent's current behavior can assist the algorithm in achieving greater benefits or better results, the agent should be encouraged to continue the current activity in order to earn the cumulative reward. The reward signal will shrink or even disappear if the agent's actual activity has a moderate or negative consequence. The agent will reorient itself and end its current path of activity. The agent will reorient itself and end its current path of activity. Therefore, it's important to use the right incentive signal. In the VNE issue, we use the ratio of long-term renewable energy usage as a reward signal. This index evaluates the efficiency of the substrate resources, notably the connection bandwidth resources. If the agent's current activity may lead to a greater revenue utilization ratio and a greater incentive signal, the agent will keep looking into the action that produces a higher revenue utilization ratio.

During the training phase, we construct a target symbol for each virtual node in the VNRs. This symbol denotes the substrate node, which contains the virtual node. The characteristic vector's jth value is 1 if the target symbol of the virtual node ni is j, while its other dimensions are all 0 for substrate node nj. The following is the way it is said:

$$N^s j = \left(0_1, 0_2, \ldots, 1_j, \ldots, 0_k\right)^T.$$ (7.14)

7.4.5   GCN–VNE algorithm

The GCN that makes up our policy network has two hidden levels. While the neighbor matrix A is fixed in each layer, the source matrices X, as well as the convolution kernel matrices, are dynamically changed. The label for each embedding result is selected at random. Additionally, we manually create an identifier vector y with n dimensions for each virtualized node, where n is the same as the total quantity of nodes in the supporting network. As the embedding label for the present virtual node, think of the node ni. The label vector y's other dimensions are all zero except for the ith dimension, which is one. The function softmax will be used to handle the output vector y of the policy network, which is an n-dimensional vector:

$$Pi = \frac{ey}{\sum_{j} eyi},$$ (7.15)

where Pi stands for the policy network outcome vector p. The resultant vector p and labeled vector y may then be used to determine the cross-entropy, which is defined as the reduction of our framework:

$$Loss(y, p) = -\sum_{i} y \log(pi)$$ (7.16)

Proposed Algorithm: GCN–VNE

Input: no. of epochs; epoch; learned rate; a; training set;
Ensure: Trained GCN parameter
1: while iter < epoch do
2: for req ∈ training set do
3:   count = 0;
4:   for vi node ∈ req do
5:       F = FMatrix();
6:       model= GCN(F);
7:       if sub.cpu ≥ vir.cpu and sec.lvl ≥ sec.req then
8:           c=model.optimizer()
9:           Gradient(c);
10:      end if
11:    end for
12:    if isMap(*all v_node ∈ req*)
13:        Floyd(req);
14:    end if
15:    if isMap(*all v_node ∈ req*)*then*
16:        if(*all v_link ∈ req*)*then*
17:            rew = rev(req)
18:        else
19:                gradient = 0;
20:        endif
21:    end if
22:    count++;
23:    if count == batch_size then
24:        update parameter;
25:        count = 0;
26:    end if
27:    end for
28:    interation++;
29: end while
30: return success;

In this work, the incentive signal is set to the revision to the cost metric. The parameter gradient update formula for policy networks may be written as follows:

$$g = a.r.g_f, \tag{7.17}$$

where r is the reward signal and g_f is the gradient computed from the loss function. The gradient of losing g_f will not participate in parameter updating, as the reward signal r cannot be produced if the link embedding fails.

Learning speed α control determines the gradient's size. If the gradient is excessively steep, the learned agent's action modification direction becomes too large, and it may overlook some of the most costly actions. The agent's training will be exceedingly time-consuming and sluggish if the gradient is too tiny. As a result, the learning rate needs to be carefully managed. We update the strategy network using a batch gradient descent technique, which not only expedites agent training convergence but also guarantees network stability. In this algorithm, the node attaching process is shown in lines 1–5, while the link attaching process is shown in lines 6–8. The Floyd method is used to determine the nearest path among all embedded nodes. In lines 9–12, the algorithm determines the gradient. The node that has the highest likelihood of appearing in the policy network's output vector is used as the embedding node during the node embedding procedure. The link embedding begins when all nodes in the request have been properly embedded. The Floyd technique is also utilized in the testing of link embedding. The embedding will be unsuccessful if any step of the process fails.

7.4.6   Analysis of computational complexity

In this part, we look at the computational complexity of the GCN–VNE training method we recommend. For the purpose of convenience, we assume that the training data contains r virtual requests and that, on average, each request has m and n virtual nodes and connections. The training algorithm's execution frequency is represented by the function $f(n)$. The forward calculation is performed by fitting the second-degree Chebyshev polynomial to the convolution kernel of the GCN network, and the computational complexity decreases to a term that remains constant. After the node embedding phase, the Floyd technique is used to find the shortest connection between nodes; during the link embedding phase, the amount of computation required may be calculated as n^3.

The updating of network characteristics, as well as the computation and storage of gradients, are hence regarded as some of the terms that remain constant with the greatest computational complexity. After simplification, the computational complexity is $O(r\,n^3)$.

7.5   Analysis and evaluation of performance

7.5.1   Setting parameter

Anaconda3 and PyCharm are used to run our simulation experiment. A real network and 1,000 VN are generated programmatically and saved as .json files. There

Table 7.2   Parameter setting

Parameter	Value
CPU Power	U[50,100]
Bandwidth Power	U[50,100]
Security level	U[0,3]
CPU requirements	U[11,40]
Bandwidth requirement	U[1,50]
Safety requirement level	U[0,3]
Virtual node connection probability	50%

Table 7.3   Algorithms idea description

Algorithm	Description
Our algorithm	To better utilize resources, the study integrates GCNN and DRL, creates a bespoke feature matrix, and calculates a security value. Model parallel training increases algorithm performance.
Node rank [32]	When mapping virtual links, order physical nodes by significance and apply the breadth-first search methodology.
MCST–VNE [12]	The MCST technique is utilized to determine the node mapping strategy, and the virtual node map is modelled as an MDP.
RL-VNE [21]	A VNE technique built on RL and CNN that trains the model using a multi-objective reward function that is specifically created.

are around 600 connections and 100 nodes in the physical network. Between 50 and 100 units, the CPU and bandwidth resource capacities are equally and randomly divided.

To approach the actual network, 1,000 VNs create a continuous VNR process that simulates the Poisson process. A freely available library for high-performance numerical computations is known as TensorFlow. Computing can be simply deployed on a range of systems thanks to its adaptable design [37].

Each virtual node has a 50% chance of connecting to another node, and each VN randomly consists of two or more virtual nodes. Virtual nodes and virtual connections' CPU and bandwidth resource demands are distributed evenly and randomly between 1 and 50 units. In particular, the VN may expand by 5 times in 100 time units, and each VNR has an exponential life cycle. Table 7.2 displays the key variables utilized in the simulation experiment.

7.5.2   Comparison algorithm

As comparative algorithms, we use the heuristic VNE method NodeRank [32], the RL–VN embedding (RL-based VNE algorithm) [21], and the MCST–VNE [12]. The performance of the algorithms is then assessed using the average profit of VNE and the accept rate of VNR. The four algorithms, including our method, are listed and described in Table 7.3.

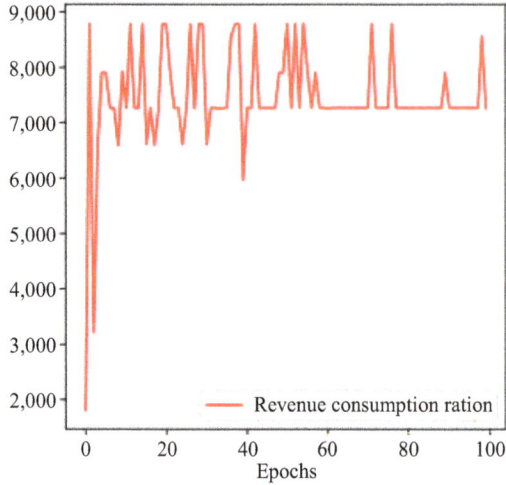

Figure 7.3   Revenue consumption ratio

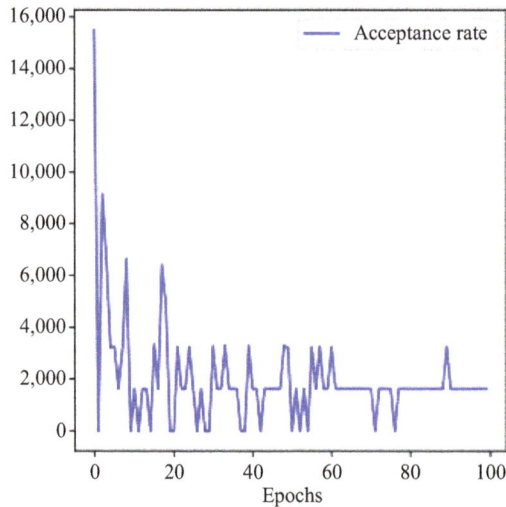

Figure 7.4   Acceptance rate

Additionally, we want to look at how the flexibility of the algorithm is affected by changing the physical network's resource capacity and VNRs' resource requirements. The algorithm's revenue, revenue-to-cost ratio, and VN accept rates are all extensively examined as shown in Figure 7.3 and Figure 7.4.

7.5.3   Result analysis

We start by evaluating the convergence of the proposed method. The loss gets diminishing as training progresses. We see that the loss quickly diminishes at the

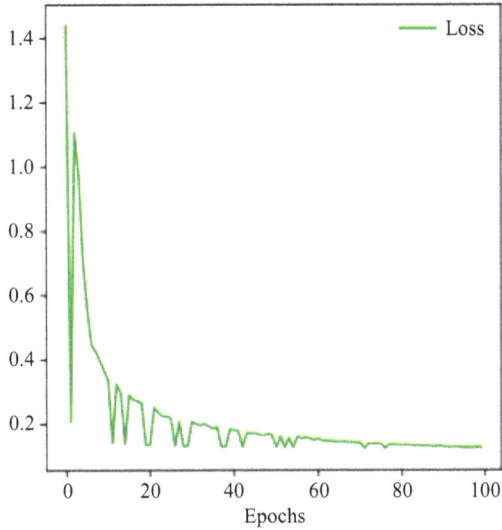

Figure 7.5   Loss

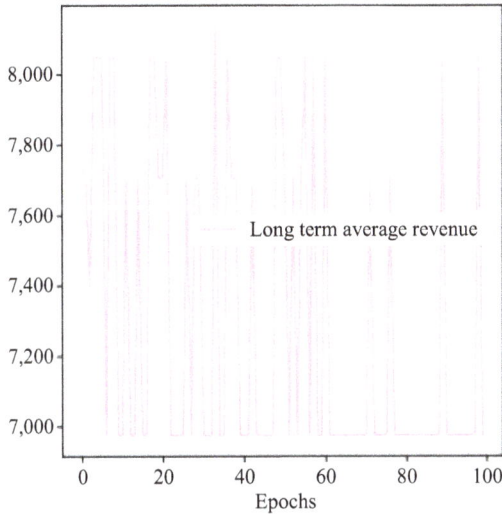

Figure 7.6   Long-term average revenue

start of training. This is brought on by the parameter randomization's creation of enough optimization space. The rate of loss declines gradually as you continue to work out. The loss is nearly consistent after 90 to 100 epochs. The outcomes of the simulation demonstrate how well as it learns, this GCN-VNE will converge on the perfect location.

Additionally, we show how three assessment metrics changed during the simulation of the proposed algorithm. From Figure 7.5 and Figure 7.6, it is clear that the proposed method achieves relatively less value in the start of train process. This is as a result of the policy network's usage of randomly initialized parameters.

The embedding process is gradually optimized alongside the training, and the metrics improve continuously. Each measure eventually reaches a stable optimal state after around 80 epochs because of the restricted substrate resources. The aforementioned tests suggest that changes in the virtual connection resource needs may also result in comparable impacts. This illustrates how effective and practical it is to investigate algorithm performance changes by changing resources characteristics of a physical networks or VNs. As a result, our approach is rather flexible. Moreover, our algorithm always ranks higher than other algorithms mentioned in the last table. As a result, we may claim that of all four of these algorithms, the GCN–VNE is probably the most efficient. Though, it has better performance when security is considered.

7.6 Conclusion

This study suggests a VNE-based algorithm approach on DRL and the GCNN that takes into account both the VNE and the method's fundamental performances. To create a powerful VNE method, we integrate the GCNN and RL algorithms using a self-defined feature matrix and evaluation criteria. There is no need to rely on any handwritten rules because the return to the agent determines the VNE choice totally. The security issue with the VNE is the primary subject of this work. Only a substrate node that fulfills the virtual node's security criteria can be mapped to it. For network businesses with strict security requirements, it is extremely important. The performance of this method is superior to other algorithms, according to simulation findings.

References

[1]   H. Cao, S. Wu, Y. Hu, Y. Liu, and L. Yang, "A survey of embedding algorithm for virtual network embedding," *China Commun.*, vol. 16, no. 12, pp. 1–33, 2019.

[2]   H. Yao, T. Mai, J. Wang, Z. Ji, C. Jiang, and Y. Qian, "Resource trading in blockchain-based industrial Internet of Things," *IEEE Trans. Ind. Informat.*, vol. 15, no. 6, pp. 3602–3609, 2019.

[3]   Z. Xu, J. Tang, J. Meng, W. Zhang, Y. Wang, C. H. Liu, and D. Yang, "Experience-driven networking: A deep reinforcement learning based approach," *in IEEE INFOCOM*, 2018, pp. 1871–1879.

[4]   J. You, R. Ying, X. Ren, W. L. Hamilton, and J. Leskovec, "*GraphRNN: Generating realistic graphs with deep auto-regressive models,*" *arXiv preprint* arXiv:1802.08773, 2018.

[5]   V. Mnih, A. P. Badia, M. Mirza, *et al.*, *"Asynchronous methods for deep reinforcement learning,"* Proceedings of the 33rd International Conference on Machine Learning, pp. 1928–1937, 2016.

[6]   C. Jiang, H. Zhang, Y. Ren, Z. Han, K. Chen, and L. Hanzo, "Machine learning paradigms for next generation wireless networks," *IEEE Wireless Commun.*, vol. 24, no. 2, pp. 98–105, 2017.

[7]   J. Wang, C. Jiang, H. Zhang, Y. Ren, K.-C. Chen, and L. Hanzo, "Thirty years of machine learning: The road to pareto-optimal wireless networks," *IEEE Commun. Surv. Tut.*, vol. 22, no. 3, pp. 1472–1514, 2020.

[8]   P. Zhang, C. Wang, Z. Qin, and H. Cao, "A multidomain virtual network embedding algorithm based on multiobjective optimization for Internet of Drones architecture in Industry 4.0," *Softw. Pract. Exper.*, vol. 52, no. 3, pp. 710–728, 2022. doi:10.1002/spe.2815.

[9]   S. Xu, "Overview of deep reinforcement learning," *Comput. Knowl. Technol.*, vol. 15, no. 3, pp. 193–194, 2019.

[10]   R. Socher, D. Chen, C. D. Manning, and A. Ng, "Reasoning with neural tensor networks for knowledge base completion," *in Advances in Neural Information Processing Systems*, 2013, pp. 926–934.

[11]   K. T. D. Nguyen and C. Huang, "An intelligent parallel algorithm for online virtual network embedding," *Proc. Int. Conf. Comput. Inf. Tele-Commun. Syst.*, 2019, pp. 1–5.

[12]   S. Haeri and L. Trajković, "Virtual network embedding via Monte Carlo tree search," *IEEE Trans. Cybernet.*, vol. 48, no. 2, pp. 510–521, 2018.

[13]   T. N. Kipf and M. Welling, *"Semi-supervised classification with graph convolutional networks,"* 2016, arXiv:1609.02907.

[14]   M. Yu, Y. Yi, J. Rexford, and M. Chiang, "Rethinking virtual network embedding: Substrate support for path splitting and migration," *Comput. Commun. Rev.*, vol. 38, no. 2, pp. 17–29, 2008.

[15]   A. Razzaq and M. S. Rathore, "An approach towards resource efficient virtual network embedding," in *Proc. Int. Conf. Evolving Internet*, Valencia, Spain, 2010, pp. 68–73.

[16]   X. Hesselbach, J. R. Amazonas, S. Villanueva, and J. F. Botero, "Coordinated node and link mapping VNE using a new paths algebra strategy," *J. Netw. Comput. Appl.*, vol. 69, pp. 14–26, 2016.

[17]   W. Hamilton, Z. Ying, and J. Leskovec, "Inductive representation learning on large graphs," *Proc. Adv. Neural Inf. Process. Syst.*, vol. 30, pp. 1024–1034, 2017.

[18]   S. Shanbhag, A. R. Kandoor, W. Cong, R. Mettu, and T. Wolf, "VHub: Single-stage virtual network mapping through hub location," *Comput. Netw.*, vol. 77, pp. 169–180, 2015.

[19]   J. Bruna, W. Zaremba, A. Szlam, and Y. Lecun, "Spectral networks and locally connected networks on graphs," *Proc. Int. Conf. Learn. Represent.*, pp. 1067–1077, 2014.

[20]   F. Tang, Z. M. Fadlullah, B. Mao, and N. Kato, "An intelligent traffic load prediction-based adaptive channel assignment algorithm in SDN-IoT: A deep

learning approach," *IEEE Internet Things J.*, vol. 5, no. 6, pp. 5141–5154, 2018.

[21] Z. Yan, J. Ge, Y. Wu, L. Li, and T. Li, "Automatic virtual network embedding: A deep reinforcement learning approach with graph convolutional networks," *IEEE J. Select. Areas Commun.*, vol. 38, no. 6, pp. 1040–1057, 2020.

[22] M. Lu, Y. Gu, and D. Xie, "A dynamic and collaborative multi-layer virtual network embedding algorithm in SDN based on reinforcement learning," *IEEE Trans. Netw. Serv. Manage.*, vol. 17, no. 4, pp. 2305–2317, 2020.

[23] Q. Li, C. Li, J. Zhang, H. Chen, and S. Wang, "A review of deep neural network compression," *Comput. Sci.*, vol. 46, no. 9, pp. 1–14, 2019.

[24] P. Zhang, C. Wang, C. Jiang, and A. Benslimane, "Security-aware virtual network embedding algorithm based on reinforcement learning," *IEEE Trans. Netw. Sci. Eng.*, vol. 8, no. 2, pp. 1095–1105, 2021. doi:10.1109/TNSE.2020.2995863.

[25] M. Dolati, S. B. Hassanpour, M. Ghaderi, and A. Khonsari, "DeepViNE: Virtual network embedding with deep reinforcement learning," *IEEE INFOCOM 2019 – IEEE Conference on Computer Communications Workshops (INFOCOM WKSHPS)*, Paris, France, pp. 879–885, 2019.

[26] Y. Yuan, Z. Tian, C. Wang, F. Zheng, and Y. Lv, "A Q-learning-based approach for virtual network embedding in data center," *Neu. Comput. Appl.*, vol. 32, no. 7, pp. 1995–2004, 2020.

[27] H. Afifi and H. Karl, "Reinforcement learning for virtual network embedding in wireless sensor networks," *2020 16th International Conference on Wireless and Mobile Computing, Networking and Communications (WiMob)*, Thessaloniki, Greece, pp. 123–128, 2020.

[28] H. Yao, X. Chen, M. Li, P. Zhang, and L. Wang, "A novel reinforcement learning algorithm for virtual network embedding," *Neurocomputing*, vol. 284, pp. 1–9, 2018.

[29] C. Zhao and B. Parhami, "Virtual network embedding through graph eigenspace alignment," *IEEE Trans. Netw. Serv. Manage.*, vol. 16, no. 2, pp. 632–646, 2019.

[30] B. Perozzi, R. Al-Rfou, and S. Skiena, "DeepWalk: Online learning of social representations," in *Proc. 20th ACM SIGKDD Int. Conf. Knowl. Discov. Data Mining*, 2014, pp. 701–710.

[31] D. Hammond, P. Vandergheynst, and R. Gribonval, "Wavelets on graphs via spectral graph theory," *Appl. Comput. Harmon. Anal.*, vol. 30, pp. 129–150, 2009.

[32] X. Cheng, S. Su, Z. Zhang, *et al.*, "Virtual network embedding through topology-aware node ranking," *Comput. Commun. Rev.*, vol. 41, no. 2, pp. 38–47, 2011.

[33] T. Ghazar and N. Samaan, "A hierarchical approach for efficient virtual network embedding based on exact subgraph matching," *2011 IEEE Global Telecommunications Conference – GLOBECOM 2011*, Houston, TX, USA, pp. 1–6, 2011.

[34] H. Yao, S. Ma, J. Wang, P. Zhang, C. Jiang, and S. Guo, "A continuous-decision virtual network embedding scheme relying on reinforcement learning," *IEEE Trans. Netw. Serv. Manage.*, vol. 17, no. 2, pp. 864–875, 2020.

[35] C. K. Dehury and P. K. Sahoo, "DYVINE: Fitness-based dynamic virtual network embedding in cloud computing," *IEEE J. Select. Areas Commun.*, vol. 37, no. 5, pp. 1029–1045, 2019.

[36] M. Thomas and E. W. Zegura, "Generation and analysis of random graphs to model internetworks," *College Comput.*, vol. 63, no. 4, pp. 413–442, 1994.

[37] M. Abadi, P. Barham, K. Chen, *et al.*, TensorFlow: A system for large-scale machine learning," *in Proc. Symp. Operating Syst. Des. Implementation*, vol. 16, pp. 265–283, 2016.

Chapter 8

Demystifying IoT, cloud, and blockchain for multimedia security – a scientific study

P.K. Paul[1], Tatayya Bommali[2], Abhijit Bandyopadhyay[3], Sanjukta Chakraborty[4], Mustafa Kayyali[5] and S.K. Sharma[6]

Abstract

In the dynamic landscape of the Internet of Things (IoT), we are on the brink of revolutionizing multimedia systems, enabling seamless real-time connectivity across essential sectors such as healthcare, entertainment, transportation, and smart cities. This expansion not only sparks thrilling possibilities but also raises critical security concerns, including data breaches, privacy challenges, and cyber threats. This chapter passionately examines the vital role of IoT in enhancing multimedia security, highlighting vulnerabilities, and emerging risks while presenting innovative solutions like encryption, blockchain, artificial intelligence (AI), and edge computing. We will delve into the significance of regulatory frameworks and explore groundbreaking trends, including post-quantum cryptography (PQC) and AI-driven threat detection. The discussion will illuminate promising research pathways aimed at strengthening multimedia security within IoT ecosystems. Moreover, we explore how cloud computing plays a pivotal role in managing multimedia content – ranging from images to videos – by offering a streamlined and efficient means of storing, processing, and accessing data online. Cloud solutions are particularly beneficial, as they can readily adapt to fluctuating user demands, such as during the premiere of a blockbuster film or a major live event. While the advantages of cloud computing are vast, it also brings forth notable security challenges that must be addressed. To enhance multimedia security across different applications, blockchain technology emerges as a powerful ally, offering decentralization, transparency, and practical benefits. With blockchain,

[1]Department of CIS and Office of the Information Scientist (Offg.), Raiganj University, India
[2]Department of Management Studies, Sanketika Vidya Parishad Engineering College, India
[3]Office of the Principal, Raniganj Institute of Computer and Information Science, India
[4]Department of Computer Science and Engineering, Seacom Skills University, India
[5]Office of the Quality Assurance & Accreditation, Maaref University of Applied Sciences, Syria
[6]College of Business, Engineering and Technology, Texas A&M University Texarkana, USA

digital media can be securely stored, verified, and tracked, effectively curbing the risks of counterfeiting, piracy, and unauthorized distribution. This chapter will thoroughly investigate the security dimensions of multimedia, particularly in the context of IoT, cloud, and blockchain systems, by focusing on their features, impacts, and the myriad challenges that lie ahead. Join us in this exciting exploration as we demystify the convergence of these technologies for a secure multimedia future!

Keywords: Multimedia security; cloud and virtualization; flexible secure computing; IoT; blockchain; digitalization; security systems

8.1   Introduction

Real-time emergencies like remote healthcare overlooking and intelligent surveillance tap into IoT-driven multimedia systems to enhance efficiency and elevate user experiences. However, this rapid proliferation of IoT devices presents significant security and privacy challenges, particularly in multimedia scenarios. The data produced by IoT frequently travels through both open and secured networks, thereby exposing it to a range of threats such as cyberattacks and unauthorized access. The limitations of IoT devices, alongside their varied architectures and protocols, make it challenging to implement adequate security measures. Guaranteeing multimedia content's confidentiality, integrity, and availability is more crucial than ever. Cloud Computing technology has transformed how multimedia is handled, offering scalable, cost-effective, and easily adaptable solutions to the needs of individuals and businesses. Cloud computing has its role in handling large-scale and dedicated multimedia applications seamlessly [1,2]. Applications such as video streaming platforms, virtual reality (VR) experiences, online gaming, and digital content creation rely heavily on the cloud. These applications demand significant computational resources and storage space, which cloud platforms can dynamically provide. Thus, there is no need to put effort and dedication into expensive hardware and similar infrastructure, as the cloud delivers resources on demand. Users can choose their requirements based on need, making it a practical and budget-friendly option. As technology continues to evolve, blockchain is poised to become a key element in the digital content landscape, ensuring that creators and users benefit from a more secure and trustworthy multimedia environment [3,4]. With further research and collaboration between industries, governments, and technology providers, blockchain can revolutionize multimedia security and reshape how digital content is protected and distributed.

8.1.1   Objective of the chapter

This work, entitled "*Demystifying* IoT, Cloud, and Blockchain for Multimedia Security – *A Scientific Study*," is theoretically mainly carried out to understand the following:

- To briefly identify the features and concerns of multimedia security in a theoretical context.
- To gain knowledge by studying existing works and papers on multimedia security concerning IoT, cloud computing, and blockchain.
- To effectively address the latest developments and issues associated with the IoT in strengthening multimedia security.
- To explore the fundamental benefits and potential advantages of cloud computing, advanced intelligent systems, and multimedia security.
- To find out the significant impact and challenges of blockchain with special reference to multimedia security.

8.1.2   Methods adopted

The work entitled "*Demystifying* IoT, Cloud, and Blockchain for Multimedia Security – *A Scientific Study*" is a theoretical chapter based on existing works, including journals, books, and a thesis related to multimedia security, *IoT*, cloud, and blockchain. The last 10 years' works are studied, and significant works are chosen while preparing this chapter. Google scholar was mainly used to collect existing research.

8.1.3   Existing works and research gap

Currently, various works are conducted in multimedia security and security aspects of IoT, cloud, and blockchain. A few selected are included in this section with major findings, issues, and other concerns.

Awadallah et al. [5] highlight the role of blockchain technology to enhance security in cloud computing. The authors offer a decentralized framework that mitigates common security vulnerabilities, including unauthorized access, by exploiting blockchain's immutable ledger and consensus procedures. The research emphasizes the capability of smart contracts to automate security protocols, ensuring data integrity and diminishing dependence on centralized cloud service providers. The article also addresses the scaling problems of blockchain within cloud systems and proposes hybrid models that reconcile efficiency and security. The authors conclude that while blockchain presents a promising solution for cloud security, further research is needed to optimize its performance and cost-effectiveness. *Habib et al.* [6] investigate blockchain on digital rights management (DRM) in online education. The researchers inform about the blockchain-based DRM system that ensures transparent and tamper-proof ownership of multimedia educational resources. By leveraging decentralized ledger technology, the system prevents illicit duplication, dissemination, and alteration of digital assets. The study also examines the advantages of smart contracts regarding license agreements and royalty sharing. While the proposed approach promotes security and fairness in content sharing, the authors note the computational cost and energy consumption problems involved with blockchain implementation in DRM. Their findings suggest a hybrid blockchain model could address these concerns while maintaining efficiency.

Jan et al. [7], this paper provides an in-depth exploration of how blockchain technology can bolster the security of the Internet of Multimedia Things (IoMT). The authors shed light on the vulnerabilities present within IoMT networks, particularly concerning data privacy and authentication issues, and integrity. They highlight blockchain's promise to safeguard multimedia transmission, enable decentralized trust management, and prevent cyberattacks. The work poses research obstacles, including latency issues, scalability restrictions, and regulatory considerations. The authors propose further research concentrating on lightweight blockchain architectures and energy-efficient consensus mechanisms suitable for IoMT applications. *Khan et al.* [8] study emphasized the IoT security and blockchain technology merger in assessing the latest advancements in secure IoT networks. The authors explore blockchain's role in combating common IoT dangers, including unauthorized access, data tampering, and distributed denial of service (DDoS) attacks. A thorough examination of various blockchain architectures – such as public, private, and consortium models – is conducted to assess their relevance and effectiveness for different IoT applications. Furthermore, the discussion includes important considerations around scalability and energy efficiency, with innovative consensus algorithms like proof of authority (PoA) and delegated proof of stake (DPoS) being proposed as potential pathways to overcome these challenges.

The authors conclude that while blockchain can considerably boost IoT security, further tuning is required to achieve seamless implementation. *Chen et al.* [26] delve into how blockchain can secure the digital forensic chain-of-custody in IoT and multimedia contexts, ensuring the integrity of critical data. The authors propose a blockchain-based system using Hyperledger Sawtooth, ensuring transparent and tamper-proof documentation of digital data. The article emphasizes the necessity of maintaining forensic integrity in cybercrime investigations, highlighting blockchain's potential to provide immutable records and cryptographic verification. The authors also investigate the problems of integrating blockchain in forensic research, including data storage limits and regulatory implications. Their findings imply that merging blockchain with forensic analysis can promote accountability and trust in digital investigations. *Liu et al.* [9] worked on blockchain-based architecture for managing multimedia data, addressing challenges such as illegal access, content validity, and piracy. The authors present a decentralized platform where multimedia assets are kept securely, and smart contracts regulate access permissions. The report illustrates how blockchain may assure content traceability, allowing content creators to keep ownership rights while enabling transparent licensing. The authors also explore the processing overhead of blockchain transactions, suggesting off-chain storage methods to boost efficiency. The research indicates that blockchain brings new-age multimedia data management, provided scalability concerns are addressed.

Additionally, *Nair and Tyagi* [10] address the dynamic intersection of artificial intelligence (AI), the IoT, blockchain, and cloud computing, illuminating their collective influence on the digital infrastructure of the future. The authors discuss how these technologies complement each other, with AI enhancing decision-making, IoT facilitating data collection, blockchain ensuring security and cloud computing providing scalable storage solutions. The chapter also showcases real-world applications, including smart cities, autonomous systems, and cybersecurity advancements. While integrating these technologies brings various benefits, the authors note the challenges of interoperability, data protection, and computational complexity. Their findings imply that future studies should focus on defining standardized standards for smooth integration. *Nguyen et al.* [11] analyze the blockchain in cloud of things (CoT) development, a paradigm that integrates IoT with cloud computing. The authors offer a blockchain-based architecture that promotes security, data integrity, and interoperability in CoT contexts. They cover applications such as decentralized data storage, secure device communication, and automated service provisioning. The report addresses important issues, including scalability, energy usage, and regulatory compliance. The authors argue that hybrid blockchain models and energy-efficient consensus techniques could address these limitations, opening the path for secure and efficient CoT implementations. *Rana et al.* [12] provide a blockchain-based secure multimedia content-sharing approach. The authors propose a method that ensures content validity and prohibits unlawful distribution through decentralized governance. A crucial innovation discussed is the integration of an ensured system update mechanism, guaranteeing that content license agreements remain up-to-date and resistant to tampering. The report emphasizes the benefits of utilizing blockchain for DRM and pirate prevention while addressing difficulties like transaction latency and storage costs.

The authors conclude that blockchain is viable for safeguarding multimedia content, but further research is needed to enhance its efficiency. *Taloba et al.* [13] analyze the function of blockchain in multimedia data processing inside IoT-enabled healthcare systems. The authors propose a hybrid blockchain paradigm that promotes data security, maintains patient privacy, and facilitates real-time data sharing across healthcare providers. The research demonstrates that, in the healthcare sector, blockchain emerges as a transformative tool for preventing data breaches and safeguarding the integrity of medical records. The authors also explore the processing cost of blockchain and present layer-two scaling strategies to boost efficiency. Their findings indicate that blockchain has considerable potential to improve IoT health care but requires additional tuning to balance security and performance. *Uddin et al.* [14] further investigate the integration of blockchain within cloud security frameworks, evaluating both the advantages and challenges associated with this integration. They illustrate how blockchain enhances security in cloud environments through mechanisms like decentralized authentication, meticulous access control, and verification of data integrity. Although this approach presents promising security advancements, it also faces hurdles such as latency concerns, vulnerabilities linked to smart contracts, and the necessity for established regulatory frameworks. The findings suggest that while blockchain holds significant potential for improving security measures, further research is imperative to enhance its efficiency and scalability.

8.1.4   *Internet of things and multimedia security*

The IoT is a game-changer across diverse fields, from health care and smart homes to industrial automation and entertainment, enabling seamless connectivity and automation. By linking billions of devices through sensors and cloud platforms, IoT redefines how multimedia content – like images, videos, and audio – is created, shared, and stored. At the same time, there are different kinds of security issues [15,16]. This section examines the vital role of IoT in multimedia security, spotlighting new threats and innovative solutions like encryption, blockchain, AI models, federated learning, and edge computing. We will also look into global security standards, regulatory policies, and the future of multimedia security research in IoT landscapes (refer to Figure 8.1).

8.1.4.1   Role of IoT in multimedia security

IoT is dedicated to revolutionizing multimedia applications to facilitate real-time data collection and analysis across diverse fields like smart cities, health care, surveillance, and entertainment. However, with the surge in interconnected IoT devices comes a critical need to address significant security concerns, especially in safeguarding multimedia content from cyber threats and unauthorized access. Protecting multimedia systems within IoT environments hinges on ensuring secure

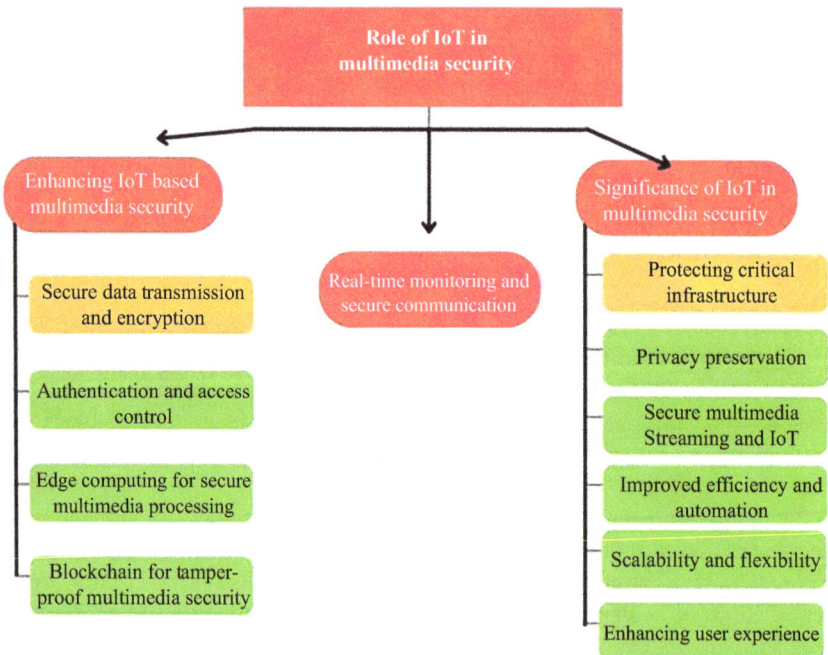

Figure 8.1   Depicts the fundamental role of the Internet of Things (IoT) in multimedia security

communication, robust authentication, strict access control, and data integrity [17,18]. This section depicts the significance of IoT in developing multimedia security, highlighting advanced strategies to mitigate risks effectively.

Secure data transmission and encryption

To maintain the integrity of multimedia data, IoT-enabled systems must implement secure transmission protocols to thwart data interception and manipulation. *End-to-end encryption (E2EE)* is essential for shielding multimedia content from sender to receiver, significantly reducing the risk of eavesdropping. Key encryption techniques include:

- Advanced encryption standard (AES): A favored symmetric encryption method widely used for securing multimedia content.
- Rivest–Shamir–Adleman (RSA): A reliable public-key encryption technique that protects data confidentiality.
- Elliptic curve cryptography (ECC): An efficient method ideal for resource-limited IoT devices.

Moreover, developing lightweight cryptographic algorithms aims to enhance security without straining computational resources, ensuring effective encryption on devices with limited power. Additionally, protected protocols like transport layer security (TLS) significantly enhance the confidentiality and integrity of multimedia data transfers.

Authentication and access control

Robust authentication and access control mechanisms are essential for securing multimedia systems within the IoT. Numerous devices rely on multimedia-based authentication techniques, such as facial recognition and voice biometrics, to verify user identities effectively. Key methods include:

- Multi-factor authentication (MFA): This combines various verification techniques (e.g., passwords, biometrics, and one-time codes) to bolster security.
- Blockchain-based access control: It provides tamper-proof authentication records, enhancing security management.
- Role-based access control (RBAC) and attribute-based access control (ABAC): These mechanisms set access permissions according to user roles and attributes, allowing for tailored security policies.
- Zero trust architecture (ZTA): A modern strategy that cautiously treats every access request, demanding continuous authentication.

These concerns are significant in preventing the unauthorized sharing of multimedia content, effectively reducing risks related to identity fraud and privilege escalation [4,19].

Edge computing for secure multimedia processing

Edge computing is transforming multimedia security by enabling data processing right at its source. This innovative strategy dramatically lowers latency and minimizes exposure to cyber threats. Moving away from traditional cloud-based systems gives us the benefits of real-time data analysis at the edge, leading to quicker response times and bolstered security. Here are some standout advantages:

- Minimized transmission vulnerabilities: Local multimedia data processing means reduced interception risks during cloud transfers.
- AI-driven anomaly detection: Machine learning algorithms operate at the edge, instantly identifying suspicious activity and addressing cyber threats.
- Secure multi-party computation (SMPC) plays a crucial role in this realm, allowing calculations to be performed on encrypted information without disclosing the original data, which greatly improves privacy.

With edge computing, IoT ecosystems can effectively and securely manage multimedia content, providing robust analysis and storage while keeping vulnerabilities at bay.

Blockchain for tamper-proof multimedia security
The advent of blockchain technology is a significant milestone for ensuring the integrity and authenticity of multimedia content generated through IoT devices. Blockchain safeguards against unauthorized changes or forgery of multimedia files by creating an unalterable, decentralized ledger. Here are the key benefits:

- Tamper-proof digital content: Each multimedia transaction is securely logged on the blockchain, ensuring authenticity.
- Smart contracts for automated security policies facilitate secure sharing and authentication of multimedia [20,21].
- Fusing public and private blockchains boosts scalability and performance in IoT-driven multimedia applications.

Furthermore, blockchain-enhanced DRM protects multimedia assets from unauthorized use, promoting fair distribution and consumption.

8.1.4.2   Real-time monitoring and secure communication
IoT is revolutionizing multimedia security through exciting advancements in real-time monitoring and communication. From intelligent surveillance to healthcare and industrial settings, IoT-enabled systems are making a significant impact. Devices like CCTV cameras, wearable health monitors, and smart home assistants produce a wealth of multimedia data essential for transmission and secure analysis [22,23]. Let's dive into some key areas:

- Real-time video surveillance: IoT security cameras are essential for monitoring public and private spaces. They demand robust encryption to thwart hacking attempts and ensure safety.
- Secure telehealth services: IoT medical devices play a vital role in healthcare by transmitting sensitive data, making end-to-end security measures critical.
- IoT-driven multimedia communication: Effective communication relies on secure protocols for voice commands, video conferencing, and immersive experiences to avoid data breaches.

IoT significantly boosts multimedia security across various applications by integrating AI analytics with secure transmission methods. Continuous innovation in IoT security frameworks is necessary with the rise of sophisticated cyber threats. As we embrace IoT, additionally, cutting-edge advancements such as PQC and federated learning are paving the way for even more robust security measures in this ever-evolving landscape, and AI-driven security models will enhance multimedia security even further, which is more dedicated to developing a safer and more robust ecosystem of the IoT.

8.1.4.3   Significance of IoT in multimedia security

IoT transforms how we create, share, and experience multimedia across diverse fields like smart cities, healthcare, industrial automation, and entertainment. This incredible synergy of IoT with multimedia technologies elevates real-time monitoring, automation, and user engagement to new heights. Yet, as these applications grow more interconnected, more dedicated security measures are required to shield sensitive information from digital dangers, unapproved access, and privacy violations. Let's dive into IoT's pivotal role in multimedia security and identify the key areas where its security innovations shine.

Protecting critical infrastructure

IoT-driven surveillance, innovative city initiatives, and industrial systems thrive on multimedia data for instant monitoring and security decisions. Protecting these streams is paramount for critical infrastructure, including:

- Smart grids: Secure multimedia monitoring in IoT power systems is crucial for preventing cyber threats and data manipulation [24,25].
- Transportation networks: Smart traffic monitoring and autonomous vehicle multimedia systems rely on secure, encrypted data to maintain safety.
- Industrial IoT (IIoT) systems: Secure multimedia solutions in manufacturing logistics are vital in monitoring operations and minimizing risks like sabotage and cyber-physical attacks.

With the proliferation of cyber-physical systems (CPS) within critical infrastructures, safeguarding against cyber threats, unauthorized access, and cyber-physical attacks is more vital than ever. To counter interruptions and enhance security, employing strong encryption, multi-layered access controls, and AI-driven anomaly detection has become essential.

Privacy preservation

The increasing prevalence of smart cameras, voice assistants, wearables, and IoT multimedia applications [26] brings significant privacy and security challenges. Issues like unauthorized data collection and surveillance breaches are serious concerns. To tackle these threats, cutting-edge privacy-preserving techniques are gaining traction, such as:

- Homomorphic encryption is dedicated to encrypting multimedia data, shielding sensitive content from exposure.

- Differential privacy: It ensures analytics provide valuable insights while protecting user privacy.
- Federated learning: AI models can now be trained without transferring raw data to centralized servers, significantly reducing privacy risks.

These innovative techniques bolster confidentiality, data sovereignty, and compliance with regulations, securing multimedia content against unauthorized access across IoT networks.

Secure multimedia streaming and IoT

IoT is revolutionizing multimedia streaming systems, which are run on a real-time basis in areas like video conferencing, remote learning, and digital entertainment. Safeguarding streaming content from cyber threats and piracy is vital for preserving data integrity. Here are some key security strategies that are making a difference:

- Secure real-time transport protocol (SRTP): This technology encrypts multimedia streams, thwarting interception and unauthorized access.
- AI-enhanced digital watermarking: Embedding encrypted metadata ensures content authenticity and deters unauthorized duplication.
- Blockchain-based content authentication: This solution offers robust provenance tracking and copyright protection, minimizing the risk of forgery.

These innovative measures strengthen the security of IoT multimedia applications, providing a seamless and secure streaming experience.

8.1.4.4   Improved efficiency and automation

IoT significantly boosts efficiency and automation across different sectors. With AI-powered video analytics, object detection, and anomaly detection, manual intervention is minimized while precision is maximized [2,27]. Noteworthy applications include:

- Intelligent surveillance: IoT cameras that autonomously detect and respond to security threats.
- Automated content moderation: Platforms that filter inappropriate content without human oversight.
- Cybersecurity threat intelligence: Real-time automated threat detection that proactively mitigates cyber risks.

These advancements reduce human error, enhance reliability, and accelerate decision-making in security-critical areas.

8.1.4.5   Scalability and flexibility

IoT networks are inherently scalable and perfect for managing expansive multimedia applications. IoT solutions efficiently handle many connected multimedia devices in smart cities, health care, and industrial sectors. Key benefits include:

- Scalable surveillance systems: Efficient management of extensive CCTV networks through IoT security frameworks.

- Flexible multimedia workflows: Dynamic adjustment of computational resources for optimal media processing.
- Cloud-edge hybrid security: Combining cloud and edge processing ensures robust multimedia security without compromising performance.

By emphasizing scalability, adaptability, and secure device integration, IoT enhances the resilience of multimedia security frameworks, paving the way for a safer digital landscape.

8.1.4.6 Enhancing user experience

IoT-powered multimedia systems' role goes far beyond security; they truly elevate our experiences with personalized, immersive, and interactive content. Whether it's enjoying smart home entertainment or exploring VR applications, robust security is vital for:

- Building user trust: Safeguarding personal multimedia content boosts consumer confidence and fosters loyalty.
- Securing AR/VR experiences: IoT enables captivating, immersive moments that depend on real-time encryption and strong authentication to guard against data leaks.
- Protecting smart entertainment: To combat cyber threats and digital piracy, streaming devices, gaming consoles, and multimedia hubs need fortified frameworks [28,29].

By embedding advanced security protocols, IoT ensures safe and smooth multimedia interactions and protects content creators and users. The importance of IoT in securing multimedia is much broader than traditional cybersecurity. It offers automated solutions, privacy safeguards, real-time monitoring, and scalable security for various sectors. As we face evolving cyber threats, integrating technologies like blockchain, harnessing AI for security analytics, leveraging federated learning, and implementing PQC are key to strengthening IoT-based multimedia ecosystems. Future explorations should aim for adaptive security models that promote resilient, intelligent, and privacy-focused multimedia solutions in the dynamic IoT realm.

8.1.5 Challenges and considerations in IoT multimedia security

The rapid growth of IoT devices in multimedia applications is revolutionizing connectivity and enabling real-time data processing, which opens up a wealth of exciting opportunities. Yet, securing multimedia content within these IoT ecosystems presents formidable challenges. Resource limitations, network vulnerabilities, cyber threats, and standardized security protocols are significant hurdles. Given the enormous amounts of sensitive data – images, videos, audio – generated by these applications, unprotected information could fall prey to malicious actors [30,31]. Let's delve into the key issues and explore potential solutions with enthusiasm!

8.1.5.1   Resource constraints

A major hurdle in securing IoT multimedia applications stems from the limited resources of IoT devices. Many smart cameras, sensors, and however, many of these devices are designed with limited processing capabilities and memory, presenting challenges in applying robust security protocols. To address these hurdles, innovators are channeling their efforts into developing lightweight security solutions that fit the unique needs of these powerful technologies.

- Lightweight encryption techniques like ECC and AES-128 strike a balance between security and efficiency.
- Hardware security modules (HSMs) that offload cryptographic tasks to dedicated security chips, boosting performance.
- AI-driven optimization, which adjusts security measures dynamically according to available resources.

Addressing these resource constraints is essential to avoid data breaches and cyber threats.

8.1.5.2   Network vulnerabilities

IoT networks are increasingly vulnerable to a wide range of cyber threats, including:

- Distributed denial-of-service (DDoS) attacks: These overwhelming assaults on networks can severely disrupt multimedia services, rendering them inaccessible.
- Man-in-the-middle (MITM) attacks: In these scenarios, cybercriminals intercept and manipulate data during transmission, posing significant risks to data integrity and privacy.
- Eavesdropping and sniffing: These tactics can compromise the confidentiality and security of multimedia streams, allowing unauthorized parties to access sensitive information.

  To strengthen our defenses, we can adopt effective measures such as:

- TLS and secure socket layer (SSL): These protocols provide encrypted communication, safeguarding data as it travels over networks.
- IoT-specific intrusion detection systems (IDS) to spot network anomalies.
- Software-defined networking (SDN) for adaptive security policies based on real-time threat intelligence [31,32].

  By addressing these vulnerabilities head-on, we can cultivate secure IoT multimedia applications and protect sensitive information from increasingly sophisticated cyber threats.

8.1.5.3   Scalability and security management issues

The rapid rise of IoT devices is reshaping the landscape of multimedia security, bringing both thrilling prospects and significant challenges. With billions of devices interconnected and generating multimedia content in real-time, ensuring reliable and secure data handling is complex. Key issues include:

- Decentralized identity management: We urgently need a scalable, robust identity verification system to protect against unauthorized access.
- Cloud-based security solutions: Merging cloud computing with edge processing is essential to boost security without excessive demands on individual IoT devices.
- Lack of universal IoT security standards: The noticeable absence of standardized security guidelines within IoT ecosystems leads to varying security practices and vulnerabilities.

Exciting initiatives like the ISO/IEC 27030 security standard aim to create essential guidelines to secure multimedia applications across diverse IoT contexts.

8.1.5.4 Data privacy and confidentiality risks

IoT devices are at the forefront of capturing and processing sensitive multimedia data – think personal images, voice recordings, and video feeds. Unauthorized access can lead to significant privacy breaches, identity theft, and regulatory issues. Some common concerns include:

- Hacked surveillance cameras that expose private spaces to harmful actors.
- Uncontrolled collection of biometric data, like facial recognition and voice authentication.
- Data leaks from poorly secured cloud storage or edge servers.

To improve privacy protection, we also have innovative techniques at our disposal:

- Homomorphic encryption: This cutting-edge method allows us to perform computations on encrypted multimedia data without the need for decryption.
- Differential privacy: Ensures that individual data points are anonymized before analysis, safeguarding personal information.
- Federated learning: Enables AI models to learn from multimedia data while keeping raw data secure and on the device, avoiding centralized data transfers.

Embracing these privacy-enhancing solutions can be transformative, significantly reducing the risk of unauthorized access while ensuring we comply with essential data protection regulations like GDPR and CCPA.

8.1.5.5 AI and deepfake threats

The rapid advancement of AI and deepfake technology [33] poses exciting yet concerning challenges for multimedia security. Deepfakes utilize generative adversarial networks (GANs) to create strikingly realistic but altered videos and audio, resulting in:

- Disinformation and fake news – The troubling proliferation of AI-generated content aimed at political or financial deception [34,35].

- Identity theft and fraud – The alarming potential for deepfake audio and video to impersonate individuals, facilitating fraudulent activities.
- Legal and ethical concerns – The difficulty in distinguishing between genuine content and synthetic forgeries is growing.

To handle these dynamic threats, innovative security measures are emerging, such as:

- Deepfake detection algorithms – Leveraging machine learning to spot manipulated content effectively.
- AI-based forensic techniques – Ensuring the authenticity of multimedia data is verifiable.
- Blockchain-powered authentication – Crafting tamper-proof digital signatures for content verification.

As AI evolves in offensive and defensive cybersecurity, proactive research and strategies are essential to safeguard IoT multimedia security.

8.1.5.6   Safeguarding data integrity and authenticity

It is essential to uphold the integrity and authenticity of multimedia content in IoT networks. There is a significant risk that malicious actors might tamper with images, videos, and audio to achieve their agendas.

- Spread misinformation.
- Deceive biometric systems.
- Alter surveillance footage, impeding investigations.

To combat these threats, we must employ robust security mechanisms such as:

- Blockchain technology – Creating immutable multimedia logs.
- AI-driven anomaly detection – Identifying suspicious or altered content.
- Digital watermarking – Embedding authentication markers within multimedia files.

IoT multimedia systems can thwart data manipulation and uphold authenticity by embracing advanced integrity verification methods. With the expanding reliance on IoT-driven multimedia applications, we face many security challenges – ranging from resource constraints and network vulnerabilities to AI-driven threats and privacy issues [36,37]. As cyber threats evolve, securing IoT multimedia ecosystems demands adaptable security frameworks, privacy-preserving techniques, and cutting-edge AI-driven detection mechanisms. Future research should zero in on standardized security protocols, quantum-resistant encryption, and decentralized security models, paving the way for resilient and next-generation IoT multimedia security solutions.

8.1.6   *Emerging trends and research pathways in IoT multimedia security*

As the landscape of the IoT continues to grow and evolve, the protection of multimedia data remains a critical challenge that demands our attention. Emerging trends and advancements in cybersecurity aim to address the vulnerabilities associated with IoT-enabled multimedia applications [38]. Future research directions focus on quantum-resistant cryptography, AI-driven security mechanisms, decentralized security models, and enhanced privacy-preserving techniques. This section highlights key future trends and research directions shaping the security landscape of IoT-based multimedia applications.

8.1.6.1 Post-quantum cryptography for IoT security

The emergence of quantum computing presents a formidable challenge to conventional cryptographic systems, such as RSA, ECC, and AES. Quantum computers are capable of undermining the encryption techniques that safeguard IoT multimedia security, which could result in serious data breaches and unauthorized access [2,7]. To mitigate this threat, efforts are underway to advance PQC, developing encryption protocols that are resilient in a future where quantum computing is prevalent. Current research directions include:

- Lattice-based cryptography [39]: This method serves as a substitute for RSA and ECC, offering protection for multimedia communications in IoT.
- Hash-based signatures: These are crucial for verifying the authenticity and integrity of multimedia content.
- Quantum key distribution (QKD): This innovative approach promises to deliver exceptionally secure encryption measures for IoT networks.

Incorporating post-quantum security features into IoT devices is vital for ensuring multimedia security remains robust against the emerging threats posed by quantum technology.

8.1.6.2 AI-driven threat intelligence and anomaly detection

As cyber threats become more sophisticated, the roles of AI and machine learning (ML) in securing IoT multimedia systems grow increasingly important. AI-powered cybersecurity strategies boost threat intelligence through:

- Real-time anomaly detection: Utilizing AI algorithms, systems can scrutinize patterns within multimedia data to spot unusual activities, such as the generation of deepfake content or instances of unauthorized access.
- Predictive threat mitigation: Machine learning models proactively identify potential security weaknesses and automatically deploy countermeasures.
- Automated behavioral analytics – AI-driven behavioral analysis improves the detection of unauthorized IoT device interactions with multimedia applications [38].

Future research will focus on explainable AI (XAI) for transparent cybersecurity solutions, ensuring trustworthy and interpretable AI-based security models for IoT environments.

8.1.6.3 Zero-trust security models for IoT

The traditional security model of implicit trust in IoT networks is no longer effective due to evolving cyber threats. The zero-trust security model (ZTSM) is emerging as a next-generation security framework that enforces strict verification and access control policies for IoT multimedia applications [8,28]. Key aspects of ZTSM include:

- Continuous authentication and authorization – Every device and user accessing multimedia data is always verified.
- Micro-segmentation – Dividing IoT multimedia networks into smaller zones to limit the spread of cyberattacks.
- Least privilege access – Ensuring IoT devices and applications only access necessary multimedia data for specific functions.

Research efforts are underway to integrate blockchain-based identity management and AI-enhanced authentication protocols to strengthen zero-trust security in IoT environments.

8.1.6.4   Utilizing blockchain for ensuring multimedia data authenticity and integrity

The potential of blockchain technology is gaining momentum as a means to uphold the integrity and authenticity of multimedia content within IoT ecosystems. Thanks to its decentralized and tamper-resistant characteristics, blockchain is well-suited for this purpose.

- Secure multimedia transactions – Preventing data manipulation in IoT-based video surveillance and streaming applications.
- Immutable content authentication – Verifying the originality of images, videos, and audio files using cryptographic hashing [27,40].
- Smart contract-based security enforcement – Automating security policies to prevent unauthorized access and data leaks.

Future research will focus on scalable and energy-efficient blockchain implementations, such as lightweight consensus mechanisms, for seamless integration into IoT-based multimedia systems.

8.1.6.5   Edge computing for secure and low-latency multimedia processing

Edge computing is revolutionizing the security of multimedia in IoT by minimizing dependence on centralized cloud services and processing data closer to the origin. The advantages of this approach include:

- Reduced latency and expedited threat detection – AI-powered security frameworks implemented on edge devices are capable of promptly detecting security breaches.
- Reduced attack surface – Storing and processing multimedia data locally prevents exposure to cloud-based cyber threats.
- Improved privacy preservation – Sensitive multimedia content remains within local IoT networks, reducing risks of unauthorized access and data interception [25,41].

Future advancements in AI-optimized edge computing security will enable real-time, decentralized protection of multimedia data in IoT ecosystems.

8.1.7   Cloud computing and multimedia security

8.1.7.1   Impact of cloud computing in multimedia

In multimedia production, cloud computing offers tools that enable real-time collaboration. This framework enables creative professionals across the globe to collaborate on projects in real time. For instance, a graphic designer in one country can edit a video while a sound engineer in another location works on its audio. This capability speeds up the production process and fosters teamwork and creativity, irrespective of geographical barriers [16,20]. Streaming services like Netflix, YouTube, and other content delivery platforms heavily rely on cloud computing to serve millions of users daily. These platforms must deliver high-quality video and audio content efficiently and without interruptions. Here, Figure 8.2 depicts the basic applications of cloud-based systems in multimedia.

Security challenges in cloud-based multimedia

The open and collaborative nature of cloud platforms makes them particularly vulnerable to cyber threats. When multimedia files like videos, images, and audio are stored or shared in the cloud, they encounter numerous significant security risks [16,42]. Key issues include:

Data breaches: Arguably the most critical threat faced in cloud-based multimedia environments, data breaches occur when unauthorized individuals gain access to sensitive multimedia assets. For instance, personal videos or photographs uploaded to the cloud might be exposed if hackers manage to bypass security protocols. Such breaches compromise privacy and can result in severe financial and reputational harm to individuals or organizations [21,43].

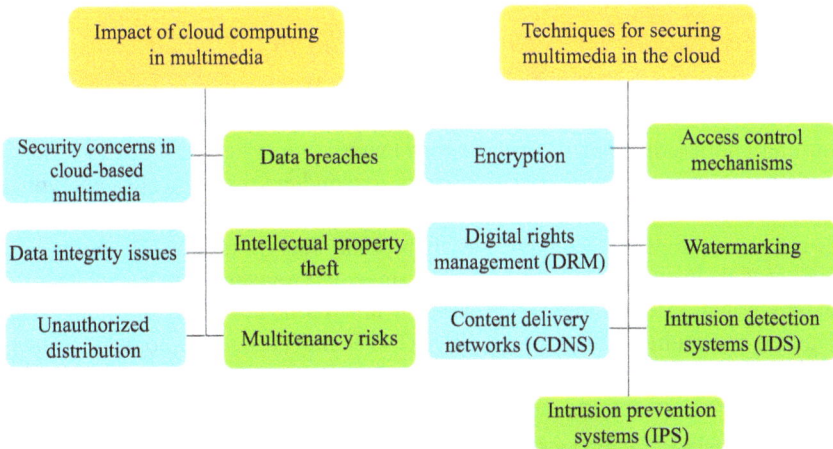

Figure 8.2   Emergence and techniques of cloud-based systems in multimedia security

Data integrity concerns: Ensuring the integrity of multimedia files within the cloud is of utmost importance. Data integrity involves confirming that files remain unchanged and reliable. However, cyberattacks, such as malware infections or unauthorized modifications, can corrupt multimedia files. For instance, a video file could be tampered with to alter its content, making it unusable or misleading. Such alterations can disrupt business operations, tarnish a brand's image, or lead to misinformation.

Intellectual property theft, in which Many multimedia files, such as movies, music tracks, and graphics, are protected by intellectual property rights. Storing these assets in the cloud increases their vulnerability to piracy [44,45]. Hackers may steal and distribute copyrighted content illegally, depriving creators and companies of rightful revenue. For example, a newly released film stored in the cloud could be accessed unlawfully and leaked online, resulting in massive financial losses.

Unauthorized distribution: Once multimedia files are uploaded to the cloud, controlling how they are distributed becomes challenging. Even authorized users may inadvertently or intentionally share content with others who should not have access. For instance, a cloud-stored presentation with sensitive images could be downloaded and shared publicly without permission. This lack of control over distribution can lead to copyright violations, loss of sensitive information, and breach of confidentiality agreements.

Multitenancy risks are also significant, as most cloud platforms use a multi-tenant architecture, where resources such as servers and storage are shared among multiple users. While this setup is efficient and cost-effective, it also raises security concerns. Data from various users is stored on the same physical servers, which heightens the risk of data leaks between users. For instance, if a security vulnerability is present in the cloud infrastructure, the data of one tenant could inadvertently be accessed by another, potentially resulting in privacy violations.

8.1.8   Techniques for securing multimedia in the cloud

Securing multimedia in the cloud involves multiple strategies to protect sensitive data and ensure its safe delivery. Each technique plays a specific role in safeguarding multimedia content from various threats. Some key methods include:

Encryption: At the core of data security in cloud environments lies encryption. This process converts multimedia files into a coded format that unauthorized parties are unable to access without the appropriate decryption key. By doing so, it safeguards multimedia data during both storage and transmission. To achieve this, robust encryption algorithms such as AES are frequently utilized. Additionally, advanced communication protocols like HTTPS and TLS are deployed to enhance security during data transfer. These protocols protect against eavesdropping and unauthorized interception, ensuring that sensitive multimedia content remains confidential and protected.

Access control mechanisms play a vital role in safeguarding multimedia by ensuring that only authorized users or systems can access or alter the data. RBAC allocates permissions based on the specific role of a user, restricting access to only the information essential for their duties. In a similar fashion, ABAC utilizes attributes like user location, device type, or the time of access to determine permission levels. Many cloud providers integrate these systems into their platforms, enabling organizations to maintain better control over who can view, modify, or distribute multimedia files, thereby minimizing the risk of unauthorized access.

DRM technologies are designed to protect the intellectual property of multimedia content creators and distributors. DRM systems enforce rules about accessing, using, and sharing digital content. For instance, they can limit the number of devices a purchased movie can be viewed on or prevent users from copying and distributing music files without permission. These restrictions safeguard content creators from piracy and unauthorized distribution, ensuring they retain control over their work and receive fair compensation.

Watermarking is a technique that embeds hidden or visible information into multimedia files, such as images, videos, or audio. This information, often in text or a logo, identifies the content's owner or source. Watermarks deter piracy because they make it easier to trace the origin of unauthorized copies. For example, if a watermarked video is leaked, the embedded data can reveal who accessed or distributed the file [46,47]. In addition to providing a layer of protection, watermarking helps content creators assert ownership and maintain credibility.

Content delivery networks (CDNs): Security is essential for efficiently delivering multimedia content to users across the globe. However, their widespread use also targets them for attacks, such as interception or unauthorized access. To secure CDNs, cloud providers use encrypted links to transmit data and deploy firewalls to block malicious traffic. Implementing these measures ensures that multimedia files are delivered to the end-users securely, without being tampered with or stolen during transit.

Furthermore, *IDS and intrusion prevention systems (IPS)* serve as advanced security mechanisms that vigilantly monitor network activity for any signs of suspicious behavior. In the context of multimedia security, these systems can detect unusual patterns, such as repeated unauthorized attempts to access files or modifications to multimedia content. Once an intrusion is identified, the system can alert administrators or automatically block the threat [48,49]. IDS and IPS help prevent potential attacks by acting as an early warning system, ensuring that multimedia data remains intact and secure in the cloud.

8.1.9 Blockchain and multimedia security

Blockchain technology has a significant role in developing multimedia security for various reasons, including decentralization, transparency, and immutability. By using blockchain, security in different digital media can be used to store, verify, and track [50,51]. Ultimately, blockchain helps address counterfeiting, piracy, and illegal distribution threats. Blockchain has a broader contribution to multimedia security, though it has some challenges, and this section illustrates such aspects. Here, Figure 8.3 shows the role of blockchain in multimedia systems.

Blockchain and multimedia security

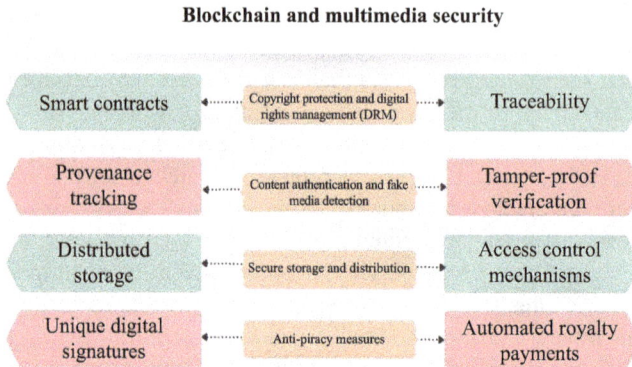

Smart contracts	← Copyright protection and digital rights management (DRM) →	Traceability
Provenance tracking	← Content authentication and fake media detection →	Tamper-proof verification
Distributed storage	← Secure storage and distribution →	Access control mechanisms
Unique digital signatures	← Anti-piracy measures →	Automated royalty payments

Figure 8.3   Significance of blockchain in multimedia security and allied concern

8.1.9.1   Understanding the essence of blockchain technology

Blockchain represents an innovative, distributed digital ledger that meticulously records transactions across a network of interconnected nodes. Operating on the principles of cryptography, it guarantees that once data is captured, it can only be amended with the collective consent of the network members. Each block within the blockchain contains a group of transactions and links to its predecessor, thus forming a continuous chain. The main features of blockchain are as follows:

- *Decentralization*: Unlike traditional databases overseen by a single entity, blockchain operates on a network of multiple nodes, significantly minimizing the risk of a centralized point of failure.
- *Immutability*: Once a transaction is recorded onto the blockchain, it cannot be altered or deleted, thereby preserving the authenticity of the information.
- *Transparency*: In public blockchains, every participant has access to the transaction history, which cultivates trust and accountability [52,53].
- *Cryptographic security*: Transactions are protected through encryption, making it difficult for unauthorized individuals to manipulate the data.
- *Consensus mechanisms*: Various algorithms, such as proof of work (PoW) and proof of stake (PoS), ensure that only valid transactions are included in the blockchain.

8.1.9.2   The contribution of blockchain to multimedia security

Blockchain can be instrumental in safeguarding multimedia content by tackling key security issues like copyright protection, verifying content authenticity, ensuring secure storage, and managing access control.

Safeguarding copyright and digital rights management (DRM)

One of the most crucial uses of blockchain technology in the realm of multimedia security is the safeguarding of intellectual property rights. Traditional DRM methods often rely on centralized databases, making them vulnerable to cyber threats and unauthorized modifications. In contrast, blockchain-driven DRM presents a decentralized solution where ownership details and licensing terms are securely recorded on an unchangeable ledger [37,54].

- *Smart contracts*: With blockchain, we can harness smart contracts that automatically uphold licensing conditions and guarantee that creators receive their due royalties.
- *Traceability*: Every transaction related to ownership and distribution of content is meticulously logged on the blockchain, ensuring that legitimate owners gain proper acknowledgment and appropriate remuneration.

Verifying content authenticity and combating fake media

As deepfake innovations and altered media proliferate, confirming the authenticity of digital assets has become a critical issue. Blockchain technology can bolster content integrity by providing a transparent and verifiable trail of its origin and any modifications it undergoes.

Provenance tracking: Multimedia files can have distinct hashes embedded within them, which are recorded on the blockchain, enabling users to confirm their authenticity and identify any changes.

- *Tamper-proof verification*: Since blockchain data cannot be changed, any modification to a multimedia file can be easily identified.

Secure storage and distribution

Blockchain provides a secure method for storing and distributing multimedia content, minimizing the chances of piracy and unauthorized access.

- *Distributed storage*: Rather than storing multimedia content on a centralized server, blockchain-based systems can utilize distributed storage technologies like the interplanetary file system (IPFS), improving security and dependability [55,56].
- *Access control mechanisms*: Only authorized users with blockchain-verified credentials can access or modify the content, reducing unauthorized distribution.

Anti-piracy measures

Digital piracy is a significant challenge in multimedia security, leading to financial losses for content creators and industries [49,57]. Blockchain can act as a deterrent by creating a transparent and trackable content distribution system.

- *Unique digital signatures*: Each multimedia file can be assigned a distinct cryptographic signature stored on the blockchain, simplifying the process of tracking unauthorized copies [58,59].
- *Automated royalty payments*: Smart contracts ensure content creators receive payments when their content is used, reducing piracy incentives.

8.1.9.3   Significance of blockchain in multimedia security

Blockchain technology provides numerous benefits that make it an effective solution for enhancing multimedia security, which includes *improved data integrity*.

By giving a tamper-proof ledger, blockchain ensures that multimedia content remains unchanged after its original creation, preserving its integrity. *Building trust and ensuring transparency*: Blockchain technology brilliantly records each transaction in a way that's easy to verify and completely transparent, which helps minimize disagreements over content ownership and distribution rights. *Improved efficiency*: Smart contracts automate content licensing and payments, eliminating intermediaries and reducing administrative overhead. *Protection against cyber threats*: Blockchain's decentralized structure makes it resistant to hacking attempts targeting centralized databases. *Empowerment of content creators*, artists, musicians, filmmakers, and other content creators can directly distribute their work using blockchain, ensuring fair compensation without reliance on intermediaries.

8.1.9.4   Obstacles and concerns in adopting blockchain for multimedia security

Although blockchain offers many benefits, its implementation in multimedia security encounters various challenges, including challenges with scalability with the blockchain networks, particularly those utilizing PoW, which encounter scalability issues because of the significant computational demands and slow transaction processing speeds. *High energy consumption*, that is, mining operations in PoW-based blockchains consume considerable energy, raising concerns about environmental impact. Meanwhile, alternative consensus mechanisms like PoS and DPoS offer smarter and more energy-efficient solutions [21,60] for maintaining network integrity. *Legal and regulatory uncertainties*: Many jurisdictions lack clear legal frameworks for blockchain-based DRM and copyright enforcement, creating uncertainty for content creators and distributors [41,61]. *Data privacy concerns* are also included. Blockchain ensures transparency, but storing sensitive multimedia metadata on a public ledger may lead to privacy concerns. Hybrid solutions combining blockchain with encryption techniques are being explored to address this issue. *Adoption barriers*: Many industries still rely on traditional security methods, and transitioning to blockchain-based solutions requires significant investment, training, and technological upgrades. *Storage limitations*: *Blockchain is unsuitable for storing large multimedia files, so* hybrid approaches using decentralized storage systems like IPFS are needed [69,70].

8.1.10   IoT, cloud, and blockchain in multimedia security: Trends, direction, and future aspects

IoT, cloud computing, and blockchain have their wider and emerging significance, and this trend is gaining (refer to Figure 8.4 and the following sections).

8.1.10.1   Future prospects and aspects of IoT and multimedia security

A crucial objective for the future is the establishment of universal IoT security standards to guarantee consistent and strong security on various platforms and devices. Creating global frameworks like ISO/IEC 27030 will significantly shape

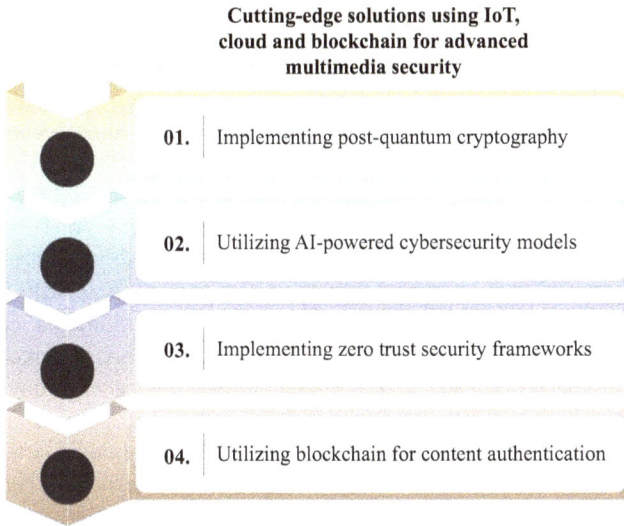

Cutting-edge solutions using IoT, cloud and blockchain for advanced multimedia security

01. Implementing post-quantum cryptography

02. Utilizing AI-powered cybersecurity models

03. Implementing zero trust security frameworks

04. Utilizing blockchain for content authentication

Figure 8.4   Cutting-edge solutions using IoT, cloud, and blockchain for advanced multimedia security

interoperable and scalable security measures. Moreover, cooperation among academia, industry, and policymakers is vital to drive innovation and implement efficient security strategies for IoT-based multimedia applications [62,63]. The secure progress of IoT multimedia ecosystems relies on continual improvements in cybersecurity technologies and proactive security systems. By adopting innovative security approaches, stakeholders can fully utilize the potential of IoT-driven multimedia applications while upholding data privacy, integrity, and user confidence.

Future research should remain flexible and adaptable to counter emerging cyber threats, ultimately shaping a resilient and secure digital future for IoT-based multimedia security. To tackle these challenges, researchers and cybersecurity experts are actively exploring cutting-edge solutions like:

- Implementing PQC to safeguard IoT multimedia security against quantum computing risks.
- Utilizing AI-powered cybersecurity models for real-time detection of abnormalities and behavioral analysis.
- Implementing zero-trust security frameworks to ensure continuous authentication and access management.
- Utilizing blockchain for content authentication prevents data tampering and improves multimedia's credibility.
- Developing secure edge computing architectures to reduce response time and enhance privacy protection.

8.1.10.2   Future prospects of cloud computing and virtualization in multimedia security

Cloud computing and virtualization are dedicated to advancing multimedia systems and are highly required for real-time cooperation. Cloud computing has a high-end requirement to allow creative teams, and it will enable multiple access to graphic software and applications. The virtualization enhances productivity, fostering teamwork, creativity, and advancing multimedia products [64,65]. The geographical barriers to multimedia-based services like Netflix, YouTube, and other content delivery can be waived. Cloud computing-based services are highly significant in drastically enhancing multimedia products and security. Numerous emerging platforms heavily rely on a new wave of multimedia content, significantly transforming the landscape. Looking ahead, services driven by multimedia are set to become increasingly efficient and sophisticated through the use of virtualization, which will facilitate adaptable resource management. This kind of flexibility is immensely advantageous for enhancing multimedia offerings while also addressing modern security needs. Today, OTT platforms are rising rapidly, and cloud-based services significantly impact multimedia products, including improving the playback experience of OTT platforms. Regarding security, using cloud-based fog can be more effective and informative in many ways [68].

8.1.10.3   Future prospects of blockchain in multimedia security

The outlook for blockchain technology in the realm of multimedia security is encouraging, as ongoing research and advancements aim to tackle current issues. Anticipated innovations include:

- *Integration with AI*: AI-driven verification systems can enhance blockchain's ability to detect fake media and authenticate content [19,61].
- *More energy-efficient consensus mechanisms:* Adopting PoS and other energy-efficient consensus models will improve blockchain sustainability.
- *Government and industry regulations*: Blockchain-based copyright protection may become standard practice as legal frameworks evolve.
- *Wider adoption in streaming services*: Major streaming platforms could implement blockchain to prevent unauthorized sharing and piracy.

Blockchain has remarkable potential to transform multimedia security by delivering decentralized, tamper-proof, and transparent mechanisms for copyright protection, content validation, secure distribution, and the fight against piracy [66,67]. However, challenges (refer to Figure 8.5) related to scalability, energy consumption, legal regulations, and adoption barriers must be addressed for widespread implementation.

8.1.11   Concluding remarks

The IoT has revolutionized multimedia applications by facilitating effortless connectivity, automation, and instantaneous data analysis. However, the swift

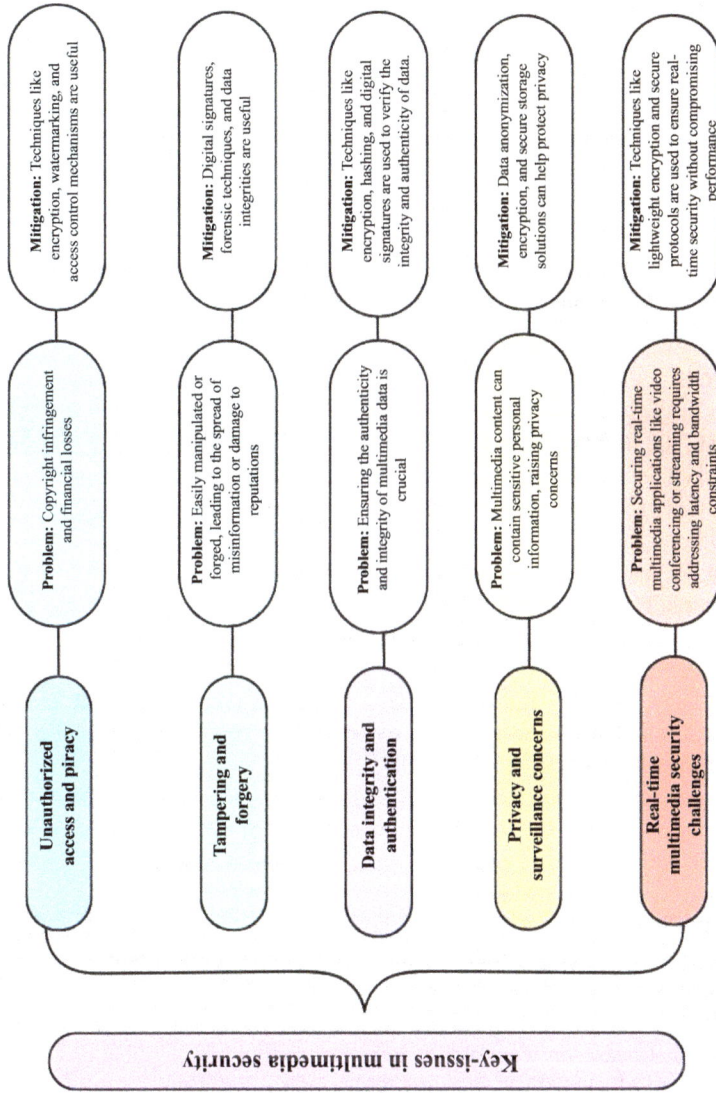

Key-issues in multimedia security

Unauthorized access and piracy

Problem: Copyright infringement and financial losses

Mitigation: Techniques like encryption, watermarking, and access control mechanisms are useful

Tampering and forgery

Problem: Easily manipulated or forged, leading to the spread of misinformation or damage to reputations

Mitigation: Digital signatures, forensic techniques, and data integrities are useful

Data integrity and authentication

Problem: Ensuring the authenticity and integrity of multimedia data is crucial

Mitigation: Techniques like encryption, hashing, and digital signatures are used to verify the integrity and authenticity of data.

Privacy and surveillance concerns

Problem: Multimedia content can contain sensitive personal information, raising privacy concerns

Mitigation: Data anonymization, encryption, and secure storage solutions can help protect privacy

Real-time multimedia security challenges

Problem: Securing real-time multimedia applications like video conferencing or streaming requires addressing latency and bandwidth constraints

Mitigation: Techniques like lightweight encryption and secure protocols are used to ensure real-time security without compromising performance

Figure 8.5   Common challenges and issues of multimedia security, emphasizing IoT, cloud, and blockchain

expansion of IoT-powered multimedia offerings has introduced intricate security challenges, including limited resources, network vulnerabilities, risks associated with deepfakes, and concerns regarding data integrity. To safeguard multimedia information in IoT environments, it's essential to adopt a holistic approach that combines sophisticated encryption methods, AI-enabled threat detection, blockchain authentication, and privacy-enhancing techniques. In today's digital landscape, ensuring the security of multimedia has emerged as a critical issue, mainly due to the growing dependence on digital content for communication, entertainment, business, and educational purposes. Multimedia assets, such as images, audio files, videos, and text, are frequently exchanged online, rendering them vulnerable to cyber threats like unauthorized access, piracy, counterfeiting, and data manipulation. While traditional security solutions like encryption and watermarking have offered some level of protection, they often fall short in addressing the new security hurdles presented by our interconnected world.

References

[1] Ajish, D. (2024). A comprehensive study on benefits and concerns of blockchain in security and compliance in banks. *International Research Journal of Modernization in Engineering Technology and Science*, 6, 2251–2265.

[2] Jain, P., and Rajesh, A. (2021, January). Security of multimedia content for ownership identification using signaling technique. In *2021 International Conference on Computer Communication and Informatics (ICCCI)* (pp. 1–5). IEEE.

[3] Al-kfairy, M., Mustafa, D., Kshetri, N., Insiew, M., and Alfandi, O. (2024). Ethical challenges and solutions of generative AI: An interdisciplinary perspective. *Informatics*, 11(3), 58–87.

[4] Bhumichai, D., Smiliotopoulos, C., Benton, R., Kambourakis, G., and Damopoulos, D. (2024). The convergence of artificial intelligence and Blockchain: The state of play and the road ahead. *Information*, 15(5), 268–300.

[5] Awadallah, R., Samsudin, A., Teh, J. S., and Almazrooie, M. (2021). An integrated architecture for maintaining security in cloud computing based on blockchain. *IEEE Access*, 9, 69513–69526.

[6] Habib, G., Sharma, S., Ibrahim, S., Ahmad, I., Qureshi, S., and Ishfaq, M. (2022). Blockchain technology: Benefits, challenges, applications, and integration of blockchain technology with cloud computing. *Future Internet*, 14(11), 341–362.

[7] Jan, M. A., Cai, J., Gao, X. C., *et al.* (2021). Security and blockchain convergence with internet of multimedia things: Current trends, research challenges and future directions. *Journal of Network and Computer Applications*, 175, 102918.

[8]   Khan, A. A., Laghari, A. A., Shaikh, Z. A., Dacko-Pikiewicz, Z., and Kot, S. (2022). Internet of things (IoT) security with blockchain technology: A state-of-the-art review. *IEEE Access*, 10, 122679–122695.

[9]   Liu, Y., Lu, Q., Zhu, C., and Yu, Q. (2021). A blockchain-based platform architecture for multimedia data management. *Multimedia Tools and Applications*, 80(20), 30707–30723.

[10]   Nair, M. M., and Tyagi, A. K. (2023). AI, IoT, blockchain, and cloud computing: The necessity of the future. In Pandey, R., Goudar, S., and Fatima, S., (eds.). *Distributed Computing to Blockchain* (pp. 189–206). Academic Press.

[11]   Nguyen, D. C., Pathirana, P. N., Ding, M., and Seneviratne, A. (2020). Integration of blockchain and cloud of things: Architecture, applications and challenges. *IEEE Communications Surveys and Tutorials*, 22(4), 2521–2549.

[12]   Rana, S., Mishra, D., and Mukhopadhyay, S. (2021). Blockchain-based multimedia content distribution with the assured system update mechanism. *Multimedia Tools and Applications*, 80(19), 29423–29436.

[13]   Taloba, A. I., Elhadad, A., Rayan, A., *et al.* (2023). A blockchain-based hybrid platform for multimedia data processing in IoT-healthcare. *Alexandria Engineering Journal*, 65, 263–274.

[14]   Uddin, M., Khalique, A., Jumani, A. K., Ullah, S. S., and Hussain, S. (2021). Next-generation blockchain-enabled virtualized cloud security solutions: Review and open challenges. *Electronics*, 10(20), 2493.

[15]   Dong, S., Abbas, K., Li, M., and Kamruzzaman, J. (2023). Blockchain technology and application: An overview. *PeerJ Computer Science*, 9, e1705–e1755.

[16]   Gad, A. G., Mosa, D. T., Abualigah, L., and Abohany, A. A. (2022). Emerging trends in blockchain technology and applications: A review and outlook. *Journal of King Saud University – Computer and Information Sciences*, 34(9), 6719–6742.

[17]   Bhushan, K., and Gupta, B. B. (2019). Network flow analysis for detection and mitigation of fraudulent resource consumption (FRC) attacks in multimedia cloud computing. *Multimedia Tools and Applications*, 78(4), 4267–4298.

[18]   Dhar, S., Khare, A., and Singh, R. (2023). Advanced security model for multimedia data sharing in Internet of Things. *Transactions on Emerging Telecommunications Technologies*, 34(11), e4621.

[19]   Natgunanathan, I., Praitheeshan, P., Gao, L., Xiang, Y., and Pan, L. (2022). Blockchain-based audio watermarking technique for multimedia copyright protection in distribution networks. *ACM Transactions on Multimedia Computing, Communications, and Applications (TOMM)*, 18(3), 1–23.

[20]   Gupta, B. B., Yamaguchi, S., and Agrawal, D. P. (2018). Advances in security and privacy of multimedia big data in mobile and cloud computing. *Multimedia Tools and Applications*, 77, 9203–9208.

[21]   Haddad, N. M., Sabah Mustafa, M., Salih, H. S., Jaber, M. M., and Ali, M. H. (2024). Analysis of the security of internet of multimedia things in wireless environment. *Journal of Cyber Security and Mobility*, 16(3), 161–192.

[22] Dwivedi, Y. K., Ismagilova, E., Hughes, D. L., *et al.* (2021). Setting the future of digital and social media marketing research: Perspectives and research propositions. *International Journal of Information Management*, 59, 59–96.

[23] Nauman, A., Qadri, Y. A., Amjad, M., Zikria, Y. B., Afzal, M. K., and Kim, S. W. (2020). Multimedia Internet of Things: A comprehensive survey. *IEEE Access*, 8, 8202–8250.

[24] Ghadi, Y. Y., Mazhar, T., Shahzad, T., *et al.* (2024). The role of Blockchain to secure internet of medical things. *Scientific Reports*, 14(1), 18422.

[25] Khan, A. A., Shaikh, A. A., and Laghari, A. A. (2023). IoT with multimedia investigation: A secure process of digital forensics chain-of-custody using blockchain Hyperledger Sawtooth. *Arabian Journal for Science and Engineering*, 48(8), 10173–10188.

[26] Chen, H., Hu, Y., and Li, Y. (2023). Blockchain-based security solutions for IoT multimedia applications. *IEEE Internet of Things Journal*, 10(5), 1234–1248.

[27] Krishna, M., Chowdary, S. M. B., Nancy, P., and Arulkumar, V. (2021, October). A survey on multimedia analytics in security systems of cyber physical systems and IoT. In *2021 2nd International Conference on Smart Electronics and Communication (ICOSEC)* (pp. 1–7). IEEE.

[28] Liu, Y., Zhu, L., and Liu, F. (2020). Design of multimedia education network security and intrusion detection system. *Multimedia Tools and Applications*, 79(25), 18801–18814.

[29] Lv, Z., Qiao, L., and Song, H. (2020). Analysis of the security of internet of multimedia things. *ACM Transactions on Multimedia Computing, Communications, and Applications (TOMM)*, 16(3s), 1–16.

[30] Rathee, G., Sharma, A., Saini, H., Kumar, R., and Iqbal, R. (2020). A hybrid framework for multimedia data processing in IoT-healthcare using blockchain technology. *Multimedia Tools and Applications*, 79(15), 9711–9733.

[31] Sathya, M., Jeyaselvi, M., Krishnasamy, L., *et al.* (2021). A novel, efficient, and secure anomaly detection technique using DWU-ODBN for IoT-enabled multimedia communication systems. *Wireless Communications and Mobile Computing*, 2021(1), 4989410.

[32] Kumar, A., Singh, R., and Sharma, P. (2023). AI-driven cybersecurity mechanisms for multimedia security in IoT. *ACM Computing Surveys*, 55(7), 234–256.

[33] Wang, Z., and Liu, Y. (2023). Deepfake detection techniques for securing multimedia content in IoT. *IEEE Transactions on Information Forensics and Security*, 18(4), 112–129.

[34] Mohammed Abdul, S. S. (2024). Navigating blockchain's twin challenges: Scalability and regulatory compliance. *Blockchains*, 2(3), 265–298.

[35] Tripathi, G., Ahad, M. A., and Casalino, G. (2023). A comprehensive review of blockchain technology: Underlying principles and historical background with future challenges. *Decision Analytics Journal*, 9–30, 100344.

[36] Guo, J., Li, C., Zhang, G., Sun, Y., and Bie, R. (2020). Blockchain-enabled digital rights management for multimedia resources of online education. *Multimedia Tools and Applications*, 79, 9735–9755.

[37] Ren, Y., Zhu, F., Zhu, K., Sharma, P. K., and Wang, J. (2021). Blockchain-based trust establishment mechanism in the internet of multimedia things. *Multimedia Tools and Applications*, 80(20), 30653–30676.

[38] Ashraf, M. U., Arif, S., Basit, A., and Khan, M. S. (2018). Provisioning quality of service for multimedia applications in cloud computing. *International Journal of Information Technology and Computer Science (IJITCS)*, 10(5), 40–47.

[39] Yang, Y., Zheng, X., Chang, V., Ye, S., and Tang, C. (2018). Lattice assumption based fuzzy information retrieval scheme support multi-user for secure multimedia cloud. *Multimedia Tools and Applications*, 77, 9927–9941.

[40] Huynh-The, T., Gadekallu, T. R., Wang, W., *et al.* (2023). Blockchain for the metaverse: A review. *Future Generation Computer Systems*, 143, 401–419.

[41] Khan, A. A., Laghari, A. A., Shaikh, A. A., Dootio, M. A., Estrela, V. V., and Lopes, R. T. (2022). A blockchain security module for brain-computer interface (BCI) with multimedia life cycle framework (MLCF). *Neuroscience Informatics*, 2(1), 100030.

[42] Idrees, S. M., Nowostawski, M., Jameel, R., and Mourya, A. K. (2021). Security aspects of blockchain technology intended for industrial applications. *Electronics*, 10(8), 951–975.

[43] Khan, D., Low, T., and Hashmani, M. (2021). Systematic literature review of challenges in blockchain scalability. *Applied Sciences*, 11–38, 9372.

[44] Khan, M. A., and Salah, K. (2018). IoT security: Review, blockchain solutions, and open challenges. *Future Generation Computer Systems*, 82, 395–411. https://doi.org/10.1016/j.future.2017.11.022.

[45] Rai, H. M., Shukla, K. K., Tightiz, L., and Padmanaban, S. (2024). Enhancing data security and privacy in energy applications: Integrating IoT and blockchain technologies. *Heliyon*, 10(19), e38917–e38943.

[46] Musa, H. S., Krichen, M., Altun, A. A., and Ammi, M. (2023). Survey on blockchain-based data storage security for Android mobile applications. *Sensors*, 23–53, 8749.

[47] Shrestha, B., Halgamuge, M. N., and Treiblmaier, H. (2020). Using Blockchain for online multimedia management: Characteristics of existing platforms. In: Treiblmaier, H., and Clohessy, T., (eds.). *Blockchain and Distributed Ledger Technology Use Cases: Applications and Lessons Learned*, 1, 289–303.

[48] Alaba, F. A., Othman, M., Hashem, I. A. T., and Alotaibi, F. (2017). Internet of things security: A survey. *Journal of Network and Computer Applications*, 88, 10–28. https://doi.org/10.1016/j.jnca.2017.04.002.

[49] Zikria, Y. B., Afzal, M. K., and Kim, S. W. (2020). Internet of multimedia things (IoMT): Opportunities, challenges and solutions. *Sensors*, 20(8), 2334.

[50] Qureshi, A., and Megias Jimenez, D. (2020). Blockchain-based multimedia content protection: Review and open challenges. *Applied Sciences*, 11(1), 1.

[51] Rangaiah, Y. V., Sharma, A. K., Bhargavi, T., Chopra, M., Mahapatra, C., and Tiwari, A. (2022, December). A taxonomy towards blockchain based multimedia content security. In *2022 2nd International Conference on Innovative Sustainable Computational Technologies (CISCT)* (pp. 1–4). IEEE.

[52] Namasudra, S., Chakraborty, R., Majumder, A., and Moparthi, N. R. (2020). Securing multimedia by using DNA-based encryption in the cloud computing environment. *ACM Transactions on Multimedia Computing, Communications, and Applications (TOMM)*, 16(3s), 1–19.

[53] Singh, I., and Lee, S. W. (2022). Self-adaptive and secure mechanism for IoT based multimedia services: A survey. *Multimedia Tools and Applications*, 81 (19), 26685–26720.

[54] Sharma, P., Jindal, R., and Borah, M. D. (2024). Blockchain-based distributed application for multimedia system using Hyperledger Fabric. *Multimedia Tools and Applications*, 83(1), 2473–2499.

[55] Lin, I.-C., Kuo, Y.-H., Chang, C.-C., Liu, J.-C., and Chang, C.-C. (2024). Symmetry in blockchain-powered secure decentralized data storage: Mitigating risks and ensuring confidentiality. *Symmetry*, 16(2), 147–162.

[56] Naidu, V. R., Bhat, A. Z., and Singh, B. (2019). Cloud concept for implementing multimedia based learning in higher education. In *Smart Technologies and Innovation for a Sustainable Future: Proceedings of the 1st American University in the Emirates International Research Conference— Dubai, UAE 2017* (pp. 81–84). Springer International Publishing.

[57] Paul, P. K., Solanki, V. K., and Kumar, R. (2019). An analytical approach from cloud computing data intensive environment to internet of things in academic potentialities. In: Peng, S. L., Pal, S., and Huang, L., (eds.). *Principles of Internet of Things (IoT) Ecosystem: Insight Paradigm* (pp. 363–381). Springer International Publishing.

[58] Smith, L., and Jones, K. (2022). Edge computing and AI-enhanced anomaly detection in IoT security. *Elsevier Journal of Network Security*, 15(6), 87–105.

[59] Zhang, P., and Lee, C. (2023). Secure multimedia streaming in IoT environments: Challenges and solutions. *Multimedia Security Journal*, 12(9), 456–478.

[60] Sultana, N. M., and Srinivas, K. (2024). Data privacy protection in cloud computing using visual cryptography. *Multimedia Tools and Applications*, 84, 1–21.

[61] Liu, J., Fan, K., Li, H., and Yang, Y. (2021). A blockchain-based privacy preservation scheme in multimedia network. *Multimedia Tools and Applications*, 80(20), 30691–30705.

[62] Paul, P. K. (2022). Wireless sensor network (WSN) vis-à-vis internet of things (IoT): Foundation and emergence. In: Chaurasiya, S.K., Dutta, J.,

Biswas, A., Dutta, G., Sarkar, M., (eds.). *Computational Intelligence for Wireless Sensor Networks* (pp. 1–16). Chapman and Hall/CRC.

[63] Verma, P., and Ram, B. (2024). Application of blockchain technology in data security. *IP Indian Journal of Library Science and Information Technology*, 1, 51–55.

[64] Paul, P. K., Chatterjee, R., Das, N., Sharma, S. K., Saavedra, R., and Bandyopadhyay, A. (2025). Multimedia security: Basics, its subfields, and allied areas. In: Deb, S., and Sahu, A.K., (eds.). *Securing the Digital World* (pp. 102–116). CRC Press.

[65] Wang, S., Zhang, Y., and Guo, Y. (2022). A blockchain-empowered arbitrable multimedia data auditing scheme in IoT cloud computing. *Mathematics*, 10(6), 1005.

[66] Tanwar, S., and Tyagi, R. (2023). Homomorphic encryption for privacy-preserving multimedia applications in IoT. *Springer Journal of Secure Computing*, 10(2), 345–369.

[67] Yang, W., Wang, S., Hu, J., and Karie, N. M. (2022). Multimedia security and privacy protection in the internet of things: Research developments and challenges. *International Journal of Multimedia Intelligence and Security*, 4 (1), 20–46.

[68] Javaid, M., Haleem, A., Singh, R. P., Khan, S., and Suman, R. (2021). Blockchain technology applications for Industry 4.0: A literature-based review. *Blockchain: Research and Applications*, 2–11, 100027.

[69] Li, J., Wang, X., and Zhang, T. (2023). Quantum cryptography: A new frontier in IoT security. *Journal of Cryptographic Research*, 8(3), 189–204.

[70] Shi, K., Zhu, L., Zhang, C., Xu, L., and Gao, F. (2020). Blockchain-based multimedia sharing in vehicular social networks with privacy protection. *Multimedia Tools and Applications*, 79, 8085–8105.

Chapter 9

Bibliometric analysis and research trends in image encryption: securing sensitive visual content (2020–2024)

Roseline Oluwaseun Ogundokun[1,2,3,9], Pius Adewale Owolawi[1], Elizabeth Mkoba[4], Abdulwasiu Bolakale Adelodun[5], Akinyomade O. Owolabi[6], Monalisa Sahu[7] and Gandharba Swain[8]

Abstract

This chapter presents a comprehensive bibliometric and content analysis of global research trends in image encryption from 2020 to 2024, using a mixed-method approach that integrates quantitative analysis, network visualization, and thematic exploration. Based on 5881 publications retrieved from the Scopus database, the study identifies substantial growth in scholarly output, with an average annual publication increase of 14.2% and a peak growth rate of 18.73% recorded in 2022. China and India emerged as the dominant contributors with 2344 and 1315 publications, respectively, while Dalian Maritime University and Quaid-i-Azam University ranked as the top institutional contributors. The leading author, Wang X., contributed 229 papers, reflecting firm individual productivity and influence. Keyword co-occurrence analysis revealed that "image encryption" (2632 mentions), "chaos" (435), and "cryptography" (348) are central to the field's intellectual structure. The study also highlights the shift toward hybrid encryption schemes, lightweight cryptography for Internet of Things (IoT), and Deoxyribonucleic acid (DNA)-based cryptographic models. Visual analytics

[1]Department of Computer Systems Engineering, Tshwane University of Technology (TUT), South Africa
[2]Department of Computer Science, Landmark University, Nigeria
[3]Department of Centre of Real Time Computer Sciences, Kaunas University of Technology, Lithuania
[4]School of Computational, Communication Science and Engineering, Nelson Mandela African Institution of Science and Technology, Tanzania
[5]ECWA College of Nursing and Midwifery, Nigeria
[6]Department of Microbiology, Landmark University, Nigeria
[7]School of Computer Science and Engineering (SCOPE), VIT-AP University, India
[8]Department of Computer Science and Engineering, Koneru Lakshmaiah Education Foundation, India
[9]Department of Computer Science, Redeemer's University Ede, Osun State, Nigeria

were employed to map publication trends and thematic evolution, including line plots, box plots, and word clouds. The most cited paper garnered 358 citations, underscoring its foundational role in the discipline. With over 300 coauthor clusters identified, collaboration emerges as a critical driver of innovation. These findings provide strategic insights for researchers, practitioners, and policy-makers, enabling informed decision-making and guiding future directions in image encryption research amid the evolving cybersecurity landscape.

Keywords: Bibliometric analysis; image encryption; multimedia security; visual data protection; digital image

9.1 Introduction

The exponential increase in digital data exchange during global interconnectedness has emphasized the necessity of robust image encryption techniques [1,38]. Visual data now constitutes a significant percentage of the information exchanged daily in various industries, including health care, commerce, communication, and national security [1]. Classical cryptographic schemes such as Rivest -Shamir -Adleman (RSA) and advanced encryption standard (AES), although effective in encrypting text data, are usually restricted when applied to image data because they are redundant and have a strong inter-pixel correlation [2]. This restriction has trig-gered a surge of studies focused on designing expert encryption schemes tailored uniquely for images.

Image encryption employs a variety of mathematical and computational fra-meworks for transforming visual information into forms that are not intelligible, hence safeguarding against unauthorized access or manipulation [3,36]. Early methods predominantly emulated conventional cryptographic practices, but more recent developments have endeavored to integrate chaos theory, DNA computing, compressive sensing, and quantum cryptography concepts ever more deeply into attempts at heightened security and efficiency [1]. Being sensitive to initial con-ditions and pseudo-random, chaotic systems has promising features for establishing secure image encryption algorithms.

Bibliometric analysis is key in exposing research development's structural and dynamic character within a given discipline [4,5]. It allows identification of key scholars, institutions, countries, networks of collaboration, and future the-matic developments. In research on image encryption, bibliometric analyses can map scientific productivity, reveal mainstream research streams, and signal future directions, providing valuable insights for both seasoned researchers and novices [5,6].

Current bibliometric research in cybersecurity has shown that image encryption studies are continuing to expand, especially with the advent of Internet of Things (IoT) devices, multimedia transmission systems, and intelli-gent surveillance systems, all of which call for lightweight yet secure encryption techniques [4,7]. The need for encryption in real-time in resource-constrained environments has also added to the focus on developing efficient chaotic and hybrid cryptosystems [8,9]. Additionally, content analysis of scholarly papers describes an interdisciplinary crossroads of cryptography, machine learning, multimedia security, and blockchain technologies to counter the high-tech threat posed by rising computational power developments, e.g., the emerging quantum computing paradigm [10,11]. Employment of DNA-based and hyperchaotic models for image encryption is a milestone in creating advanced and secure encryption methods [12].

India and China have been noted as the highest countries with regard to research production in this field due to substantial government investment in research and development [6]. Dalian Maritime University and Quaid-i-Azam University are some of the institutions that have become high-ranking producers [4], with top-productivity authors such as Wang X., Zhang Y., and Singh H. have shaped the field extensively by enormous writings. Keyword co-occurrence analysis indicates that the most significant areas of research over the past five years include "chaos," "image encryption," "security analysis," "cryptography," and "chaotic maps" [13,14]. Interestingly, the research on chaos-based encryption schemes continues to expand because of their robustness in generating unpredictable and sensitive encryption mechanisms. Additionally, emerging fields of interest such as lightweight encryption for IoT devices, privacy-preserving surveillance, and DNA-based cryptographic models are expected to guide future research [15].

In the wake of dynamism and rapid technological progress in this field, a detailed bibliometric and trend analysis is essential to trace the development patterns systematically, the intellectual structure, and frontiers of research in image encryption. This study applies a mixed-method approach incorporating quantitative bibliometric analysis with qualitative content analysis to explore international research activity between 2020 and 2024. Based on Scopus-indexed information and Python-based data analytics tools, this study performs statistical analysis, network visualization, and keyword mapping to extract major findings in the scholarly domain.

This work consistently identifies leading authors, institutions, principal contributor countries, influential papers, and emerging research themes, providing in-depth insight for researchers, practitioners, and policymakers. The results seek to guide strategic decision-making, develop collaborations, and indicate prospective areas for novel research in protecting image data against increasingly developing cybersecurity challenges [16,17].

9.2 Background and motivation

The extensive application of computer technologies and ubiquitous utilization of multimedia information have significantly increased the significance of image security in modern society. From telemedicine and smart surveillance to safe communication and cloud computing, images are being used to carry private and confidential information [5,15]. However, the ease with which digital images can be intercepted, altered, or illegally distributed poses significant threats to data privacy and information integrity. Traditional encryption algorithms, designed primarily for alphanumeric data, often fail to address the specific structural and statistical properties of image data, necessitating the development of specialized image encryption techniques. In the last 20 years, imaging encryption research accelerated through the demands of developing speed-efficient, low-weight, but highly secure algorithms to withstand continually advancing cyberattacks [18]. Chaos theory, DNA computing, and compressive sensing methods have gained popularity based on their potential to enhance confusion and diffusion properties necessary for successful encryption. Moreover, the emergence of interdisciplinary fields – integrating cryptography with machine learning, blockchain, and quantum computing – has introduced new paradigms for securing image data [11].

Despite such advances, a limited large-scale understanding of the development trajectory, influential contributors, common research trends, and emerging patterns in the study of image encryption exists. Even though previous bibliometric reviews have mapped out broader image processing and cybersecurity fields [4,6], the specialized analysis targeted toward image encryption alone, particularly within the recent transformation period from 2020 to 2024, is limited. The impetus behind this research is twofold. First, by carrying out an exhaustive bibliometric and trend analysis, this work seeks to shed light on the composition, dynamics, and progression of the field of image encryption research. Identifying who the key authors are, which organizations and nations dominate, and where thematic trends are developing can strategically inform future research agenda and funding options [7,9].

Second, with the dynamic technology landscape and constant new security threats, it is imperative to address current research voids and probable avenues for innovation. This study fulfills these necessities by exploring variations in keyword focus, technological advances from conventional cryptographic methods to chaos- and DNA-based cryptography schemes, and the advent of interdisciplinary research collaborations. Through this comprehensive approach, the research aims to develop actionable findings that can inform and guide researchers, business executives, and policymakers in improving next-generation image encryption system design [10,13].

Finally, this research is motivated and backed by the imperative need to secure visual information in an increasingly networked digital world. Through an intensive, evidence-led critique of past and ongoing research activity, this research aims to guide the strategic evolution of secure, effective, and strong image encryption methods.

9.3   Materials and methods

Following the outlined motivations and research needs, this research uses a repeatable and systematic methodological strategy to analyze worldwide research trends in image encryption. Through bibliometric analysis, content analysis, and sophisticated data visualization techniques, the research attempts to identify important information regarding the evolution and future directions of the field. The following section describes the materials, data sources, methodological procedures, and analytical methods to achieve these research objectives. This chapter describes the methodology for conducting a comprehensive bibliometric and content analysis of research trends in image encryption. A mixed-methods approach was employed, integrating quantitative methods like descriptive statistical analysis and network visualization with qualitative methods like content interpretation. The study uses evidence from the Scopus database, processed in the Jupyter Notebook environment using Python libraries specifically optimized for data wrangling, analysis, and visualization. The research design is transparent, reproducible, and rigorous, and covers data collection, cleaning, descriptive and inferential analysis, and graphical presentation of results. Each iteration is thoughtfully designed to offer strong insights into the publication trends, thematic changes, leading authors, and emerging lines of research in the image encryption domain.

9.3.1   Research methodology

This research employs a mixed-methods methodology by employing both quantitative and qualitative analyses in providing a comprehensive evaluation of the dataset. Conducted within the Python-assisted Jupyter Notebook, the study leverages its reproducible and interactive nature for fast data analysis. As evident from Figure 9.1, the methodological framework begins with establishing the environment via importing necessary libraries such as NumPy, Matplotlib, and Pandas. The dataset, acquired in comma-separated values (CSV) mode from Scopus, was read into the notebook for processing. Subsequent data wrangling was performed to remove duplicate records and correct missing values, enhancing quality. Descriptive statistical analysis then ensued to determine variable distribution and uncover important trends. Finally, visualization techniques were employed to support result interpretability and derive usable information from the dataset.

9.3.2   Data collection

The data employed in this study are bibliometric publication data from the Scopus database with quantitative and qualitative characteristics. For the qualitative components – title, abstract, authors' affiliations, and keywords –

Figure 9.1   The research methodology

systematic content analysis was applied to identify recurring themes, emerging research trends, and co-authorship patterns. The qualitative components of the data, publication titles, author keywords, and participant quotations, were examined utilizing qualitative data analysis methods. Descriptive statistics were used to report and illustrate salient aspects of the data, such as measures of central tendency and dispersion. However, inferential statistical methods allowed explorations of associations, trends, and patterns within the broader literature context.

9.3.3   Data wrangling

Data wrangling is an important data management procedure involving data detection, extraction, cleaning, and integrating data to analyze [19,20]. Raw data needs to be prepared from diverse sources for further analysis and decision-making [21]. Moreover, big datasets need to be prepared for decision support systems and business intelligence [22].

Python is now an excellent tool for data wrangling and analysis. It offers numerous libraries and packages for data manipulation, including NumPy, SciPy, TensorFlow, and Keras. The versatility of Python is also expressed in cloud computing applications, which can handle effective data processing and machine learning processes. To ensure efficient data wrangling, researchers have developed systems like wrangle search, where data wrangling functions from previous Python codes are mined automatically so that analysts can leverage existing work [23]. For data science novices, comprehensive resources such as "Minimalist Data Wrangling with Python" provide a novice's guide to significant concepts and techniques, from data cleaning, feature selection, exploratory analysis, dimensionality reduction, clustering, and pattern modeling [24]. Such enhancements to Python-based data wrangling packages and tools contribute to more efficient and convenient data preparation and analysis workflows.

9.3.4   Data analyses

Data analyses are crucial for extracting useful information from raw data to facilitate decision-making in various fields [15,25]. It involves multiple stages such as cleaning, transformation, and modeling of data with the help of different techniques, including data mining, text analytics, and data visualization [15]. Appropriate selection of the appropriate analysis techniques is necessary because various techniques have differing strengths and applications [26]. The recent advances in artificial intelligence, machine learning, and deep learning have significantly enhanced data analyses capacity [25].

9.3.4.1   Descriptive statistics

Descriptive statistics are good data summarization and presentation tools. They provide qualitative or quantitative summaries of datasets, typically presented in tables or graphs [27]. Descriptive statistics include frequency, central tendency, dispersion, and position [28]. Descriptive statistics include minimum and maximum values, range, percentiles, mean, median, mode, standard deviation, variance, skewness, and kurtosis [29]. In spatial statistics, they are adjusted to accommodate geographic space, that is, mean center and standard deviation ellipse [29]. Descriptive statistics are relevant to initial data analyses in research and should precede inferential statistical comparisons [28]. Descriptive statistics aid in the description of relationships among variables in samples or populations [28] and may reveal patterns in data [16]. However, descriptive statistics cannot be used to make inferences outside the data or to test hypotheses [16].

9.3.4.2 Content analysis

Content analysis is a widely used methodology for systematically studying human communication in various forms, including written texts, images, and audio [17,30]. It involves breaking down content into smaller units for analysis, ranging from qualitative approaches like grounded theory to quantitative techniques using concept dictionaries [11,30]. The method has roots in positivism and is often used in combination with other research methods [17]. Content analysis offers flexibility in addressing research questions where quantitative data are unavailable and provides an unobtrusive way to study socio-cognitive concepts in context [11].

9.3.5 Visualization

Data visualization means the graphical representation of information, which makes it easier to comprehend complex datasets through the utilization of visual elements like charts, graphs, and maps [8]. Data visualization helps identify patterns, trends, and correlations that may not be apparent in textual data [8]. Data visualization is crucial for decision-makers since it allows them to see analytics in a graphical format and make complex concepts easier to grasp [7]. Data visualization tools provide easy ways of analyzing big data, and therefore, it is useful in numerous areas, including education. Data visualization enables storytelling by recognizing trends and outliers within organizations [31]. Computer graphics in recent times have significantly impacted current visualization techniques. While data visualization is most directly useful to data teams, its application extends beyond data analysts and scientists as a tool for conveying data-driven insights across organizational levels [10].

9.3.5.1 Line plot

Line plots are widely used graphical representations in scientific and business contexts to illustrate trends and correlations. Automated analysis of line plots in technical documents has been developed to extract relevant information, including text and plot data, with high accuracy for axes extraction and moderate success in label extraction and curve tracing [32]. A novel approach, the line mosaic plot, uses lines instead of tiles to represent cell sizes in contingency tables, offering a more intuitive visualization of data.

9.3.5.2   Bar chart

Bar charts are widely used graphical representations of qualitative data, employing rectangular bars which correspond proportionally to the represented values [33]. Bar charts have the benefits of faster interpretation of data and esthetic appeal over text or table use [33]. However, their effectiveness is design-dependent. Cleveland and McGill have found that the aligned bar charts are better than stacked ones, and comparisons between adjacent bars are more accurate than between distant bars [34].

9.3.5.3   Word Cloud

Word clouds are visual representations of text data, indicating keywords in words of proportional size based on frequency or importance [9,14]. Word clouds have evolved from organizational tools to effective visualization techniques for summarizing documents and analyzing text [12]. Lohmann *et al.* [13] built Concentri Cloud, a multi-text comparison word cloud design involving words in concentric circles with document-specific vocabulary on the outside circle and overlapping vocabulary on the inside circles. These new word cloud techniques expand beyond the random traditional designs, covering semantic relationships and time information [12]. Word clouds have been used in various fields, with mathematics education utilizing them as classroom organizational and communication tools [14].

9.3.5.4   Box plot

Box plots make excellent visual aids for appreciating and comparing data distributions. Created by John Tukey, they give a graphical depiction of key statistical measures like median, quartiles, and potential outliers [35,36]. The box displays the interquartile range, while whiskers extend to the minimum and up to a maximum, thus giving a summary of data level, spread, and symmetry. Box plots are primarily applied to compare many data groups simultaneously and are better than histograms for several distribution plots. Box plots can be easily hand-drawn or through statistical software and are therefore accessible to researchers [37]. Box plots are more than mere table substitutes; they are tools that complement quantitative thinking. Due to their ability to detect underlying patterns and enrich data interpretation, box plots must be utilized more frequently in exploratory data analyses.

9.4   Result

This chapter presents the results of the comprehensive bibliometric and content analyses of image encryption research works published from 2020 to 2024. The findings are organized to uncover striking characteristics of the dataset, including publication trends and growth rates, authors and institutions making contributions to the field, highly cited papers, keyword co-occurrence, and emerging research topics. Quantitative results are complemented by visualizations such as line plots, bar charts, box plots, and word clouds to facilitate interpretability and enable a more nuanced understanding of the evolving research landscape. The findings provide a sound foundation for a discussion of the intellectual structure, collaboration networks, and future directions in the research domain of image encryption.

9.4.1   Experimental setup

The computer runs on Windows 11, a 64-bit operating system, an x64-based processor equipped with Python 3. The system's processor is Intel(R) Core (TM) i3-3110M CPU @ 2.40 GHz 2.40 GHz with a RAM of 16.00 GB. All the code used in this study was written in the Jupyter notebook environment, and all relevant Python libraries were installed using pip install and correctly imported.

The statistical summaries provided in Table 9.1 offer insights into the dataset's numerical aspects. The dataset spans publications from 2020 to 2024, with an average publication year of approximately 2022.27. The median citation count is 4, with 25% of publications having 1 or fewer citations and 75% having 15 or fewer. This highlights a skewed distribution with a long tail of highly cited works, as shown in Figure 9.2.

Table 9.1   Statistical summary of the numerical data

	Year	**Cited by**
Count	5881	5881
Mean	2022.270532	13.951539
Std	1.403021	26.198176
Min	2020	0
25%	2021	1
50%	2022	4
75%	2024	15
Max	2024	358

Box plot of citation counts

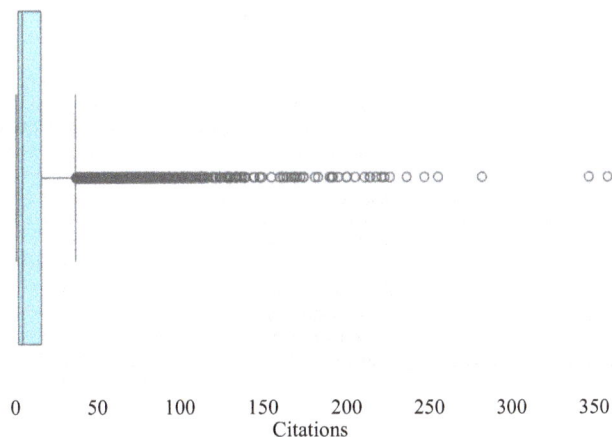

Figure 9.2   Box plot of citation count

The statistical summaries in Table 9.2 provide valuable insights into the dataset's non-numerical aspects. The dataset includes 5364 unique authors, with "Yadav S.; Singh H." one of the most frequent author pairs, appearing in 6 publications. There are 5866 unique titles, with the most common title occurring only 3 times, indicating a high degree of originality. Similarly, the dataset contains 1389 unique source titles, with "Multimedia Tools and Applications" being the most frequently cited, accounting for 510 publications.

Affiliations also show diversity, with 5179 unique institutions represented. The "School of Information Science and Technology" appears most frequently, with 58 occurrences. The dataset further highlights 5876 unique abstracts, demonstrating a strong research uniqueness, as the most repeated abstract appears only twice. Regarding publishing entities, there are 274 unique publishers, with the Institute of Electrical and Electronics Engineers (IEEE) being the most prominent, responsible for 1239 publications. Additionally, the dataset includes 236 unique sponsors among 362 entries, with IEEE being the most common, sponsoring 37 studies. Regarding conference participation, there are 1122 unique names, with the "2023 IEEE International Conference on Memristi …" being the most frequently mentioned, appearing 7 times. Conference dates and locations vary significantly, though "Virtual, Online" is the most frequent location. The dataset is heavily composed of articles, which comprise 4322 of the 5881 entries, emphasizing a strong focus on traditional journal publications. Additionally, 5789 publications are

Table 9.2   Statistical summary of the nonnumerical data

Items	Count	Unique	Top	Freq
Authors	5881	5364	Yadav S.; Singh H.	6
Title	5881	5866	Optical image encryption using non-separable v …	3
Source title	5879	1389	Multimedia tools and applications	510
Affiliations	5879	5179	School of Information Science and Technology, …	58
Authors with affiliations	5880	5710	Wang Q., National and Local Joint Engineering …	5
Abstract	5881	5876	Due to the significance of image data over the …	2
Publisher	5876	274	Institute of Electrical and Electronics Engine …	1239
Sponsors	362	236	IEEE	37
Conference name	1403	1122	2023 IEEE International Conference on Memristi …	7
Conference date	1401	897	23 September 2022 through 25 September 2022	9
Conference location	1402	509	Virtual, online	97
Document type	5881	4	Article	4322
Publication stage	5881	2	Final	5789
Open Access	1758	7	All open access; Gold open access	1041

Table 9.3 Publication growth rate from 2020 to 2024

Year	Count	Grow rate (%)
2020	901	–
2021	977	8.435072
2022	1160	18.730809
2023	1316	13.448276
2024	1527	16.033435

Publication volume over the past 5 years

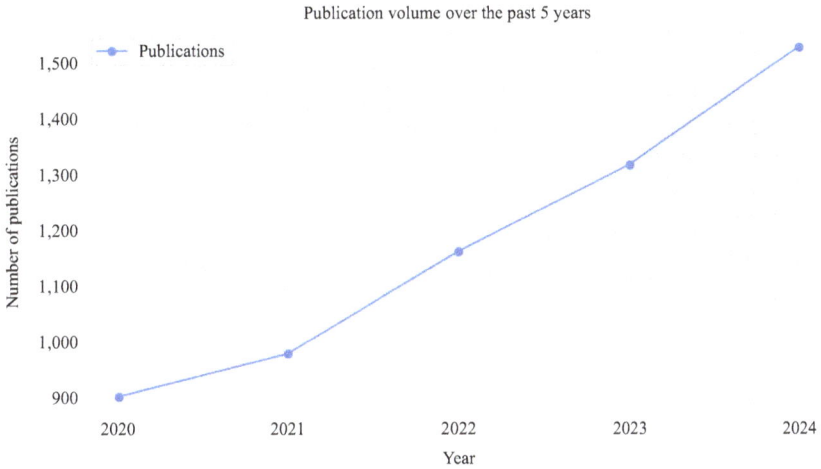

Figure 9.3 Publication volume over the past five years

classified as being in the final stage, indicating that most entries are completed works. Finally, there are 7 categories of open-access publications, with "All Open Access: Gold Open Access" being the most common, appearing 1041 times.

Table 9.3 and Figure 9.3 provide an overview of the publication growth rate from 2020 to 2024, highlighting the annual increase in publications within the dataset. In 2020, the dataset began with 901 publications. As this is the starting point, no growth rate has been calculated for this year. In 2021, the number of publications increased to 977, representing a growth rate of approximately 8.44%. In 2022, the publication count rose to 1160, marking a significant growth rate of 18.73%. In 2023, the growth continues with 1316 publications, although the growth rate slightly decreases to 13.45%. In 2024, the dataset records 1527 publications, with a growth rate of 16.03%.

Table 9.4 highlights the top contributing authors in the dataset, showcasing those with the most publications. Wang X. leads the list with 229 publications, indicating a significant contribution to the research output. Zhang Y. follows with 146 publications, and Zhang X. with 117 publications. Wang Y. and Zhang J. also appear prominently, with 112 and 96 publications, respectively. The list continues with Wang Z. (89 publications), Li J. (85 publications), Li X. (81 publications), Liu Y. (77 publications), and Zhang H. (75 publications).

Table 9.5 presents an overview of the top contributing countries in the dataset, offering insights into the geographical distribution of research output. These data are essential for understanding the global research landscape and identifying key regions driving scientific advancements.

China leads with 2344 publications, followed by India with 1315. Iraq and Egypt contribute 244 and 203 publications, respectively, while Saudi Arabia accounts for 163. The United States and Pakistan have comparable outputs, with 120 and 119 publications, respectively. Turkey, Iran, and Morocco complete the list with 97, 87, and 79 publications, respectively.

Table 9.6 presents the top affiliating institutions contributing to the dataset, offering insights into the academic and research centers that are most active in producing scholarly outputs. Identifying these key institutions helps highlight the leading research hubs and their potential influence on scientific advancements within the dataset's scope.

Table 9.4   Top contributing authors

Authors	Count
Wang X.	229
Zhang Y.	146
Zhang X.	117
Wang Y.	112
Zhang J.	96
Wang Z.	89
Li J.	85
Li X.	81
Liu Y.	77
Zhang H.	75

Table 9.5   Top contributing countries

Country	Count
China	2344
India	1315
Iraq	244
Egypt	203
Saudi Arabia	163
United States	120
Pakistan	119
Turkey	97
Iran	87
Morocco	79

Table 9.6   Top affiliating institution

Affiliations	Count
School of Information Science and Technology, Dalian Maritime University, Dalian 116026, China	58
Department of Mathematics, Quaid-i-Azam University, Islamabad, Pakistan	17
Electronic Engineering College, Heilongjiang University, Harbin 150080, China	16
Faculty of Mathematics and Computer Science, Guangdong Ocean University, Zhanjiang 524088, China	13
School of Information and Control Engineering, China University of Mining and Technology, Xuzhou 221116, China	13
School of Electrical and Automation Engineering, East China Jiaotong University, Nanchang 330013, China	11
School of Software, Nanchang University, Nanchang 330031, China	11
College of Computer Science and Electronic Engineering, Hunan University, Changsha 410082, China	10
College of Software, Taiyuan University of Technology, Jinzhong, 030600, China; College of Information and Computer, Taiyuan University of Technology, Jinzhong 030600, China	10
College of Computer Science and Technology, Hengyang Normal University, Hengyang 421002, China; Hunan Provincial Key Laboratory of Intelligent Information Processing and Application, Hengyang 421002, China	9

The School of Information Science and Technology, Dalian Maritime University, China, leads with 43 publications. It is followed by the Department of Mathematics, Quaid-i-Azam University, Pakistan, with 17 publications. The Electronic Engineering College, Heilongjiang University, China, contributed 16 publications. Both the Faculty of Mathematics and Computer Science, Guangdong Ocean University, China, and the School of Information and Control Engineering, China University of Mining and Technology, China, have 13 publications each.

The School of Electrical and Automation Engineering, East China Jiaotong University, China, and the School of Software, Nanchang University, China, each account for 11 publications, while the College of Computer Science and Electronic Engineering, Hunan University, China, and the College of Software, Taiyuan University of Technology, China, each contribute ten publications.

Table 9.7 presents the frequently cited papers within the dataset, highlighting those that have garnered significant attention in the research community. The most highly cited publication in the dataset is "Image encryption algorithm for synchronously u …" by Wang X. and Gao S., which has 358 citations. Following closely is "Fractal sorting matrix and its application on …" by Xian Y. and Wang X., with 347 citations. The paper "Image encryption algorithm based on the matrix …" by Wang X. and Gao S. has 282 citations, while "A Comprehensive Review on Image Encryption Tec …" by Kaur M. and Kumar V. has 255 citations.

Other notable contributions are "A novel chaos-based optical image encryption u …" by Farah M.A.B. *et al.*, with 247 citations, "Cross-plane colour image encryption using a tw …" by Hua Z. *et al.*, with 236 citations, "An Efficient Image

Table 9.7   Frequently cited papers

S/ No	Title	Authors	Cited by
1	Image encryption algorithm for synchronously u ...	Wang X.; Gao S.	358
2	Fractal sorting matrix and its application on ...	Xian Y.; Wang X.	347
3	Image encryption algorithm based on the matrix ...	Wang X.; Gao S.	282
4	A Comprehensive Review on Image Encryption Tec ...	Kaur M.; Kumar V.	255
5	A novel chaos-based optical image encryption u ...	Farah M.A.B.; Guesmi R.; Kachouri A.; Samet M.	247
6	Cross-plane colour image encryption using a tw ...	Hua Z.; Zhu Z.; Yi S.; Zhang Z.; Huang H.	236
7	An Efficient Image Encryption Scheme Based on ...	Lu Q.; Zhu C.; Deng X.	226
8	A novel one-dimensional sine powered chaotic m ...	Mansouri A.; Wang X.	223
9	Design and Analysis of Multiscroll Memristive ...	Lai Q.; Wan Z.; Zhang H.; Chen G.	221
10	An image encryption scheme based on a new hybr ...	Farah M.A.B.; Farah A.; Farah T.	218

Encryption Scheme Based on ..." by Lu Q. *et al.*, with 226 citations, "A novel one-dimensional sine powered chaotic m ..." by Mansouri A. and Wang X., with 223 citations, "Design and Analysis of Multiscroll Memristive ..." by Lai Q. *et al.*, with 221 citations and "An image encryption scheme based on a new hybr ..." by Farah M.A.B. *et al.*, with 218 citations.

Table 9.8 and Figure 9.3 present an overview and visualization of the top keyword co-occurrences in the dataset. This data provides insights into the key research themes and focus areas, reflecting the interests and priorities of the research community. "Image Encryption" is the dominant keyword, with the highest count (2632). "Encryption" (465) and "Chaos" (435) appear as closely related keywords. "Chaotic Systems" (360) and "Chaotic Maps" (276) demonstrate a strong relationship between chaos theory and cryptography.

"Cryptography" (348), "Security" (268), and "Decryption" (150) are also key to the keyword co-occurrence analysis. "Logistic Map" (178) and "S-Box" (164), "Security Analysis" (128) and "Information Security" (114), "Color Image Encryption" (126), "Reversible Data Hiding" (124), and "Compressive Sensing" (108) are also notable. "Hyperchaotic Systems" (106) emerges as an important area, while "Multimedia Security" (109) and "Image Security" (109) are also present.

Figure 9.4 is the visualization of the keyword trends over the specified periods and provides insights into the shifting focus and interests within the research

Table 9.8   Top keyword co-occurrence

Count	Author keywords
Image encryption	2632
Encryption	465
Chaos	435
Chaotic system	360
Cryptography	348
Chaotic map	276
Security	268
Logistic map	178
S-box	164
Chaotic maps	163
Diffusion	152
Decryption	150
Security analysis	128
Color image encryption	126
Reversible data hiding	124
Information security	114
Image security	109
Multimedia security	109
Compressive sensing	108
Hyperchaotic system	106

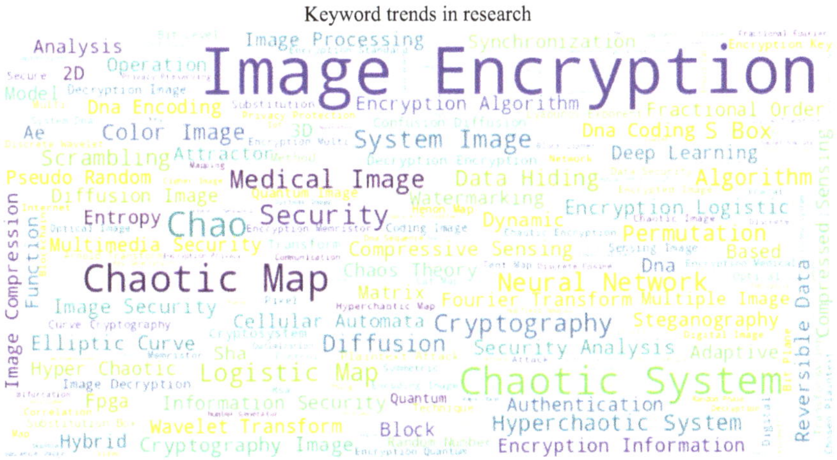

Figure 9.4   Word Cloud of keyword trends

community. From 2020 to 2022, During this period, "hybrid cryptograph," "image encryption," "modulus operation," "random key," and "substitution cipher" were the most popular keywords. Keywords such as "2-d cellular automata," "block cipher," "IoT," and "lightweight" also followed a similar trend, peaking in 2021.

Table 9.9   Author cluster

Research team	Number of publications
Yadav S., Singh H.	6
Kumar A., Dua M.	6
Roy M., Chakraborty S., Mali K.	6
Chuman T., Kiya H.	5
Wang Q., Zhang X., Zhao X.	5
Ahmad I.; Shin S.	5
Ahmad I.; Shin S.	5
Bai B., Wei H., Yang X., Gan T., Mengu D., Jarrahi M., Ozcan A.	4
Sabir S., Guleria V.	4
Ravi R.V., Goyal S.B., Djeddi C.	4

The themes "image encryption," "Internet of Things," "privacy," "secure surveil-lance," and "security" also peaked in 2021. The interest in these topics increased from 2020, reaching a peak in 2021.

Between 2022 and 2024, this period witnessed the emergence of "chaos," "chaotic system," and "image encryption" as the leading new trends. Additionally, keywords such as "Arnold transform," "DNA," and "image encryption" evolved. These themes reached their peak in 2023, reflecting a concentrated period of research and development. However, they began to fall out of trend toward the end of 2024. "Chaos-based encryption," "cryptanalysis," and "image encryption schemes" began gaining traction in 2022 and are expected to continue growing through 2025.

The information about author clusters in Table 9.9 provides insights into the collaborative nature of research within the dataset. There are over 300 author clusters, each comprising at least 2 researchers. Some clusters span multiple institutions and departments. Three clusters have produced six publications, while four clusters have published five.

9.5   Discussion

9.5.1   Interpretation of the findings

This study employs a mixed-method approach, utilizing Python in the Jupyter Notebook environment to integrate quantitative and qualitative analyses, perform data wrangling, conduct descriptive statistical analysis, and apply data visualization techniques for comprehensive dataset evaluation.

In Table 9.1, the standard deviation of 1.40 indicates a moderate spread around the mean, suggesting that most publications are concentrated around the years 2021 to 2023. The quartiles further confirm this, with 25% of publications from 2021, 50% from 2022, and 75% from 2024. The average number of citations per pub-lication is approximately 13.95, with a high standard deviation of 26.20, indicating significant variability in citation counts. The minimum citation count is 0, while the maximum is 358, suggesting that a few publications are highly cited, skewing the average. Table 9.2 indicates a total of 5364 unique authors, reflecting a diverse range of contributors, with only a few authors having multiple publications. The dataset includes 5866 unique titles, with the most common title appearing only 3 times, highlighting a high diversity in research topics. Among the 1389 unique source titles, *Multimedia Tools and Applications* is the most frequent, accounting for 510 publications, suggesting a concentration of research in specific journals or conference proceedings. With 5179 unique affiliations, the dataset represents a broad spectrum of institutional contributors. Additionally, 5710 unique author-affiliation combinations indicate extensive collaboration across institutions. The dataset also contains 5876 unique abstracts, demonstrating a high level of originality in research content. Of the 362 entries with sponsorship details, 236 unique sponsors are recorded, with IEEE being the most frequent, appearing 37 times, indicating a variety of funding sources, with IEEE playing a significant role. Conference dates and locations vary widely, with "Virtual, Online" being the most common, reflecting the growing impact of virtual events.

Journal articles dominate the dataset, comprising 4322 of the 5881 entries, underscoring a strong focus on traditional academic publishing. Furthermore, 5789 publications are in the final stage, indicating that most entries represent completed works. The dataset includes 7 categories of open access, with "All Open Access; Gold Open Access" being the most prevalent, appearing 1041 times, signifying that a substantial portion of the research is freely accessible. Table 9.3 and Figure 9.3 present an overview of the publication growth rate from 2020 to 2024, illustrating the annual increase in research output within the dataset. A moderate rise in 2021 indicates steady growth, suggesting a growing interest or capacity in the relevant research fields. During 2022, a sharp boost can be accounted for by heightened research endeavors, augmented expenditure, or working up to the maturity of ongoing investigation. Growth continues in 2023, yet more slowly, perhaps indicating the plateauing of the peak of the previous year. By 2024, the pace of growth will recuperate, corresponding to renewed activity generated by newly emerging technologies, intensified cooperation, or enhanced finance opportunities.

In general, the dataset shows a consistent increasing trend in publications generated within a span of five years, but with varying growth rates. The consistent increase in 2021 is followed by a peak in 2022, then a slight drop in 2023 and a rebound in 2024. This trend may be an expression of mass changes taking place in the research world as a result of global events, technological advancements, or shifting research directions. Long-term growth indicates a dynamic and expanding research environment, with significant implications for the transmission of knowledge and evolution of the subject matter within the dataset. The prevalence of several authors bearing the surname Zhang in Table 9.4 suggests that the surname is not uncommon, and perhaps it suggests a concentration of research activity where it is prevalent, e.g., China. In addition, the rate of occurrence of surnames like Wang, Zhang, and Li shows significant input by Chinese researchers since these are some of the most commonly occurring Chinese surnames.

The statistics reveal that a few authors contribute to a high percentage of the papers, perhaps reflecting their dominance in specific research areas, large-scale collaborative networks, or involvement in big collaborative research projects. The surname over-representation also reflects the distribution of research activity across regions and suggests a high representation of Chinese researchers in the sample. This concentration of contributions can influence the scope and direction of research in these areas. Table 9.5 indicates China as the lead contributor, marking its high research and development input and growing role in the international world of science. India comes second, indicating its important interest in research. Iraq and Egypt are also included in the list among leading contributors, indicating active research communities and perhaps an indication of concentrated research activities or international collaboration.

The database indicates contributions from a combined 98 countries and suggests an international and richly diverse research environment. The dominance of Chinese and Indian journals suggests these countries are major contributors to output from research, understandably due to their large populations, expanding bases of academics, and official promotion of research endeavors. The inclusion of countries from the Middle East and North Africa, such as Iraq, Egypt, and Saudi Arabia, suggests regional hubs of research activity, possibly sustained by strategic investment and collaboration. The numbers point toward a globalized research environment with deep insights by both the hitherto dominating powers and the newer research nations. Globalization brings vitality to the scientific world and inspires inter-border associations leading to an increasingly integrated and universal science development.

School of Information Science and Technology, Dalian Maritime University, China, is the top affiliating institution from Table 9.6, which indicates a high research activity and output of this institution in the fields of information science and technology. Department of Mathematics, Quaid-i-Azam University, Pakistan, is second with 17 publications. This indicates a high contribution of this department toward mathematical research as well as its applications. The figures reveal a high proportion of Chinese institutions among the top contributors, which reflects China's emphasis on research and development in technology and engineering fields. The inclusion of Quaid-i-Azam University from Pakistan reveals its contribution to mathematical research. Table 9.7 illustrates the highly cited papers in the dataset, which reveal those that have attracted a lot of attention within the research community. The citation pattern of these papers is quite uneven, with some of the classic papers receiving far more citations than others. This suggests that these papers have had a significant influence in their respective fields, which has guided later research and development.

The keyword co-occurrence data in Table 9.8 and Figure 9.4 highlight a clear emphasis on image encryption and the application of chaotic systems and maps in the development of secure encryption methods. The application of security, cryptography, and specific techniques such as S-Boxes and logistic maps highlights the technical richness and creativity in research communities. The application of encryption and security is an expression of the increasing demand for safeguarding digital information in a connected world.

Figure 9.5 indicates an evolving research landscape with evolving priorities and emerging technologies. The initial focus on chaos-based cryptography and IoT security concerns indicates the demand for malleable and efficient encryption mechanisms. The emergence of chaos-based methods subsequently is a search for more secure and creative methods. The continued interest in cryptanalysis and encryption schemes is an indication of continued interest in secure mechanisms against evolving challenges. These tendencies confirm the worth of adaptability and creativity in cryptography and information protection. The author cluster information presented in Table 9.9 provides an insight into the collaborative nature of research work among the collection.

There are over 300 author clusters, implying that there exists a dense and extensive network of co-authorship among researchers. All the clusters contain at least two researchers, which means that collaboration is the way of the day for this research team. With such a stipulation, it guarantees that all the clusters are genuine collaborations rather than single contributions. The clusters are sometimes cross-institutional and interdepartmental, indicating that they have taken place between

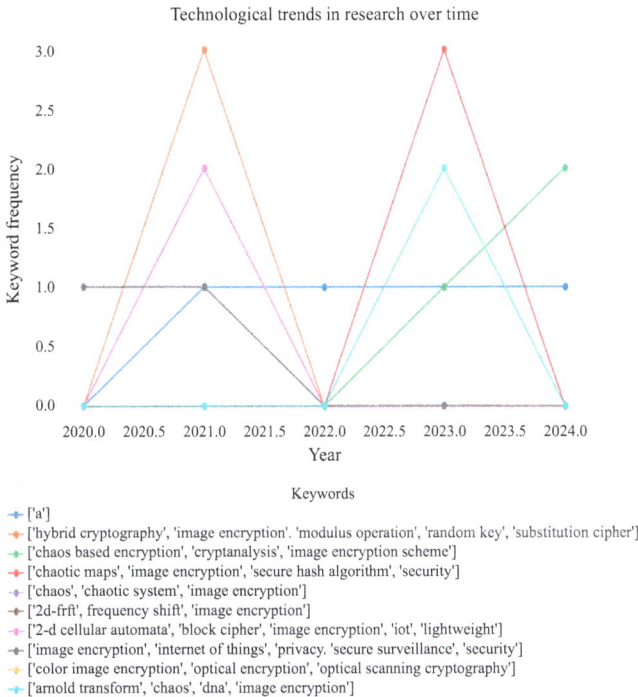

Keywords

['a']
['hybrid cryptography', 'image encryption'. 'modulus operation', 'random key', 'substitution cipher']
['chaos based encryption', 'cryptanalysis', 'image encryption scheme']
['chaotic maps', 'image encryption', 'secure hash algorithm', 'security']
['chaos', 'chaotic system', 'image encryption']
['2d-frft', frequency shift', 'image encryption']
['2-d cellular automata', 'block cipher', 'image encryption', 'iot', 'lightweight']
['image encryption', 'internet of things', 'privacy. 'secure surveillance', 'security']
['color image encryption', 'optical encryption', 'optical scanning cryptography']
['arnold transform', 'chaos', 'dna', 'image encryption']

Figure 9.5   Line plot of the technological trends in research over time

institutions and departments, and they exhibit interdisciplinary and interinstitutional collaborations. This can ensure a diverse sharing of methodologies and ideas, and hence the quality and potential of the research.

There is more than one cluster with distinct publication outputs, which gives evidence in support of collaborative cooperation in being able to produce groundbreaking research. Cooperation can lead more easily to greater inclusiveness, merging disciplines and approaches across clusters. The incidence of clustering, which transcends institutions and departments, has possibilities for interdisciplinary research to the extent that breakthrough research might emerge. Varying publications among the different clusters indicate disparate productivity. Effective clusters may also serve as centers of innovation and expertise, with the potential to influence research priorities and trends.

Author clustering suggests an active and networked research community. The diversity in cluster size and composition suggests a rich combination of established and emerging research groups, each contributing to knowledge advancement in a particular area. This collaborative environment is likely to remain innovative and discover.

9.5.2   Evaluating research goals

9.5.2.1   Publication volume and growth rate over the past five years

We found the growth rate in 2022 was 18.73%, dropped in 2023, but again increased in 2024. This indicates a vibrant research environment with flurries of activity, likely as a result of technology changes and good levels of funding. The same pattern was found within publications of image research studies in the years from 2013 to 2017 [4,18].

9.5.2.2   Top contributing authors, countries, and institutions

Wang X. is the leading productive author with 229 papers, indicating a high contribution to the field. Other leading authors include Zhang Y. and Zhang X., indicative of a group research effort in which these last names are strong, that is, China. China leads at 2344 papers, seconded by India at 1315. This is reflective of significant spending on research and development in such countries. The Dalian University of China leads in being the top contributor with 43 publications, which attests to its high research focus on information science and technology. This is in line with another study of bibliometric analysis on image processing from 2014 to 2019, when China is among the top productivity nations [6].

9.5.2.3   Frequently cited papers and landmark research

The most cited paper, "Image encryption algorithm for synchronously u ..." authored by Wang X. and Gao S., has 358 citations. This is a measure of its contributory foundation in image encryption research. Other highly cited papers are those of Xian Y. and Wang X., and Kaur M. and Kumar V., showing their high contribution to the discipline.

9.5.2.4 Keyword co-occurrence to reveal evolving research themes

The pattern of the dataset is to emphasize "image encryption," "chaos," and "cryptography." The use of keywords like "chaotic system" and "security" shows ongoing demand for data security. Use of keywords like "DNA" and "Arnold transform" points toward a shift in the direction of encryption. This is contrary to more research studies of images between 2013 and 2017 [4,18] and deep learning utilization [5].

9.5.2.5 Technological shifts from traditional cryptography to AI- and blockchain-enhanced encryption

While the dataset predominantly focuses on traditional cryptographic methods, the presence of keywords like "chaos" and "DNA" suggest the direction toward advanced and interdisciplinary approaches. Despite the absence of AI and blockchain, their applications in cryptography represent an upcoming trend within the broader research landscape.

9.5.2.6 Interdisciplinary collaborations between multimedia security, cryptography, and machine learning communities

The presence of over 300 author clusters, some spanning institutions and departments, reflects the widespread collaboration. This bodes well for possibilities of interdisciplinary research, particularly in multimedia security and cryptography domains. The use of machine learning techniques in these topics is likely an emerging area of interest.

9.5.2.7 Future directions

The focus of the dataset on efficient and scalable encryption methods strikes a chord with the need for IoT lightweight encryption methods. The advent of chaos-based methods signals possible applications in quantum image encryption. Even though homomorphic encryption is not handled as a special category, its use for multimedia content cloud storage is a crucial area of study in the future because of increased demand for cloud data processing in secure ways.

9.6 Conclusion

This article critically analyses research trends in image encryption, cryptography, and security using a mixed-method to evaluate a heterogeneous dataset from 2020 to 2024. The findings show an active and dynamic state of research with tremendous growth in publication rate, especially in the years 2022 and 2024. This is a positive trend, reflecting global concern for safeguarding digital data through innovative methods of encryption driven by technological growth, increased funding, and interdisciplinarity.

Key contributors, particularly from China and India, reflect the enormous investment by these countries in research and development, placing them at the forefront of the world's scientific community. Due to the ever-increasing demand for increased data protection in the interconnected world, the dataset reflects a concentration of research on specific areas, such as chaotic encryption systems, hybrid cryptography, and multimedia security. Seminal articles and prolific writers such as Wang X. and Zhang Y. also refer to the powerful impact of a small set of influential works in shaping the course of the research. The prevalence of collaboration, as reflected in the large number of author clusters and institutional partnerships, is a major driver of innovation. Collaborative research yields cross-disciplinary work, particularly at the interfaces among cryptography, multimedia security, and machine learning, which converge to tackle increasingly complex encryption challenges. The increasing popularity of chaos-based encryption and potential applications of blockchain and AI technology signal areas of future research in cryptography.

The research points out several key areas of research for the future, such as cloud-based data protection, quantum image encryption, and lightweight IoT encryption. There are many opportunities to advance the field because of the growing need for efficient, scalable, and secure encryption solutions.

This study underscores the need for ongoing innovation in cryptography and data protection, fueled by cross-disciplinary research, technological advances, and global research efforts. The findings presented herein will guide subsequent research studies, ensuring that encryption techniques evolve to keep up with the digital world's growing complexity.

References

[1] Abikoye, O. C., and Ogundokun, R. O. (2021). Efficiency of LSB steganography on medical information. *International Journal of Electrical and Computer Engineering (IJECE)*, *11*(5), 4157–4164.

[2] Akande, N. O., Abikoye, C. O., Adebiyi, M. O., Kayode, A. A., Adegun, A. A., and Ogundokun, R. O. (2019). Electronic medical information encryption using modified Blowfish algorithm. In *Computational Science and Its Applications – ICCSA 2019: 19th International Conference, Saint Petersburg, Russia, July 1–4, 2019, Proceedings, Part V 19* (pp. 166–179). Springer International Publishing.

[3] Jimoh, R. G., Awotunde, J. B., Ogundokun, R. O., and Adeoti, D. S. (2024). A randomized encryption algorithm for the MPEG-DASH digital rights management. *Information Security Journal: A Global Perspective*, *33*(5), 473–485.

[4] Chen, S., Wang, Y., and Qiu, S. (2018a). Bibliometric trend analysis on global image processing research. *Other Conferences*, 134. https://doi.org/10.1117/12.2505675.

[5] Gokhale, A., Mulay, P., Pramod, D., and Kulkarni, R. (2020). A bibliometric analysis of digital image forensics. *Science & Technology Libraries*, *39*(1), 96–113. https://doi.org/10.1080/0194262X.2020.1714529.

[6]   Khan, U., Khan, H. U., Iqbal, S., and Munir, H. (2022). Four decades of image processing: a bibliometric analysis. *Library Hi Tech*, *42*(1), 180–202. https://doi.org/10.1108/LHT-10-2021-0351.

[7]   Gupta, V. K. (2019). An analysis of data visualization tools. *International Journal of Computer Applications*, *178*(10), 4–7. https://doi.org/10.5120/IJCA2019918811.

[8]   Islam, M., and Jin, S. (2019). An overview of data visualization. *2019 International Conference on Information Science and Communications Technologies (ICISCT)*. https://doi.org/10.1109/ICISCT47635.2019.9012031.

[9]   Padmanandam, K., Bheri, S. P. V. D. S., Vegesna, L. H., and Sruthi, K. (2021). A speech recognized dynamic word cloud visualization for text summarization. *International Congress on Information and Communication Technology*, 609–613. https://doi.org/10.1109/ICICT50816.2021.9358693.

[10]   Agam Sinha. (2024). Data visualization pitfalls: a systematic review. *Darpan International Research Analysis*, *12*(3), 149–159. https://doi.org/10.36676/dira.v12.i3.62.

[11]   Reger, R. K., and Kincaid, P. A. (2021). Content and text analysis methods for organizational research. In *Oxford Research Encyclopedia of Business and Management*. https://doi.org/10.1093/ACREFORE/9780190224851.013.336.

[12]   Rajan, V. M., and Ramanujan, A. (2021). *Architecture of a semantic word cloud visualization*. 95–106. https://doi.org/10.1007/978-3-030-49500-8_9.

[13]   Lohmann, S., Heimerl, F., Bopp, F., Burch, M., and Ertl, T. (2015). Concentri cloud: Word cloud visualization for multiple text documents. *International Conference on Information Visualisation*, 114–120. https://doi.org/10.1109/IV.2015.30.

[14]   Nickell, J. V. (2012). Word clouds in math classrooms. *Mathematics Teaching in the Middle School*, *17*(9), 564–566. https://doi.org/10.5951/MATHTEACMIDDSCHO.17.9.0564.

[15]   Islam, M. (2020). Data analysis: Types, process, methods, techniques and tools. *International Journal on Data Science and Technology*, *6*(1), 10. https://doi.org/10.11648/J.IJDST.20200601.12.

[16]   Downie, N. M., and Starry, A. R. (2019). Descriptive and inferential statistics. *Companion Encyclopedia of Psychology*, 75–95. https://doi.org/10.4135/9781473920446.N5.

[17]   Neuendorf, K. A., and Kumar, A. (2015). Content analysis. *The International Encyclopedia of Political Communication*, 1–10. https://doi.org/10.1002/9781118541555.WBIEPC065.

[18]   Chen, S., Wang, Y., and Qiu, S. (2018b). Bibliometric trend analysis on global image processing research. In *Optical Sensing and Imaging Technologies and Applications*, *10846*, 854–865. SPIE. https://doi.org/10.1117/12.2505675.

[19]   Furche, T., Gottlob, G., Libkin, L., Orsi, G., and Paton, N. W. (2016). Data wrangling for big data: Challenges and opportunities. *Advances in Database Technology – EDBT*, *2016-March*, 473–478. https://doi.org/10.5441/002/EDBT.2016.44.

[20]   Konstantinou, N., Koehler, M., Abel, E., *et al.*, (2017). *The VADA Architecture for Cost-Effective Data Wrangling.* In *Proceedings of the 2017 ACM International Conference on Management of Data.* 1599–1602. https://doi.org/10.1145/3035918.3058730.

[21]   Azeroual, O. (2020). Data wrangling in database systems: Purging of dirty data. *Data*, *5*(2), 1–9. https://doi.org/10.3390/data5020050.

[22]   Patil, M. M., and Hiremath, B. N. (2018). A Systematic Study of Data Wrangling. *International Journal of Information Technology and Computer Science*, *10*(1), 32–39. https://doi.org/10.5815/IJITCS.2018.01.04.

[23]   Cambronero, J., Fernandez, R., and Rinard, M. (2021). *Wrangle search: Mining data wrangling functions from Python programs [Under submission].*

[24]   Gagolewski, M. (2022). *Minimalist data wrangling with Python.* arXiv preprint arXiv:2211.04630.

[25]   Nnachi, A. B., Arinze, E. D., and Uchechukwu, A. J. (2024). Exploring the frontiers of data analysis: A comprehensive review. *INOSR Applied Sciences*, *12*(1), 62–68. https://doi.org/10.59298/INOSRAS/2024/12.1.62680.

[26]   Bihani, P., and Patil, S. (2014). *A Comparative Study of Data Analysis Techniques, International journal of emerging trends & technology in computer science*, 3(2), 95–101.

[27]   Sharma, S. K., Kanchan, T., and Krishan, K. (2018). Descriptive Statistics. *The Encyclopedia of Archaeological Sciences*, 1–8. https://doi.org/10.1002/9781119188230.SASEAS0165.

[28]   Kaur, P., Stoltzfus, J., and Yellapu, V. (2018). Descriptive statistics. *International Journal of Academic Medicine*, *4*(1), 60. https://doi.org/10.4103/IJAM.IJAM_7_18.

[29]   Lee, J. (2020). Statistics, descriptive. *International Encyclopedia of Human Geography*, 13–20. https://doi.org/10.1016/B978-0-08-102295-5.10428-7.

[30]   Baxter, J. (2020). Content analysis. *International Encyclopedia of Human Geography*, 391–396. https://doi.org/10.1016/B978-0-08-102295-5.10805-4.

[31]   Gerela, P., Mishra, P. N., and Vipat, R. (2022). Study on data visualization. *International Journal of Health Sciences*, 6298–6305. https://doi.org/10.53730/ijhs.v6ns3.7393.

[32]   Nair, R. R., Sankaran, N., Nwogu, I., and Govindaraju, V. (2015). Automated analysis of line plots in documents. *IEEE International Conference on Document Analysis and Recognition,* 796–800. https://doi.org/10.1109/ICDAR.2015.7333871.

[33]   Kasmana, K., and Adipraja, F. M. (2019). The benefits of using bar charts in company websites. *IOP Conference Series: Materials Science and Engineering*, *662*(3), 1–6. https://doi.org/10.1088/1757-899X/662/3/032003.

[34]   Talbot, J., Setlur, V., and Anand, A. (2014). Four experiments on the perception of bar charts. *IEEE Transactions on Visualization and Computer Graphics*, *20*(12), 2152–2160. https://doi.org/10.1109/TVCG.2014.2346320.

[35]   Liu, Y. (2008). Box plots: use and interpretation. *Transfusion*, *48*(11), 2279–2280. https://doi.org/10.1111/J.1537-2995.2008.01925.X.

[36] Nuzzo, R. L. (2016). The box plots alternative for visualizing quantitative data. *PM&R*, 8(3), 268–272.
[37] Junaidi, J. (2014). *Deskripsi Data Melalui Box-Plot*.
[38] Ogundokun, R. O., and Abikoye, O. C. (2021). A safe and secured medical textual information using an improved LSB image steganography. *International Journal of Digital Multimedia Broadcasting*, *2021*(1), 8827055.

Chapter 10

Privacy protection of medical data using NTRU-based post-quantum cryptography

B.V.S.S. Praneeth[1], Rupa Ch[1], Ch. N. Manikanta[1], D. Pavan Kumar[1], Monalisa Sahu[2] and Aditya Kumar Sahu[3]

Abstract

In the medical field, images play a crucial role in patient monitoring, treatment planning, and diagnosis. Since the digital healthcare revolution, a large number of medical images are being transmitted and stored, making security essential. Any breach in data confidentiality can lead to misdiagnosis of patients. The traditional encryption techniques, like McEliece, have been used to shield medical data. However, the rise of quantum computers poses a threat to the prevailing encryption techniques, as they can easily break and expose medical data to quantum attacks. To overcome these attacks, post-quantum cryptography (PQC) has been introduced. Nth degree truncated polynomial ring unit (NTRU) is a lattice-based PQC algorithm that uses complex mathematical concepts that cannot be solved by quantum computers. Harnessing NTRU's lattice-based security ensures the robustness and resilience of medical data. The proposed system implements the PQC-based NTRU algorithm to protect medical data against quantum attacks.

Keywords: Medical images; quantum attacks; post-quantum cryptography; NTRU

10.1 Introduction

Healthcare digitization has revolutionized medical imaging as a key component of contemporary diagnostics, by providing real-time patient monitoring and accurate treatment planning. The sharing and storage of confidential medical information on

[1]Department of Computer Science and Engineering, Siddhartha Academy of Higher Education Deemed to be University, Vijayawada, India
[2]School of Computer Science and Engineering (SCOPE), VIT-AP University, India
[3]Department of Computer Science and Engineering, SRM University-AP, Andhra Pradesh, India

interconnected networks will make them prone to unparalleled security threats [1]. Traditional encryption schemes, like the McEliece cryptosystem, depend on mathematical challenges that are becoming more susceptible to quantum algorithms such as Shor's [2]. In addition to the classical algorithms, other approaches like fuzzy logic have been applied so far to improve data security [3,4]. The encryption systems designed using lightweight ciphers for IoMT data protection have elevated the level of securing medical data [4]. However, when these approaches are faced with quantum threats, they lack the resilience needed to ensure medical data security. These vulnerabilities may violate patient confidentiality and result in dangerous misdiagnoses, particularly as the technology of quantum computing continues to improve. These imminent threats and attacks emphasize the need to embrace quantum-resistant solutions since encrypted medical data might be decrypted with the advantages of quantum abilities.

In this demanding scenario, the PQC algorithms stand out as a strong contender for security because of their resilience [5]. NTRU is one of the PQC algorithms that continues to be a strong alternative that takes advantage of the shortest vector problem (SVP) that resists classical and quantum attacks [6]. The NTRU structure enables efficient key generation and encryption, which are important for the intensive workflows involved in medical imaging. As opposed to conventional schemes, NTRU's lattice-based design avoids the risks of quantum attacks and hence enjoys a robust safeguard against the quantum computing era.

This paper is dedicated to applying the NTRU algorithm to effectively secure medical images. By leveraging NTRU's lattice-based complexity, the system under consideration will resist quantum decryption efforts while allowing fast encryption and decryption operations without compromising the integrity of the diagnostic images [7,8]. Through the incorporation of NTRU in medical imaging systems, this new solution not only protects patient information from today's and tomorrow's quantum attacks but also retains the precision and confidentiality required by contemporary health care.

10.1.1   Problem statement

Development of a robust and secure medical data privacy protection system by the deployment of PQC-based NTRU algorithm to overcome the vulnerabilities of conventional cryptographic algorithms, to withstand against evolving quantum attacks.

10.1.2   Motivation

In the present digital age, the rapid growth of image communication, particularly in sensitive areas such as health care, necessitates stringent data protection [9]. Conventional encryption algorithms such as RSA and ECC are susceptible to quantum computing, which creates a dangerous threat to data security. As a counterpoint, we suggest the implementation of the NTRU post-quantum cryptographic algorithm for secure image encryption [10]. NTRU provides high-speed performance and robust resistance against both classical and quantum attacks [11]. By using NTRU to encrypt RGB image data, we provide confidentiality, integrity, and protection against brute-force decryption attempts [12]. This method not only protects sensitive visual information but also future-proofs systems against future cryptographic challenges. This project is in line with the increasing demand for quantum-resilient security in the changing digital world.

10.1.3   Research contributions

The proposed work contributions are as follows:

- To ensure protection and security from quantum computing attacks by leveraging the PQC-based NTRU algorithm for sensitive medical data.
- To strengthen data confidentiality and integrity by addressing the weaknesses of traditional cryptographic algorithms.
- To develop a practical encryption system to ensure the protection and security of medical data, making it appropriate for real-world healthcare applications. The remaining part of the paper consists of a literature survey in Section 10.2, followed by the preliminaries in Section 10.3, the proposed methodology with a detailed explanation of the working of the device in Section 10.4, results and analysis of the proposed smart device, including performance evaluation in Section 10.5, conclusion in Section 10.6, followed by references.

10.2   Literature survey

Shahbaz Khan *et al.* [13] proposed a quantum image encryption method that follows a two-step chaotic process. This model consists of four modules that perform a qubit and pixel-level encryption on both sides. This encryption method contains special quantum circuits that use chaotic transformations and a shuffling operation to enhance security. Even if half of the image is changed, it is able to recover the original image. It takes multiple encryption steps; processing images may take time. Karthika and Priya [14] proposed an image processing and cloud storage system using the NTRU algorithm. This system creates a 128-bit private key to encrypt and verify images, protecting them from unauthorized access. This encrypted image is stored in the cloud so that only an authorized user can decrypt the image. This system achieved a PSNR of 5.24 dB and an MSE of 82.10 dB, indicating its effectiveness. But it could be too slow for real-time image processing. Kuznetsov *et al.* [15] proposed a biometric-based cryptographic key generation system using deep learning. This study provides a new method that uses biometric data to create cryptographic keys instead of traditional passwords or PINs. The process is more efficient because it avoids the need for complex storage and distribution of keys. It specifically uses convolutional neural networks (CNNs) to extract unique features from human facial images. Then it was processed by using code-based cryptographic extractors. However, Biometric data may vary due to conditions like lighting and aging. Arjona *et al.* [16] proposed a biometric authentication system using homo- morphic encryption, which allows biometric data to be compared while staying encrypted. Only the final result is revealed when the private key is known, which enhances security. This system is used for non-device-centric authentication architectures with clients. This system uses the Classic McEliece encryption algorithm, a strong algorithm that is selected by NIST. It reduces storage needs by over 90.5% compared to other post-quantum homomorphic encryption methods, making it highly efficient. It requires higher processing power compared to traditional methods.

Hoe *et al.* [17] proposed an isogeny-based cryptography method for secure communication in post-quantum cryptography (PQC). It mainly uses Montgomery curves because they offer fast elliptic curve arithmetic and isogeny computation. CSIDH commonly uses Montgomery curves because they allow fast calculations, but they become slow and inefficient when handling large-degree isogenies. To overcome this, a new optimization method enhances CSIDH while still using Montgomery curves. Because of this, the CSIDH method is 6.4% faster than the original, proving the effectiveness of Montgomery curves. Even though the method improves CSIDH speed by 6.4%, it remains slower than other existing approaches. Park *et al.* [18] proposed an end-to-end encryption (E2EE) of video conferencing systems. The E2EE protocols are used in Zoom and Secure Frame, two popular video conferencing systems. It uses a different user authentication method than zoom and SFrame in a government public key infrastructure (GPKI)-based video conferencing system. In E2EE-related protocols, a key encapsulation mechanism (KEM) is utilized in the user key exchange process. The different encryption methods used by Zoom and SFrame make it hard to apply the same security upgrade to both systems. Malina *et al.* [19] proposed an intelligent infrastructure (II) service integrated with the Internet of Things (IoT) to enhance security and privacy. This work provides privacy-enhancing techniques and PQC methods to secure billions of connected devices from future quantum attacks. This system provides PET schemes using post-quantum cryptographic techniques and showcases their real-world applications. The study uses lattice-based cryptographic schemes as the best choice for IoT/II systems. But implementation remains challenging due to limited processing power and energy constraints. Cultice *et al.* [20] have introduced a hierarchical tree-structured solution for a reliable security framework that supports PQC for controller area network-based additive manufacturing devices. They have used subnet hopping in the middle of isolated CAN buses. The framework introduced can employ a third-party device in plug and play fashion while reducing the attack. It decreases the message cost by 25% over lightweight and 90% over post-quantum. However, in the case of denial-of-service attacks, the attackers can attack up to k subnet nodes. If the number of subnet nodes is large, the attack may be significant, which could be a major drawback.

Rupa and Avadhani [21] proposed a novel message encryption scheme that employs the concept of "cheating text." The core idea is to embed decoy or misleading information within the encrypted message to confound potential attackers. This approach aims to enhance the security of the encrypted data by introducing ambiguity, making it more challenging for unauthorized entities to decipher the actual content. The authors detail the methodology for integrating cheating text into the encryption process and discuss the potential benefits and drawbacks of the

Table 10.1   Summary of literature survey

Refs.	Input	Repository	Algorithm	Complexity	Cryptographic approach
Shahbaz Khan *et al.* [13]	Image	Cloud	QIE with chaotic confusion-diffusion	Exp.	Quantum-based
Karthika and Priya [14]	Image	Cloud	NTRU-based encryption with cloud storage	Poly. (NTRU)	Lattice-based
Kuznetsov *et al.* [15]	Bio data	Secure DB	Code-based biometric key Gen. with deep learning	Linear	Code-based
Arjona *et al.* [16]	Bio data	HSM	Homomorphic encryption (McEliece)	Exp.	Code-based
Heo *et al.* [17]	Text	On-premise	CSIDH with Montgomery curves	Sub-Exp.	Isogeny-based (CSIDH)
Park *et al.* [18]	Video	Bulletin	End-to-end PQC encapsulation	Poly.	Lattice-based
Malina *et al.* [19]	Text	IoT storage	Post-quantum KEM	Varies	Lattice-based
Cultice *et al.* [20]	Text	Secure TPM	PQC, hierarchical CAN security	Depends	Hybrid PQC
Rupa and Avadhani [21]	Text	N/A	Modified message digest algorithm	Polynomial	Hash based
Bhaskar and Rupa [22]	Text/ image	N/A	Corrected block TEA	Polynomial	Chaotic system-based

technique. Bhaskar and Rupa [22] introduce an advanced symmetric block cipher that leverages chaotic systems to enhance encryption strength. Chaotic systems are known for their sensitivity to initial conditions and pseudo-random behavior, which are utilized to generate complex key sequences that improve the confusion and diffusion properties of the cipher. The authors provide a comprehensive analysis of the cipher's architecture, detailing how chaotic maps are integrated into the encryption and decryption processes. They also present security analyses demonstrating the cipher's robustness against common cryptographic attacks.

The summary of several existing post-quantum works is listed in Table 10.1. It provides the details of input, repositories, algorithms, complexities, and cryptographic approaches employed in each work.

10.3   Preliminaries

This section provides the prerequisite knowledge about the PQC-based NTRU cryptography algorithm. The NTRU concept is a lattice-based cryptosystem that utilizes the hardness of structured lattices to provide robust security against quantum attacks. The basic definitions and notations to understand the background of NTRU are as follows.

10.3.1   Polynomial ring unit

It refers to the set of polynomials with the coefficients in a finite field, combined with addition and multiplication. The polynomial ring is denoted as $Zq[x]$ where q denotes integer modulo for the coefficients that results in a set of polynomials with coefficients always less than q. If an element has its inverse within the same polynomial ring, it is called a polynomial ring unit.

10.3.2   Nth degree truncation

It is the process of reducing a polynomial such that the degree of the truncated polynomial is always less than N. An Nth degree truncation applied to the polynomial ring is denoted as follows:

$$Z[x]/(xN - 1).$$

10.3.3   Cyclic convolution function

The result of the cyclic convolution multiplication of two polynomials, F and G is:

$$F * G = \sum_{i=0}^{k} F_i G_{k-i} + \sum_{i=k+1}^{N-1} F_i G_{N+k-i} \tag{10.1}$$

where Fi, Gi denote the ith term coefficient of F, G.

10.4   Proposed methodology

The architecture diagram of the proposed system is shown in Figure 10.1. It highlights the methodology of the proposed system, mainly consisting of two modules, namely post-quantum encryption and post-quantum decryption. The post-quantum encryption module describes the process of securing medical images by encrypting them. It consists of two submodules, namely key generation and encryption. The key generation module handles the generation of the public key and the private key. The encryption module describes the procedure of encrypting medical images with the help of the public key. The encrypted image is constructed at the end of the encryption module. The decryption module describes the procedure of decrypting the encrypted image using the private key to extract the original medical image. The main modules of the proposed system are module 1: key Generation, module 2: NTRU encryption, and module 3: NTRU decryption.

10.4.1   Key generation

The NTRU algorithm has three parameters: N, p, and q. Here, N refers to the maximum degree of the polynomials $(N - 1)$, p is a small positive integer acting

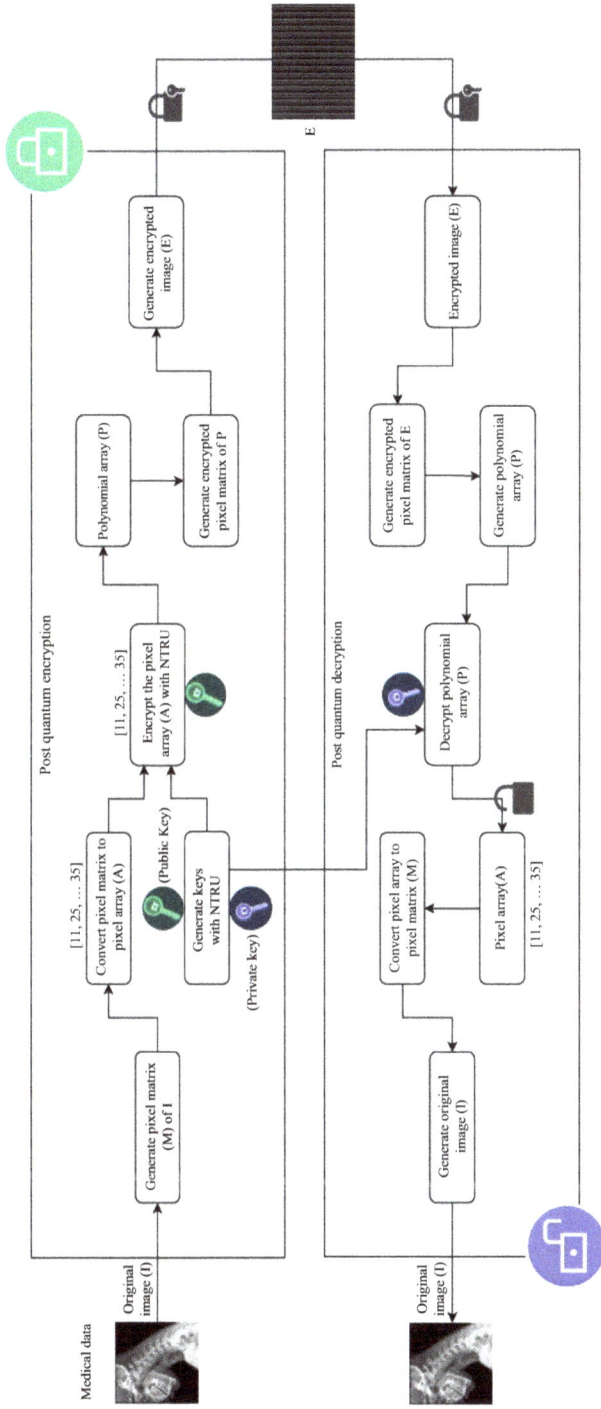

Figure 10.1 Proposed system architecture

as a small modulus and q is a larger positive integer acting as a modulus. The key generation process starts with choosing two random polynomials, f and g, in the polynomial ring unit $Rq(x)$ in variable x. The reason for choosing polynomials is that the coefficients of the polynomials must be less than or equal to q. It is important that f must have an inverse modulo p and q to exist in the ring R_a (i.e., f is invertible in the ring R_a). To compute these inverses, the Euclidean algorithm is used. The following function *NTRU generates keys (f, g, N, p, q)*, which is used to generate a public and private key pair (K_{pub}, K_{priv}). At this step, K_{pub} is generated using the polynomial f cyclic convolution product with g. K_{priv} is the polynomial containing both f and fp.

$$f_p * f \equiv 1 \pmod{p} \tag{10.2}$$

$$f_q * f \equiv 1 \pmod{q} \tag{10.3}$$

$$K_{pub} \equiv g * f_q \pmod{q} \tag{10.4}$$

$$K_{priv} = (f_p, f) \tag{10.5}$$

The inverse of a random polynomial f is computed with respect to p and q using the Euclidean algorithm. The public key is derived by multiplying f_q and $g (moduloq)$ and the private key consists of a polynomial pair (f, f_p).

10.4.2 NTRU encryption

The medical image taken as input will be converted into a pixel matrix (M) consisting of pixel values, which is further flattened into a one-dimensional array (A). NTRU encryption is applied to the elements of array A to create the polynomial array (P). Every pixel value in A is converted into binary format and taken as a string b. Every individual element of the polynomial is converted into the polynomial form with its coefficients less than q. The characters of the string s are taken as $(s_0, s_1, \ldots, s_{N-1})$. The pixel polynomial P will be generated by using the above set of values as coefficients, as follows:

$$P = s_0x^0 + s_1x^1 + \cdots + s_{N-1}x^{N-1} \tag{10.6}$$

$$C = p \times r \times K_{pub} + P \tag{10.7}$$

$$C = C \bmod q \tag{10.8}$$

Each cipher polynomial is added to an array C_p. With the coefficient values present in the C_p the encrypted image will be generated.

Algorithm 1: Working of the NTRU for securing medical data.

Input: Medical image, N, p, and q.
Output: Medical image.
Step 1: Read the medical image as a 2D matrix $I[][]$.

Step 2: Flatten the $I[][]$ into 1D array $A[]$.
Step 3: Calculate the inverse of the polynomial f with respect to q and p.

$$f_q * f \equiv 1 \ (modulo \ q)$$

$$f_p * f \equiv 1 \ (modulo \ p)$$

Step 4: Generate public key and private key

$$K_{pub} \equiv f_q * g (modulo \ q)$$

$$K_{priv} = (f_p, f)$$

Step 5: for i in $A[]$:
Transform i into the binary format string s having length n and construct a pixel polynomial P as follows:

$$s = tobinary(i)$$

$$P = \sum_{j=0}^{n-1} s_j x^j$$

Generate a random polynomial r and Cipher C.

$$C \equiv p \times r * K_{pub} + P \ (modulo \ q)$$

Store the cipher polynomial C in $B[][]$.
Step 6: Generate the encrypted image using $B[][]$.
Step 7: Read the encrypted image as the pixel matrix $E[][]$.
Step 8: Rearrange the pixel matrix $E[][]$ into a 2D array containing cipher polynomials.
Step 9: for C in $E[][]$:
Compute intermediate polynomials a using private key K_{priv} and extract the original pixel value

$$a \equiv f * C \ (modulo \ q)$$

Using the centering procedure, transform coefficients of a into the range of

$\left[-\frac{q}{2}, \frac{q}{2}\right]$.

$$a = centering(a, q)$$

Compute the original pixel polynomial P

$$P \equiv f_p * a (modulo \ p)$$

Using the centering procedure, transform the coefficients of b into the range of $\left[\frac{-p}{2}, \frac{p}{2}\right]$

$$P = centering(P, p)$$

Extract the coefficients of P to convert them into a binary string s.
Compute the original pixel value i.

$$i = to\_integer(s)$$

Store i in the decrypted image array $D[][]$.
Step 10: Construct the decrypted image from $D[][]$.

10.4.3   NTRU decryption

In the decryption part, the encrypted image is taken as input. The encrypted image is converted and re-arranged into a two-dimensional matrix, where each one-dimensional matrix stores a pixel polynomial. Each pixel polynomial is taken and decrypted to retrieve the original pixel value as follows:

$$a \equiv f * c \;(\mathrm{mod}\ q) \tag{10.9}$$

$$b \equiv a \;(\mathrm{mod}\ p) \tag{10.10}$$

$$P \equiv f_p * b \;(\mathrm{mod}\ q) \tag{10.11}$$

The coefficients of P are extracted as $(a_0, a_1, a_2, \ldots, a_{N-1})$. The coefficient set is converted into a string to get the original pixel value in binary format. This binary format is converted into an integer, which gives the original pixel value. Then the pixel values are rearranged to a matrix to get the original image.

Algorithm 1 provides a detailed explanation of the proposed methodology, focusing on reading the medical image in array form and flattening the image array. Then it explains the process of NTRU Encryption of the pixels and generating the encrypted image. The remaining flow illustrates the NTRU decryption to retrieve the original pixel values and formation of the original image.

10.5   Results and analysis

The analysis outcomes are derived through the development of a Java-based framework on a Windows 11 64-bit operating system, utilizing hardware specifications including a 12th Gen Intel(R) Core (TM) i5-1240P processor, 512 GB SSD storage, and 8 GB of RAM.

10.5.1   Performance evaluation

The transformations of medical data through encryption and decryption using the NTRU scheme are illustrated in Table 10.2. It includes test cases along with the original image, the encrypted image, and the reconstructed image. The response time for the process of encryption and decryption, based on the sizes of the input images in powers of 2, is shown in Table 10.3. The input image size is taken as the primary variable and measured in kilobytes (KB), and the response time is measured in milliseconds (ms).

Table 10.2   Test cases with original, encrypted, and reconstructed images

Test case	Original image	Encrypted image	Reconstructed image
Test case 1			
Test case 2			
Test case 3			
Specifications test case 4			

Table 10.3   Encryption and decryption time for different data sizes

Data size (in KB)	Encryption time (in ms)	Decryption time (in ms)
2	376	198
4	453	233
8	498	287
16	536	342
32	597	390
64	650	466
128	702	542
256	755	619

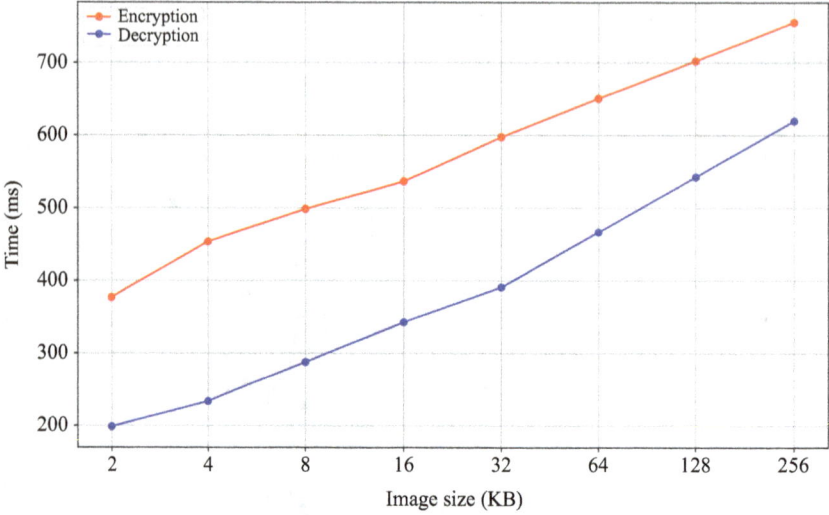

Figure 10.2   Encryption and decryption times vs image size

Table 10.4   Performance comparisons with classical algorithms

Author	Algorithm	Category	Structure	Time complexity	Latency
Arab *et al.* [23]	AES	Classical	Symmetric block cipher	$O(n)$	Low
Du and Ye [24]	RSA	Classical	Asymmetric algorithm	$O(n^3)$	High
Singh and Manglem Singh [25]	ECC	Classical	Asymmetric algorithm	$O(n^2)$	Moderate
Li *et al.* [26]	DES	Classical	Symmetric algorithm	$O(n^2)$	Low
Arjona *et al.* [16]	McEliece	Post-quantum	Asymmetric code-based algorithm	$O(n^3)$	Moderate
Proposed system	NTRU	Post quantum	Asymmetric lattice-based algorithm	$O(n \log n)$	Low

The response times observed by applying the PQC algorithm on various image sizes provided valuable insights about the performance of the algorithm. It is obvious that as the image size is increasing, the response times of encryption and decryption also increase, making it evident that there exists a linear relationship between the input size and response time, as shown in Figure 10.2. Various image encryption systems that are built on classical cryptographic algorithms are compared with the proposed encryption system based on attributes like type of algorithm, their category, structure, time complexity to implement those algorithms, and their latency as described in Table 10.4.

10.6   Conclusion

The proposed work offers a robust approach for shielding medical data against quantum threats by employing the lattice-based NTRU algorithm. The proposed system provides solid encryption and decryption capabilities to secure medical data against both traditional and quantum security challenges. The proposed system's quick encryption and higher performance capabilities ensure the confidentiality and integrity of medical data. The performance of the proposed system showcases the efficiency and reliability of the NTRU algorithm in securing medical data. NTRU's performance in key generation, encryption, and decryption makes it ideal for securing medical data.

References

[1]   Singh, A., Sharma, V. S., Basheer, S., and Chowdhary, C. L. "A Deep Cryptographic Framework for Securing the Healthcare Network from Penetration." *Sensors* 2024, 24, 7089. https://doi.org/10.3390/s24217089.

[2]   Zhang, X., Mu, D., Zhang, W., and Dong, X. "Encryption Algorithm MLOL: Security and Efficiency Enhancement Based on the LOL Framework." *Cryptography* 2025, 9, 18. https://doi.org/10.3390/cryptography9010018.

[3]   Devi Sree, G., Rupa, C., and Gayathri, U. "Multimedia Encryption Over Wireless Communication Using Fuzzy Logic based Approach." *2023 International Conference on Integrated Intelligence and Communication Systems (ICIICS), Kalaburagi, India,* 2023, pp. 1–5, https://doi.org/10.1109/ICIICS59993.2023.10421322.

[4]   Rupa, C. "Meaningful Cipher Data Generation: Novel Encipher using Fuzzy Logic For Reducing Cyber Data Attacks." *2023 Innovations in Power and Advanced Computing Technologies (i-PACT), Kuala Lumpur, Malaysia,* 2023, pp. 1–6, https://doi.org/10.1109/i-PACT58649.2023.10434566.

[5]   Drzazga, B. and Krzywiecki, L. "Review of Chosen Isogeny-Based Cryptographic Schemes." *Cryptography* 2022, 6, 27. https://doi.org/10.3390/cryptography6020027.

[6]   Ortiz, J. N., de Araujo, R. R., Aranha, D. F., Costa, S. I. R., and Dahab, R. "The Ring-LWE Problem in Lattice-Based Cryptography: The Case of Twisted Embeddings." *Entropy* 2021, 23, 1108. https://doi.org/10.3390/e23091108.

[7]   Wu, Q., Zhang, J., and Li, Z. "Generalized NTRU Algorithms on Algebraic Rings." *Electronics* 2024, 13, 4293. https://doi.org/10.3390/electronics13214293.

[8]   Lee, J., Ryu, H., Lee, M., and Park, J. "Cryptanalysis on 'NTRU+: Compact Construction of NTRU Using Simple Encoding Method'." *IEEE Transactions on Information Forensics and Security,* 2024, 19, 9508–9517. https://doi.org/10.1109/TIFS.2024.3471074.

[9]   Memon, Q. A., Al Ahmad, M., and Pecht, M. "Quantum Computing: Navigating the Future of Computation, Challenges, and Technological

Breakthroughs." *Quantum Reports* 2024, 6, 627–663. https://doi.org/10.3390/quantum6040039.

[10]   Shojaei, P., Vlahu-Gjorgievska, E., and Chow, Y.-W. "Security and Privacy of Technologies in Health Information Systems: A Systematic Literature Review." *Computers* 2024, 13, 41. https://doi.org/10.3390/computers13020041.

[11]   Kamarposhti, M. S., Ng, K.-W., Chua, F.-F., Abdullah, J., Yadollahi, M., Moradi, M., and Ahmadpour, S. "Post-quantum healthcare: A roadmap for cybersecurity resilience in medical data." *Heliyon* 2024, 10, e31406. https://doi.org/10.1016/j.heliyon. 2024.e31406.

[12]   Karthika, K. and Priya, R. "Image Processing-Based Protection of Privacy Data in Cloud Using NTRU Algorithm." *Signal, Image and Video Processing* 2024, 18, 1–16. https://doi.org/10.1007/s11760-024-03008-4.

[13]   Shahbaz Khan, M., Ahmad, J., Al-Dubai, A., *et al.*, "Chaotic Quantum Encryption to Secure Image Data in Post Quantum Consumer Technology." *IEEE Transactions on Consumer Electronics* 2024, 70, 4, pp. 7087–7101. https://doi.org/10.1109/TCE.2024.3415411.

[14]   Karthika, K. and Priya, R. D. "Image Processing-Based Protection of Privacy Data in Cloud Using NTRU Algorithm." *SIViP* 2024, 18, 4003–4018. https://doi.org/10.1007/s11760-024-03008-4.

[15]   Kuznetsov, O., Zakharov, D., and Frontoni, E. "Deep Learning-Based Biometric Cryptographic Key Generation with Post-quantum Security." *Multimedia Tools and Applications* 2024, 83, 56909–56938. https://doi.org/10.1007/s11042-023-17714-7.

[16]   Arjona, R., López-González, P., Román, R., and Baturone, I. "Post-Quantum Biometric Authentication Based on Homomorphic Encryption and Classic McEliece." *Applied Sciences* 2023, 13, 757. https://doi.org/10.3390/app13020757.

[17]   Heo, D., Kim, S., Yoon, K., Park, Y.-H., and Hong, S. "Optimized CSIDH Implementation Using a 2-Torsion Point." *Cryptography* 2020, 4, 20. https://doi.org/10.3390/cryptography4030020.

[18]   Park, Y., Yoo, H., Ryu, J., Choi, Y.-R., Kang, J.-S., and Yeom, Y. "End-to-End Post-Quantum Cryptography Encryption Protocol for Video Conferencing System Based on Government Public Key Infrastructure." *Applied System Innovation* 2023, 6, 66. https://doi.org/10.3390/asi6040066.

[19]   Malina, L., Dzurenda, P., Ricci, S., *et al.*, "Post-Quantum Era Privacy Protection for Intelligent Infrastructures." *IEEE Access*, 2021, 9, pp. 36038–36077. https://doi.org/10.1109/AC-CESS.2021.3062201.

[20]   Cultice, T., Clark, J., Yang, W., and Thapliyal, H. "A Novel Hierarchical Security Solution for Controller-Area-Network-Based 3D Printing in a Post-Quantum World." *Sensors* 2023, 23, 9886. https://doi.org/10.3390/s23249886.

[21]   Rupa, C. and Avadhani, P. S. "Message Encryption Scheme Using Cheating Text." *2009 Sixth International Conference on Information Technology: New Generations, Las Vegas, NV, USA*, 2009, pp. 470–474. https://doi.org/10.1109/ITNG.2009.232.

[22] Bhaskar, C. U. and Rupa, C. "An Advanced Symmetric Block Cipher Based on Chaotic Systems." *2017 Innovations in Power and Advanced Computing Technologies (i-PACT), Vellore, India*, 2017, pp. 1–4. https://doi.org/10.1109/IPACT.2017.8244891.

[23] Arab, A., Rostami, M. J., and Ghavami, B. "An Image Encryption Method Based on Chaos System and AES Algorithm." *Journal of Supercomputing* 2019, 75, 6663–6682. https://doi.org/10.1007/s11227-019-02878-7.

[24] Du, S. and Ye, G. "IWT and RSA Based Asymmetric Image Encryption Algorithm." *Alexandria Engineering Journal* 2023, 66, 979–991.

[25] Singh, L. D. and Manglem Singh, K. "Image Encryption Using Elliptic Curve Cryptography." *Procedia Computer Science* 2015, 54, 472–481.

[26] Li, S.-Y., Gai, Y., Shih, K.-C., and Chen, C.-S. "An Efficient Image Encryption Algorithm Based on Innovative DES Structure and Hyperchaotic Keys." *IEEE Transactions on Circuits and Systems I: Regular Papers,* 2023, 70, 10, 4103–4111. https://doi.org/10.1109/TCSI.2023.3296693.

Chapter 11

Advances in image steganography: a survey of methods, metrics, and emerging trends across domains

Biswajit Patwari[1], De Rosal Ignatius Moses Setiadi[2], Debosree Ghosh[3], Srishti Dey[4], Utpal Nandi[5], Sudipta Kr Ghosal[6] and Monalisa Sahu[7]

Abstract

Steganography, the art of concealing information within a non-suspicious medium, has emerged as a critical tool in modern information security. This chapter explores the fundamental principles of steganography, its evolution over time, and its applications in digital communication. It categorizes and analyzes various techniques, including spatial domain, frequency domain, and other methods, highlighting their strengths, limitations, and practical use cases. Key metrics such as imperceptibility, capacity, and robustness are discussed to evaluate the effectiveness of steganographic systems. The chapter also examines deep learning-based steganography and steganalysis, which leverage artificial intelligence to improve embedding strategies and detect hidden data, respectively. Applications in fields such as secure communication, copyright protection and data integrity verification etc. are reviewed, alongside challenges like trade-offs between capacity and imperceptibility, and vulnerabilities to steganalysis. This chapter also highlights recent advancements and evolving trends, underscoring its critical role in addressing emerging cybersecurity threats. It explores how steganography can help mitigate security risks in the cyber-physical systems and examines its potential impact in the era of quantum computing. By providing a comprehensive overview of

[1]Department of Computer Science, Panihati Mahavidyalaya, India
[2]Research Group for Quantum Computing and Materials Informatics, Faculty of Computer Science, Dian Nuswantoro University, Indonesia
[3]Department of Computer Science and Technology, Swami Vivekananda Institute of Science and Technology, India
[4]Department of Computer Science and Engineering, University of Kalyani, India
[5]Department of Computer Science, Vidyasagar University, India
[6]Department of Cyber Forensics and Information Security, Behala Government Polytechnic, India
[7]School of Computer Science and Engineering (SCOPE), VIT-AP University, India

steganography's current landscape, this chapter highlights its critical role in safe-guarding modern digital communication and outlines future research directions for enhancing its security and efficiency.

Keywords: Steganography; capacity; imperceptibility; robustness; steganalysis

11.1 Introduction

With the advancement of communication systems, sharing digital content such as text, images, audio, and video over the internet has become effortless. However, since these transmissions occur over public networks, sensitive data is at risk of being intercepted, altered, duplicated, or erased by unauthorized entities. This necessitates strong security measures to safeguard confidential information, including private messages, corporate data, and personal records, against potential cyber threats. Given the increasing volume of digital data and the growing concerns over cybersecurity, secure communication methods have become a major focus of research. Among the various security techniques, cryptography [1,2] and information hiding [1,2] have gained significant attention. Cryptography protects messages by converting them into an unreadable format, while information hiding conceals the existence of secret data within cover media, ensuring minimal distortion. Unlike cryptography, which obscures message content, information hiding aims to make the data invisible to unauthorized viewers.

Two major branches of information hiding are steganography [3,4] and digital watermarking [5–7]. Steganography embeds confidential data within ordinary-looking media, such as images, audio, or video, making the presence of hidden information undetectable. Digital watermarking, on the other hand, incorporates identifiable information, such as copyright details, into digital content to assert ownership and protect intellectual property. While both techniques fall under information hiding, their objectives and implementation methods differ. Steganography ensures secrecy, whereas watermarking aims to make embedded information detectable for authentication and copyright protection. Digital watermarks can be either visible, such as logos imprinted on images, or invisible, subtly embedded within the media and accessible only to authorized users. In contrast, steganography ensures that hidden data remains completely undetectable to avoid suspicion. While digital watermarking is designed to endure common modifications like compression, resizing, and format conversion, steganography prioritizes secrecy over resilience to alterations. Digital watermarking is classified into three main types: fragile [5], semi-fragile [6], and robust [7], depending on how they react to changes in the media. Steganography, on the other hand, operates within two domains: the spatial domain, where data is embedded directly into pixel values, and the transform domain, where it is hidden within frequency components of the media. The primary applications of digital watermarking include copyright protection, where ownership data is embedded to prevent unauthorized reproduction; content authentication, which detects any modifications to media; broadcast monitoring, which tracks content distribution using unique watermark signatures; and image forensics, which helps verify authenticity and detect tampering.

Steganography is categorized into different types, arranged in a hierarchical structure as illustrated in Figure 11.1.

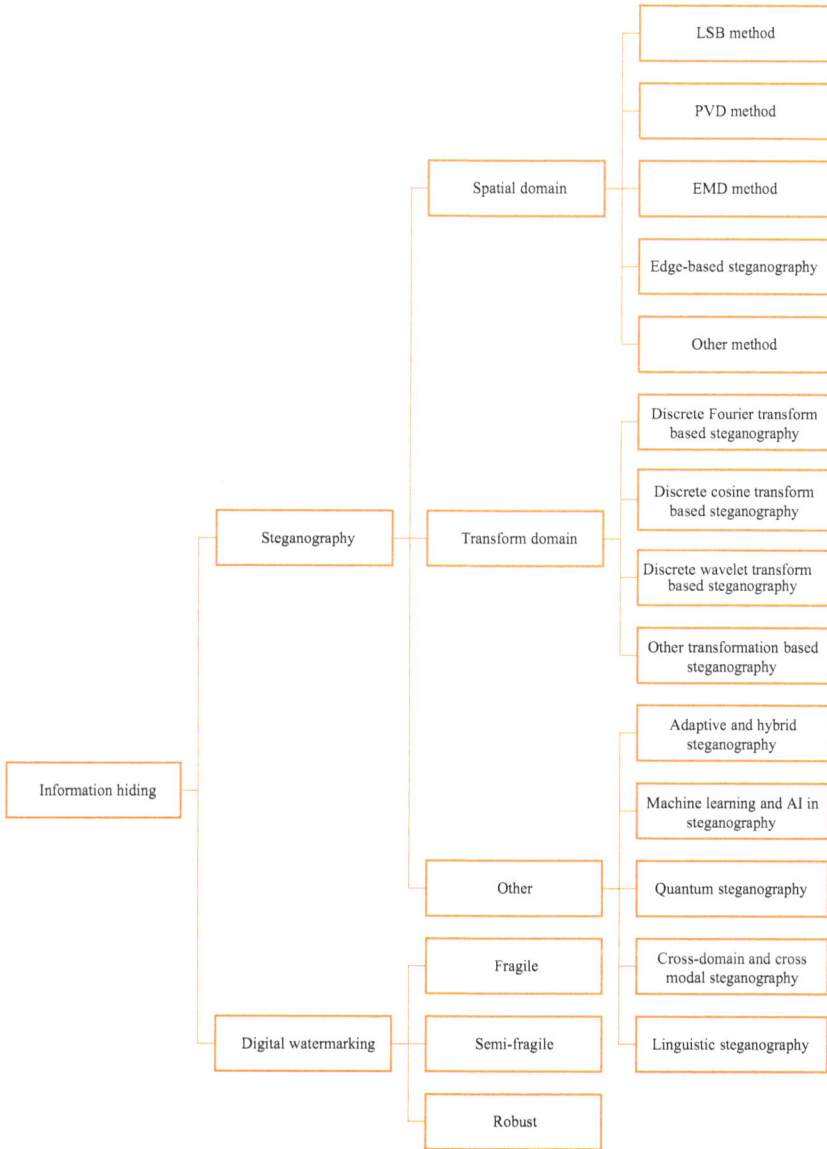

Figure 11.1   Classification of information hiding schemes

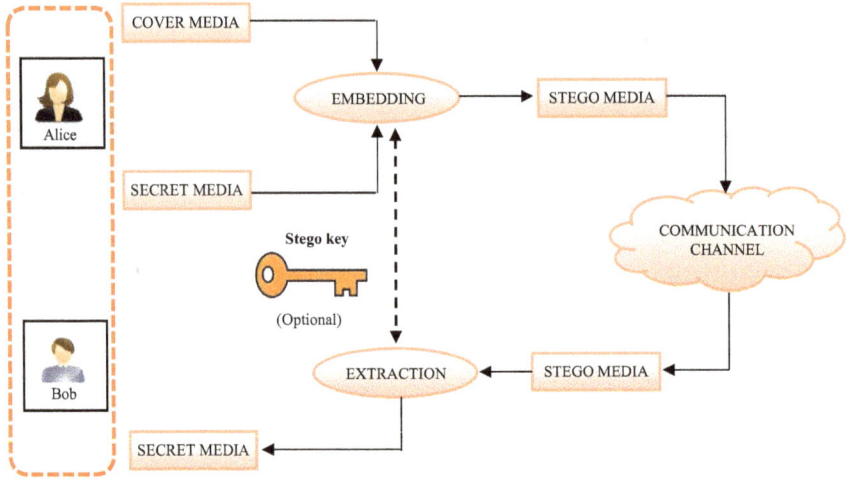

Figure 11.2   Block diagram of steganography system

Meanwhile, steganography is widely used for confidential communication by hiding messages within regular files, espionage by embedding classified data within inconspicuous media, digital forensics to uncover hidden information for investigative purposes, and medical imaging to securely store patient details within diagnostic images. The block diagram of a traditional steganography system is shown in Figure 11.2.

In this chapter, we emphasize on steganography [3,4] instead of digital watermarking [5–7] as it provides a discreet and reliable method for embedding information, ensuring both confidentiality and effective covert communication. Its ability to ensure secrecy without drawing attention from unintended recipients makes it an ideal choice for sending confidential information through digital media such as images, audio files, or text from one place to another over the internet seamlessly. It enables us to accomplish covert communication, which is essential in areas like espionage, secure data exchange, and digital forensics. Its adaptability makes it even more effective, as it can be implemented in both the spatial and transform domains, allowing users to select the most suitable method based on security needs and application requirements. Furthermore, steganography is especially useful in situations where encryption alone might raise suspicion. Encrypted messages clearly indicate the presence of sensitive information, which could attract scrutiny. In contrast, a well-executed steganographic approach conceals data within seemingly ordinary media, making it less likely to be detected by unauthorized parties.

11.2   Techniques in digital steganography

Digital steganography [3,4,8–20] includes some scientific techniques to hide information within digital media, such as images, audio, video, and text, allowing

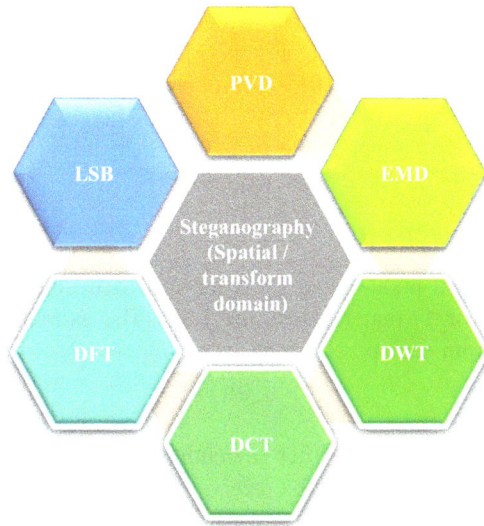

Figure 11.3   Popular steganography methods

for covert communication. These approaches fall into two primary domains: spatial and transform (frequency). Popular spatial domain methods include least significant bit (LSB) [9,11,14], pixel value differencing (PVD) [12,15], and exploiting modification direction (EMD) [16], and most of the spatial domain methods are inspired by this trio. On the other hand, transform domain methods are primarily based on the discrete Fourier transform (DFT) [21], the discrete cosine transform (DCT) [22,23], and the discrete wavelet transform (DWT) [24]. Each type presents different trade-offs concerning capacity, imperceptibility, and robustness.

The key spatial and transform domain steganography methods are illustrated in a hexagonal radial layout, as shown in Figure 11.3.

11.2.1   Spatial domain techniques

Spatial domain techniques [4,14–20] are the most straightforward and most common methods of steganography due to their simplicity, high embedding capacity, and low computational complexity. They are also suitable for real-time computation and are very flexible. The spatial domain techniques avoid the computational overhead and loss of precision in the frequency domain transformations. They deal with the pixel intensity values of an image, adjusting them in such a manner that no perceptible distortion is created.

11.2.1.1   LSB technique

The LSB technique [9,11,14] is a widely used approach in spatial domain steganography because it is simple to perform, has a high embedding capacity and will not lead to large differences in perception. It operates by replacing the LSB of pixel values in a digital image with secret data bits, ensuring that the modification remains visually imperceptible. In a grayscale image, each pixel is typically represented using 8 bits (e.g., 11010110 for a pixel value of 214). By modifying the last bit, such as changing 11010110 to 11010111, the pixel value changes from 214 to 215, a difference that is negligible to human perception. Mathematically, the embedding process can be represented as follows:

$$p'_i = p_i - (p_i \bmod 2) + b \tag{11.1}$$

where P_i is the original pixel value, P_i' is the modified pixel value, and b is the secret bit (either 0 or 1) that replaces the LSB. The extraction process simply retrieves the LSB from each pixel using:

$$b = p'_i \bmod 2 \tag{11.2}$$

The stego image is generated after embedding, which is visually similar to the cover image almost. At extraction, the receiver scans the stego image, extracts the LSB of a specified pixels and recreates the original message.

Variations in LSB techniques [9,11,14] involve the specific bits modified during embedding. The basic method, LSB replacement (LSBR) [10,11], alters bits from the upper-left corner of the image and proceeds horizontally and vertically. Conversely, the stego-key-directed LSB method, or random LSB, uses a user-defined stego-key to distribute the secret message throughout the image. Another variant, LSB matching (LSBM) [11], minimizes changes by aligning binary data with pixel values, altering as few pixels as possible.

11.2.1.2   PVD technique

PVD [12,15] is an advanced steganographic technique that improves upon the LSB method [9,11,14]. PVD enhances the imperceptibility of hidden data, often achieving higher peak signal-to-noise ratio (PSNR) [1–20,25] values than LSB, which makes detection more challenging. It works well in textured or edge regions but can be less efficient in smoother areas. The method's complexity also raises computational demands. For any arbitrary grayscale cover image, the difference value d_i is obtained from every non-overlapping two-pixel block, p_i and p_{i+1}. If the grayscale values of the respective pixels are g_i and g_{i+1}, then d_i is computed as $g_i + 1 - g_i$, which ranges from −255 to 255. The number of bits n that can be embedded in block B with index i is determined by the function $= \log_2 u - l + 1$, where the lower and upper bound values of the range r_{i+1} ($i = 1, 2, \ldots, w$) are defined. A bit stream of embedded data S with n bits is selected on the next step, and a new difference $d'_i = |l_i + b|$ is computed, where b is the integer value of the sub-stream S for the computed n value. Since the value b is in the range from 0 to $u_i - l_i$, the value of d_i is in the range from l_i to u_i. After calculating d_i, embed b by performing an inverse calculation from g_i to yield the new gray (g'_i, g'_{i+1}) values for the pixels in the corresponding two-pixel block (p'_i, p'_{i+1}) of the stego-image.

For instance, consider a pair of pixels of the cover image CI as $(p_1, p_2) = (123, 220)$.

We compute the difference value $d = |p_2 - p_1| = |220 - 123| = 97$.
Here, $d \in R5$ as per the range table provided in [12]

$$w_5 = u_5 - l_5 + 1 = 27 - 64 + 1 = 64$$

Since number of bits $t = \log_2 w_5 = 6$, let us consider a 6-bit secret data
s = 010000 whose decimal equivalent $b = 16$
$$d' = l_5 + b = 64 + 16 = 80$$

$$m = |d' - d| = |80 - 97| = 17$$

Stego-pixel pair $(p'_1, p'_2) = (p_1 + \frac{m}{2}, p_{i+1} - \frac{m}{2}) = (123 + 9, 220 - 8)$
$= (132, 212)$.
At the recipient end, $d' = |p'_1 - p'_2| = |132 - 212| = 80$

$$b' = d' - l_5 = 80 - 64 = 16$$

Therefore, the example clearly demonstrates that the embedded data and the
extracted data have identical values.

11.2.1.3 Exploiting modification direction (EMD) technique

Zhang *et al.* [13] introduced an enhanced modification direction (EMD)
[13,16,26] scheme for information hiding, which efficiently embeds secret data
into grayscale images. This approach represents each secret digit using an $(2n +
1)$-ary notation, where n pixels from the cover image form a single embedding
unit. A key feature of this method is its ability to conceal n bits of secret data
within n cover pixels, modifying at most one pixel per group from $(2n + 1)$
possible combinations. The modification is restricted to either an increment or a
decrement by one, preserving image quality while embedding information. The
technique partitions the carrier image into n adjacent pixels following a row-major
order. The extraction function is defined as a weighted sum modulo $(2n + 1)$,
expressed as follows:

$$f(g_1, g_2, \ldots, g_n) = \left[\sum_{i=1}^{n} g_i \cdot i\right] \mod (2n + 1) \tag{11.3}$$

Secret digit d in the $(2n + 1)$-ary notational system is embedded into pixels
$(g_1, g_2, \ldots, g_n)$ with three conditions. Firstly, if $d = f$, then no modification is
needed. Secondly, if $d \neq f$, then compute $s = d - f \mod (2n + 1)$. If $s < n$, then
increase the value of g_s by 1; otherwise, follow the third case by decreasing
g_{2n+1-s} by 1. If f is greater than d, which yields s to be a negative number, then s
is updated as $(2n + 1) + s$. The decoder extracts the fabricated secret data by
taking the n embedded pixels to calculate the same extraction function. The
technique has two major limitations, one of which is its limited payload capacity
(1.16 bpp) and another is the characteristic of dealing with only odd base secret
digits $(2n + 1)$.

For instance, consider a pair of pixels of the cover image CI as,

$$(g_1, g_2) = (220, 117)$$
$$f = \{(220 \times 1) + (117 \times 2)\}\%5$$
$$= (220 + 234)\%5$$
$$= 454\%5 = 4$$
$$d = 3$$
$$s = (d - f)\%(2n + 1)$$
$$= (3 - 4)\%5 \ldots [\text{as} \, n = 2]$$
$$= -1$$

Since s is negative

$$s = (2n + 1 + s)\%(2n + 1)$$
$$= (5 - 1)\%5$$
$$= 4\%5$$
$$= 4$$

As $s > n$,

$$g_1' = g_{2n+1-s} - 1$$
$$= g_{5-4} - 1$$
$$= g_1 - 1$$
$$= 220 - 1$$
$$= 219$$
$$g_2' = g_2 = 117$$

At the recipient end,

$$(g_1', g_2') = (219, 117)$$

To extract embedded data, we compute $f' = \{(219 \times 1) + (117 \times 2)\}\%5 = 3$

Therefore, the example clearly demonstrates that the embedded data and the extracted data have identical values.

In addition to the three methods mentioned above, several advanced spatial domain techniques are presented in Table 11.1, structured with the following columns: Reference, Method, Dataset, Major Advantages, and Major Challenges.

Table 11.1   *Comparative analysis of various spatial domain steganographic techniques*

Reference	Method	Dataset	Major advantages	Major challenges
Cheng and Chan [14]	OPAP-LSB	Lena, Baboon, Jet, Scene	Improved image quality using OPAP	Not much secured
Grajeda-Marín *et al.* [15]	TWPVD	Airplane, Barbara, Boat, Goldhill, Lena	• This method overcomes the overflow and underflow issues. • Utilized the embedding space fully.	• Trade-off between payload and quality • Not optimized for resilience against steganalysis attacks
Hajizadeh *et al.* [16]	High-capacity EMD	Lena, Baboon, F16, Pepper, Boat, Tiffany.	Balanced payload and visual quality	Low security
Gaurav and Ghanekar [17]	Canny edge-based	SIPI image database	Minimal modification with high-quality and imperceptible image steganography	Embedding capacity is not significantly high
Ghosal *et al.* [18]	Kirsch edge-based	SIPI image database	Payload varies and can sometimes exceed 3 bpp, ensuring high capacity	Both embedding capacity and image quality can be further enhanced effectively
Sultana *et al.* [19]	Hybrid edge based on prediction error space	Ten 512 × 512 images were used from the BOSS dataset	Optimized data embedding capacity with enhanced stego-image quality and fidelity	The proposed method highly depends on edge detection and has high computational complexity
Jan *et al.*[20]	Hybrid edge based	Lena, Baboon, Peppers, Airplane, Sailboat, Barbara, Cameraman	Integrating chaos functions for encryption enhances the security	Robustness can be further enhanced

11.2.2   Transform domain techniques

Transform domain techniques [6–8,21–24,27–30] in steganography are advanced methods used to embed secret data in multimedia files like images, audio, and video. Unlike spatial domain techniques, which change pixel values directly, transform domain methods manipulate the frequency components, enhancing the robustness and security of hidden data against detection and attacks. These techniques address the limitations of spatial methods, which, while simple and capable of high embedding capacities, often lack robustness against steganalysis. Transform domain techniques [6–8,21–24,27–30] in image steganography [3,4,8–20] employ orthogonal and non-sinusoidal transforms to embed data with distinct mathematical and operational characteristics. In orthogonal transforms, basis functions are sinusoidal, and the results contain only real numbers, whereas non-sinusoidal basis transforms include complex numbers. As discussed, the widely used transforms include the DFT [21], DCT [22,23], and DWT [24], which embed data in ways that are less perceptible and more resistant to compression and noise. While they generally maintain high quality of stego media, evidenced by better PSNR values, transform domain techniques can involve complex computations and may have lower embedding capacity. Nonetheless, their enhanced security makes them a preferred choice for applications requiring data integrity and confidentiality.

11.2.2.1   Discrete Fourier transform (DFT)

DFT [21] is an essential tool in image steganography, allowing secret messages to be embedded within the frequency domain of an image. It transforms a spatial image into its frequency components, where higher-frequency coefficients are preferred for embedding because they are less perceptible to the human eye. An inverse discrete Fourier transform (IDFT) [21] is used to restore the image to its initial form after the data has been embedded, with only minor alterations that are imperceptible to the observer. DFT [21] is defined by the following equation:

$$X[u, v] = \sum_{x=0}^{N-1}\sum_{y=0}^{N-1} f(x,y)e^{-j2\pi\left(\frac{ux}{M}+\frac{vy}{N}\right)} \tag{11.4}$$

where $f(x,y)$ is the grayscale image intensity at coordinate (x,y) and $X(u,v)$ is the transformed frequency-domain representation. DFT-based steganography is computationally expensive as it has a computing complexity of $O(n^2)$ and its coefficients are typically expressed as complex numbers. This method is robust against common processing operations like compression and noise addition, as the embedded data is spread across various frequency components. However, DFT [21] is susceptible to steganalysis attacks, which can identify statistical anomalies from the embedding process. Current advancements include adaptive techniques that selectively modify DFT coefficients based on the image's characteristics and the combination of DFT with other transform methods, such as the DCT [22,23] and wavelet transform [24], to improve robustness and imperceptibility.

11.2.2.2   Discrete cosine transform (DCT)

DCT [22,23] is a popular orthogonal transform in image steganography that converts spatial domain image data into frequency domain coefficients, allowing for imperceptible data embedding. Due to high energy packing capability it is widely used in secure communication, digital watermarking and image compression. In DCT-based image steganography [22,23], cover image is divided into 8×8 pixel blocks, applies DCT to each block, compresses them using a quantization table, and embeds the secret message into the DCT coefficients. The forward 2D-DCT of an image signal $f(m, n)$ is given by:

$$F[K, L] = \alpha(k)a(l) \sum_{m=0}^{N-1} \sum_{n=0}^{N-1} f(m, n)\cos\left(\frac{(2m+1)\pi k}{2N}\right) \times \cos\left(\frac{(2n+1)\pi l}{2N}\right)$$

(11.5)

where $f(m, n)$ is input image pixel intensity at position (m,n), $F[K,L]$ is DCT coefficient [22,23] at frequency position (K,L) and $\alpha(k)$, $\alpha(l)$ are normalization factors. Even though DCT-based steganography is highly imperceptible and resistant to some image processing techniques, steganalysis attacks that identify statistical anomalies related to embedding can still affect it. Recent advancements focus on adaptive embedding techniques that dynamically adjust the embedding strength based on image characteristics and the use of machine learning (ML) algorithms [31] to enhance imperceptibility and robustness against sophisticated steganalysis methods.

11.2.2.3 Discrete wavelet transform (DWT)

DWT [24] is a non-sinusoidal technique that has become essential in image steganography due to its ability to balance imperceptibility and robustness. Unlike sinusoidal transforms, such as the Fourier transform, which represent signals as a combination of infinite-length sinusoids, DWT decomposes signals into different frequency bands with varying resolutions. DWT operates in the frequency domain by breaking down an image into multi-resolution sub-bands: LL (approximation), LH (vertical details), HL (horizontal details), and HH (diagonal details). High-frequency sub-bands, like HH, are preferred for embedding secret data because the human visual system is less sensitive to changes in edge regions, which minimizes perceptible distortions. In comparisons, DWT performs better than both DCT [22,23] and DFT [21] in terms of time-frequency resolution, compression efficiency, and edge preservation. The computational complexity of DWT is $O(n)$, making it faster than DCT and DFT. DWT-based image steganography offers significant advantages, including robustness, high imperceptibility, and improved payload capacity. However, it also faces challenges related to computational costs, sensitivity to compression, potential information loss, and trade-offs between capacity and imperceptibility. Hybrid approaches, such as combining DWT with DCT or cryptography, can help mitigate these issues.

Apart from the three methods discussed earlier, Table 11.2 presents various advanced spatial domain techniques, organized under the columns: Reference, Method, Dataset, Major advantages, and Major challenges.

11.3   Evaluation metrics

The effectiveness of a steganographic system is determined by three key factors: capacity, imperceptibility and robustness. These parameters influence the overall performance and usability of the technique.

Table 11.2   Comparative analysis of various transform domain steganographic techniques

Reference	Method	Dataset	Major advantages	Major challenges
Shehada, D., and Bouridane, A.	DCT [23]	ALASKA2, BOSSBase	Produces stego images with enhanced quality and increased PSNR values	Susceptible to attacks
Seyyedi, A., and Ivanov, N.	Integer wavelet transform (IWT) [27]	Barbara, Peppers, Baboon, Lena, Airplane and Boat	Balanced Payload and PSNR	Floating point to integer wavelet coefficients shifting may cause error
Atawneh *et al.*	DWT [24]	SIPI, CVG RSP	Enhance the efficiency of embedding and effectively addresses the overflow and underflow issues	Payload is limited
Mandal *et al.*	Hartley transform [28]	The USC-SIPI image dataset	Complex computational process	PSNR is suboptimal
Subhedar, M. S., and Mankar, V. H.	Singular value decomposition (SVD) [29]	Seven USC-SIPI Images	Improved stego image imperceptibility	Higher computational complexity
Ghosal *et al.*	Binomial transform [30]	The USC-SIPI image dataset	Improved computation complexity due to the uses of integer sequences	Payload is fixed, not variable

11.3.1   Capacity

Capacity refers to the maximum amount of information that can be embedded within a cover medium while maintaining acceptable quality. High-capacity techniques allow for larger data embedding but may compromise security and imperceptibility.

- *Measurement metrics*:
    - o Bits per second (bps) for audio and video.
    - o Total bits that can be embedded without noticeable degradation.
    - o Bits per pixel (bpp) for images.

Out of all the metrics, *bpp* (bits per pixel) is the most commonly used, as images are a popular choice for concealing secret information. To grasp this concept, let's think of it in an analogy. *Bpp* is a measure of how many binary bits can be embedded into the cover media (image), and it is typically expressed in bits per pixel.

Consider that we want to embed a $\lambda \times \lambda$ secret image within a cover image of size $X \times Y$ to achieve a payload of *bpp*. The dimension $\lambda \times \lambda$ of the hidden image must satisfy the following condition:

$$\lambda \times \lambda \times 8 \leq X \times Y \times B \tag{11.6}$$

For the maximum payload, λ should be:

$$\lambda_{\max} = \text{floor}\left(\sqrt{\frac{X \times Y \times B}{8}}\right) \tag{11.7}$$

- *Factors influencing capacity*:
  o Characteristics of the cover medium (e.g., textures in images, frequency variations in audio).
  o Embedding domain (spatial vs. transform-based techniques).
  o Compression techniques (lossy formats reduce capacity).

11.3.2 Imperceptibility

Imperceptibility ensures that modifications made to the cover medium remain undetectable to human perception. If alterations are noticeable, the presence of hidden information may be suspected.

- *Measurement metrics*:
  o *Mean squared error (MSE)*: It is the total squared error between the CI and the stego-image (SI). The following is the mathematical representation of MSE:

$$\text{MSE} = \frac{1}{MN} \sum_{p=0}^{P-1} \sum_{q=0}^{Q-1} [CI(p,q) - SI(p,q)]^2 \tag{11.8}$$

  This formulation makes it clear that, in order to maximize accuracy, the MSE must always be minimized. In this case, the SI and CI have $P \times Q$ dimensions.
  o *PSNR*: Measures signal quality relative to noise introduced by embedding. In other words, PSNR is nothing but the ratio of signal power to noise power (measured in dB) is known as the PSNR [1–20,25], and it can be represented as follows:

$$\text{PSNR} = 20\log_{10}\left(\frac{MAX_{CI}}{\sqrt{MSE}}\right) \tag{11.9}$$

  In this instance, the signal is denoted by MAX_{CI}, and for an 8-bit grayscale cover image (CI), its maximum intensity value is 255.
  o *Structural similarity index (SSIM)*: Assesses perceptual similarity between original and modified media. The structural similarity between cover images and stego-images is represented by the SSIM [12–20]. Stated differently, SSIM can be used to quantify the structural correlation between the stego-image and the cover image. The range $[-1, 1]$ is the allowed

range for SSIM values. The following is the mathematical representation of SSIM:

$$\text{SSIM}(CI, SI) = \frac{(2\mu_{CI}\mu_{SI} + c_1)(2\sigma_{CI.SI} + c_2)}{(\mu_{CI}^2 + \mu_{SI}^2 + c_1)(\sigma_{CI}^2 + \sigma_{SI}^2 + c_2)} \quad (11.10)$$

where the variances are represented by σ_{CI}^2 and σ_{SI}^2, the means by μ_{CI} and μ_{SI}, and the covariance of CI and SI, respectively, by $\sigma_{CI.SI}$. Two variables are used in this case c_1 and c_2 to stabilize the division operation using a weak divisor.

- *Challenges*:
  o Maintaining visual and auditory quality while embedding high amounts of data.
  o Avoiding perceptible distortions caused by embedding techniques.

11.3.3   Robustness

Robustness [7,22,32,33] determines the resilience of hidden data against intentional and unintentional modifications such as compression, noise addition, and signal processing. In other words, it is the ability of the stego-image to resist the external attacks (both visual and geometrical) is known as robustness. Several tests are conducted to summarize the performance of robustness against attacks such as primary set, Chi-square, RS analysis, etc.

- *Threats to robustness*:
  o Lossy compression (JPEG for images, MP3 for audio).
  o Noise addition and filtering.
  o Cropping, scaling, and re-encoding.

- *Enhancing robustness*:
  o Embedding in transform domains (DCT, wavelet) to resist compression artifacts.
  o Incorporating error correction codes (e.g., Reed–Solomon) to recover lost information.
  o Using adaptive embedding techniques based on cover medium properties.

11.4   Recent trends and advances in steganography

11.4.1   Adaptive and hybrid techniques

Adaptive and hybrid techniques in steganography have become significant trends in research to address classic challenges, such as balancing hiding capacity, imperceptibility, and resistance to steganalysis. New methods integrate the spatial domain, the transformation domain, and adaptive logic approaches to create more robust solutions. By leveraging complex features within images, these approaches aim to maximize embedding effectiveness while minimizing detectable visual distortions. Research conducted by Shafi *et al.* [34] proposed a combination of fuzzy logic and wavelet transformation to enhance hiding capacity and imperceptibility. This approach begins with a bit reduction algorithm applied to each byte of secret data to be hidden within the cover image, reducing memory usage and increasing hiding capacity. The wavelet transform is then used to shift the image into the frequency domain, allowing data embedding in high-frequency coefficients less sensitive to human visual perception. After embedding, an optimal pixel adjustment algorithm is applied to minimize visual distortions between the cover and stego images.

Laishram and Tuithung [35] introduced J-HMMSteg, an adaptive JPEG steganography technique utilizing a Hidden Markov model (HMM) for embedding data with minimal distortion. The process consists of three main stages: first, block analysis of the image to construct statistical features by examining intra- and inter-block correlations of quantized DCT coefficients. Second, these features are used to build an HMM capable of capturing complex characteristics of the image, such as smoothness and regularity. Finally, data embedding is performed using a maximum likelihood embedder controlled by a threshold based on Kullback–Leibler divergence (KLD), ensuring that only acceptable distortion is introduced. J-HMMSteg demonstrates high imperceptibility, resistance to RS steganalysis, and enhanced security against ensemble-based steganalysis.

Setiadi *et al.* [36] introduced MaxMiPOD, an advancement of the MiPOD method designed to enhance the security and imperceptibility of content-adaptive steganography. This method employs image sharpening techniques to highlight complex features such as textures and edges, combined with fuzzy-based edge detection that prioritizes embedding areas in regions with complex features. Additionally, "complexity-first" and "spreading" rules are applied by optimizing residuals through weighting that accounts for edge and non-edge areas differently. MaxMiPOD significantly outperforms MiPOD and other state-of-the-art cost-function-based methods, especially at low payloads, where detection rates approach random levels (PE near 0.5) against SRM and maxSRMd2 steganalysis attacks. Sondas and Erturk [37] proposed the multi-pixel-pair (MPP) approach to enhance data hiding [38] capacity in the hybrid near maximum histogram (HNMH) method. In this approach, the peak histogram value of the cover image is used as a reference to determine pixel pairs where secret data will be hidden. Embedding is performed using the LSB on relevant pixels, while image partitioning is applied to extend hiding capacity dynamically. The MPP approach enables broader pixel utilization within the cover image, minimizing visible changes and improving resistance to steganography detection. Consequently, the MPP approach significantly enhances the performance of the classic HNMH method.

Adaptive and hybrid approaches have proven effective in improving visual quality and embedding capacity while maintaining resilience against modern steganalysis [39]. The studies above demonstrate that integrating adaptive algorithms, such as fuzzy logic and sharpening techniques, with transformation domains can significantly minimize trade-offs and optimize steganography security. Advances in this technology pave the way for further exploration, particularly in applying hybrid algorithms that leverage the strengths of various techniques to enhance system performance.

11.4.2   Machine learning and AI in steganography

Applying ML and artificial intelligence (AI) in steganography presents a more sophisticated adaptive method. With the ability to optimize embedding and increase resistance to steganalysis, this technology offers a safer and more effective solution than traditional methods. For example, a study conducted by Yang *et al.* [40] proposed a hybrid ML-based algorithm to improve hiding capacity and security. Their approach uses adaptive segmentation to separate images into foreground and background regions, where embedding is performed with high capacity in the foreground and low capacity in the background. Before embedding, the secret information is encrypted using hybrid ML, while the embedding process is performed with the breadth-first search (BFS) algorithm. This method significantly improves the quality of visual perception compared to traditional methods and provides high security against steganalysis.

Dhawan *et al.* [41] integrated hybrid fuzzy networks and fuzzy logic to improve security, payload capacity, and visual quality of hidden images. This method combines LSB substitution, PVD, and EMD techniques for embedding. By applying backpropagation learning in a hybrid fuzzy neural network (HFNN), this method adaptively improves the quality of hidden images by optimizing embedding parameters. Li *et al.* [42] proposed a steganography method based on GAN and DWT to improve the imperceptibility and security of hidden data. This approach utilizes the high-frequency sub-band of DWT as the main embedding location, which is more difficult to detect visually. The GAN generator is designed to optimize the embedding through a hybrid loss function that combines spatial loss and transformation domain. At the same time, the discriminator detects the presence of hidden data to improve the embedding performance. Significant improvement in PSNR with better robustness to steganalysis detection.

Li *et al.* [43] introduced StegoFormer, combining a U-shaped CNN network [44] with a transformer block to integrate high-resolution spatial and global self-attention features. In addition, the authors proposed a shifted window local loss (SWLL) mechanism to reduce errors in the center area of the steganographic image. This method enables the encoder to generate images with higher visual quality. Data augmentation is performed with Gaussian mask augmentation (GMA) to improve security. StegoFormer outperforms DL-based steganography algorithms in terms of anti-steganalysis capability and steganography effectiveness. Chen *et al.* [45] proposed a novel approach for color image steganography with a hybrid local texture descriptor-based cover selection algorithm. The study used a local binary pattern (LBP) adjusted for color channel correlation, especially utilizing the green channel as an important component in enhancing steganography security. This local descriptor is combined with local phase quantization (LPQ) to create a hybrid texture descriptor that integrates spatial and frequency domain information. Experiments show that this algorithm significantly improves the efficiency of cover image selection and embedding security. This approach is unique because it targets color images, which were often neglected in previous studies, mainly focusing on grayscale images.

Li *et al.* [46] also proposed a cover selection method for nature images and metaverse. This approach evaluates each candidate image based on its resistance to steganalysis using steganalysis features extracted and processed through an ensemble-based steganalysis classifier. This method provides significant improvement in undetectability metrics compared to conventional cover selection methods, with an average increase in PE (minimum total error) of 2.37–4.21% depending on the steganalytic tool used, such as SPAM, SRMQ1, maxSRMd2, and TLBP. The use of ML and AI in steganography has shown promising results regarding security and imperceptibility. Algorithms such as StegoFormer, HFNNs, GAN-based steganography models, and even cover selection steganography provide innovative solutions to overcome classic challenges in steganography. AI-based approaches improve embedding performance and provide opportunities for exploring new, more adaptive methods.

11.4.3   Quantum steganography

Quantum steganography is an emerging field that combines the principles of quantum mechanics, such as superposition, entanglement, and non-cloning, to enhance data-hiding techniques. Unlike classical steganography, which embeds data into digital media, quantum steganography allows embedding in quantum states, providing significantly higher security. Related studies, such as Jiang *et al.* [47] introduced two quantum-based steganography algorithms designed using the novel enhanced quantum representation (NEQR). The first algorithm, plain LSB, embeds secret data directly into the LSB bits of quantum cover image pixels, while block LSB divides the image into pixel blocks and performs segmented embedding. The NEQR representation enables encoding pixel color information on a quantum computational basis using superposition. Experimental results show that these algorithms offer good imperceptibility and flexibility in managing embedding capacity through block size adjustments, making them more resistant to manipulation. However, this approach has limitations regarding maximum capacity for larger images and inefficiency in real-time processing. Joshi *et al.* [48] developed a modified quantum key distribution (QKD) protocol to support steganography. This approach integrates continuous modulation and backward communication, enabling senders and receivers to embed secret data during the key exchange process. The technique reduces vulnerability to detection thanks to the inherently secure QKD against third-party attacks. Simulations indicate that this method can maintain the security of hidden data while ensuring resistance to detection by modern steganalysis algorithms. However, the approach still requires further development to improve implementation efficiency.

Khan and Rasheed [32] introduced a steganography algorithm based on the improved flexible representation of quantum images (IFRQI), a quantum representation model that allows embedding secret data in quantum cover images. This technique divides the color pixel information of the secret image into four image planes, each with a specific bit depth. It encodes them using controlled rotations with a chaotic laser system. This approach leverages quantum superposition to store data in highly compact quantum states. Experimental results show high resistance to steganalysis, with an average PSNR value above 50 dB, demonstrating excellent imperceptibility. The algorithm excels in embedding capacity, making it a top choice for applications where data hiding space is a priority. In another study, Khan and Rasheed [49] developed the multi-channel effective quantum image representation (MCEQI) model, an extension of IFRQI for color images. This scheme divides the red, green, and blue color channels into four planes, each with two-bit depth, and encodes them using controlled rotations based on chaotic keys. This technique enables embedding secret data into the MCEQI states of quantum cover images with significantly higher capacity. The embedding process also includes key sensitivity analysis to ensure maximum security. Evaluation results indicate that this method achieves highly accurate extraction capabilities and strong resistance to entropy-based steganalysis attacks.

Maurya *et al.* [50] proposed a DCT-based steganography scheme utilizing a quantum substitution box (S-Box) to hide grayscale secret images into color cover images. The process involves applying DCT to the cover image and embedding secret bits using the quantum S-Box into random image channels based on the DCT frequency order in a zigzag pattern. This method enhances security through key sensitivity analysis and attack resistance while demonstrating superior visual quality. Rijati *et al.* [51] introduced a watermarking technique based on a combination of DWT–DCT transformations optimized using quantum variational circuits (QVC). Although technically more focused on watermarking than steganography, this method provides insights into how quantum-inspired approaches can optimize embedding positions and intensities. The technique employs quantum-inspired annealing (QIA) to select optimal DCT coefficients, while QVC dynamically adjusts embedding intensity parameters. The results show that this approach minimizes the trade-off between imperceptibility and robustness.

The above review highlights how quantum steganography evolves through pure, hybrid, and quantum-inspired approaches. Fully quantum-based approaches rely entirely on quantum mechanics, such as quantum circuits or states for embedding and extraction, as seen in [47,32]. Study by Joshi *et al.* [48] also employs a pure quantum method but focuses on modifying QKD techniques to make them suitable for steganography. Hybrid quantum-classical approaches combine quantum techniques with classical steganography, improving imperceptibility, and capacity through quantum methods in specific stages, such as chaos-based embedding in MCEQ I [49], quantum S-Box for embedding [50] and QVC for embedding optimization [52]. Quantum-inspired techniques are also applied in [51], utilizing QIA to determine embedding coefficients. The use of quantum mechanics remains highly promising for further development, considering that quantum technology is still in its early stages. The future of quantum steganography will depend on its scalability and the implementation of practical systems, especially in secure data communication.

11.4.4 Cross-domain and cross-modal steganography

The increasing need for secure and covert communication has driven advancements in cross-domain steganography [33,53], where secret data are hidden across different modalities, such as image-to-audio, image-to-video, and multimodal embedding. Traditional steganographic techniques often rely on a single-domain carrier, such as images or text, which limits flexibility, robustness, and resistance to steganalysis. Single-domain approaches are vulnerable to domain-specific detection techniques, as steganalysis models can be trained to detect anomalies in pixel patterns, statistical distributions, or linguistic inconsistencies [53,44]. Cross-domain steganography [33,53] mitigates this risk by distributing hidden data across different media types, making detection significantly more challenging [54]. However, this also introduces technical complexities, such as maintaining synchronization between modalities, ensuring embedding reversibility, and optimizing capacity without degrading perceptual quality.

Song *et al.* [55] introduced INRSteg, a novel framework based on implicit neural representations (INRs), which facilitates cross-modal steganography across image, audio, video, and 3D shapes. Unlike traditional steganographic methods that rely on deep neural networks [56] for training, INRSteg [55] transforms data into a continuous representation, eliminating high computational costs and domain adaptation issues. Using layer-wise permutation encoding, INRSteg [55] enhances security and makes secret data harder to detect. The ability to embed multiple secret data types in a single cover medium demonstrates INRSteg [55]'s potential for multi-domain and multi-secret hiding, making it more robust than existing deep learning-based steganography methods.

Taremwa *et al.* [33] explored audio-image steganography using cross-modal transformer (CMT) models [33], integrating vision transformers (ViT) [33] and audio spectrogram transformers (AST) [33] to optimize feature extraction and alignment between auditory and visual data. By leveraging cross-modal attention mechanisms, the proposed framework significantly improves embedding capacity by 58.7% and PSNR by 57.5%, outperforming traditional LSB and ViT-based steganography. Deep learning-based audiovisual synchronization enhances security, making this approach promising for secure communication and multimodal data hiding [38].

Zhao *et al.* [57] introduced THInImg, a cross-modal steganography approach capable of hiding lengthy audio sequences in images, which can later be decoded into talking head videos. Unlike conventional image-based steganography, THInImg exploits human facial features as an embedding space, allowing up to 80 s of video and audio to be hidden in a single 160×160 image. This method provides iterative embedding for multiple levels of access control, offering secure and efficient covert communication. By employing non-uniform compression aligned with human auditory perception, THInImg reduces the bit size of audio while preserving high-quality recovery. Compared to GAN-based or STFT-based approaches, THInImg achieves superior compression efficiency and hiding capacity, making it suitable for covert video transmission and copyright protection.

Cui *et al.* [58] presented a multi-stage residual hiding framework for image-into-audio steganography designed to encode and decode image data into audio carriers while maintaining imperceptibility and security. By employing multi-stage networks, this approach gradually encodes residual errors into different audio subsequences, ensuring that the secret image can still be partially recovered even if part of the audio carrier is lost. Compared to traditional GAN-based or frequency-domain steganography methods, this approach enhances robustness against detection and improves payload flexibility. The results indicate that modifications to the audio carrier remain undetectable by human listeners, proving the feasibility of high-capacity cross-modal steganography.

The advancements in AI-driven cross-modal steganography have paved the way for more secure and robust data-hiding techniques. Approaches such as INRSteg [55], THInImg, and multi-stage residual hiding have significantly improved embedding performance and resistance to steganalysis. Introducing transformer-based architectures and INRs further expands the potential of cross-domain steganography, enabling seamless integration across various data modalities. These innovations open new opportunities for secure communication, digital watermarking, and multimodal information protection.

11.4.5   Linguistic steganography

The practice of concealing information within an innocuous carrier has traditionally relied on images, audio, and video as mediums for embedding secret messages. However, linguistic steganography, or text-based steganography, has gained attention due to its applicability in covert communication through natural language. Unlike other modalities, text lacks redundant bits, making it more challenging to manipulate without altering meaning or grammatical correctness [59]. Traditional linguistic steganography methods include synonym substitution, word reordering, and format-based techniques such as spacing manipulations or font variations. These approaches, however, suffer from low embedding capacity, poor robustness against linguistic analysis, and susceptibility to detection via statistical steganalysis. With the advent of deep learning and large-scale neural language models, a new generation-based linguistic steganography has emerged, offering higher imperceptibility, adaptability, and increased payload capacity [60,61]. Recent advances in AI-driven linguistic steganography, focusing on neural text generation, arithmetic coding, and multimodal approaches, have revolutionized text-based data hiding, enabling more secure and undetectable steganography techniques.

Some related studies, such as those by Ziegler *et al.* [62] proposed a neural linguistic steganography technique leveraging arithmetic coding and large-scale neural language models. Unlike traditional synonym substitution or word reordering methods, their approach generates text from scratch while encoding hidden messages. By ensuring the generated text statistically aligns with natural language distributions, the method achieves high imperceptibility and resists detection by linguistic steganalysis tools. This work represents a significant step in generation-based linguistic steganography, allowing higher embedding capacity than previous methods while maintaining natural readability.

Wen *et al.* [63] introduced a carrier-less steganographic method that generates image descriptions as stego text. Using CNN–LSTM networks, their framework encodes images into steganographic text under two different embedding schemes: word-by-word hiding (WWH) and sentence-by-sentence hiding (SSH). Compared to traditional text steganography, this method achieves higher embedding capacity and better reversibility without requiring a predefined cover text. This paper is unique as it combines image-to-text translation with steganographic embedding, making it highly effective for platforms like social media, video streaming, and online image-sharing services.

Gupta Banik and Bandyopadhyay [64] explored a novel approach to text steganography by using natural language processing (NLP) and part-of-speech (POS) tagging. Their method focuses on semantic and syntactic transformation rather than simple word substitution, ensuring the stego text remains contextually meaningful. The proposed method preserves linguistic structure by leveraging POS alignment while embedding secret information in text. Compared to earlier rule-based linguistic steganography techniques, this approach is more resistant to detection and offers greater flexibility in embedding strategies.

A formal evaluation framework for AI-driven linguistic steganography was proposed in [65], employing artificial neural networks (ANNs), RNNs, and LSTMs. The study compared two recurrent neural network-based steganography methods (RNN-stega vs. RNN-generated lyrics) and analyzed their embedding rate, security, and robustness. Their key contribution lies in benchmarking AI-based text steganography, emphasizing the importance of statistical language models (Markov chains) and deep learning models (LSTMs, Huffman coding). This research provides critical insights into how AI models influence steganographic security, undetectability, and payload optimization.

Steinebach [66] explored how large language models (LLMs) like ChatGPT can be utilized for linguistic steganography and watermarking. Using controlled prompt engineering, the study demonstrates how ChatGPT can generate steganographic cover texts seamlessly integrating hidden messages. The key innovation lies in leveraging AI-generated stego covers that follow specific linguistic patterns, making them difficult to distinguish from normal text. This work raises important questions about AI-generated steganography's role in security applications and potential misuse, highlighting the dual-use nature of LLM-based steganographic techniques.

The advent of AI-powered linguistic steganography has transformed the field from low-capacity detectable methods to sophisticated generation-based approaches with high levels of insensitivity and adaptability. The studies surveyed demonstrate how deep learning models [67–69], arithmetic coding, and neural representations improve security, linguistic coherence, and embedding efficiency in text-based steganography. However, as these techniques advance, so do the challenges of detecting, preventing, and ethically regulating their use. Future research should focus on improving security, developing anti-steganalysis methods, and exploring ethical frameworks to ensure the responsible application of AI in stega-nography applications.

11.5   Challenges and open issues

11.5.1   Trade-off among evaluation metrics

It is already discussed in Section 3 that the performance of a steganographic system depends on three important metrics: capacity [12,16,19,37], imperceptibility [12,14–20,24], and robustness [7,22,32,33]. However, it is examined from the state-of-the-art methods that it is always a big challenge to keep a balance among the SEKEY evaluation metrics to optimize performance, as shown in Figure 11.4.

- *Capacity vs. imperceptibility*: Higher data embedding reduces imperceptibility, making alterations more noticeable. Therefore, if you aim to maximize the capacity, the alteration introduced to the cover medium becomes more sig-nificant, making it easier for an observer or detection system to identify the presence of hidden information. On the other hand, reducing the embedded data increases imperceptibility, making the system harder to detect, but at the cost of hiding less information.
- *Capacity vs. robustness*: High capacity typically weakens robustness since embedded data is spread thinly across the medium. As you increase the *capacity* [12,16,19,37], the amount of available space in the cover medium for embedding the data becomes more stretched. To maintain high capacity, the data must often be spread thinly across the medium. This can make the system more vulnerable to data loss or corruption during attacks or transformations, weakening its *robustness*. Essentially, a larger hidden payload may reduce the ability of the system to withstand changes or attacks because the data is more dispersed and more likely to be altered or removed.

Selecting the right balance depends on the specific application requirements, whether it is maximizing capacity for a large payload or increasing robustness to resist applications such as tampering, copyright infringement, etc.

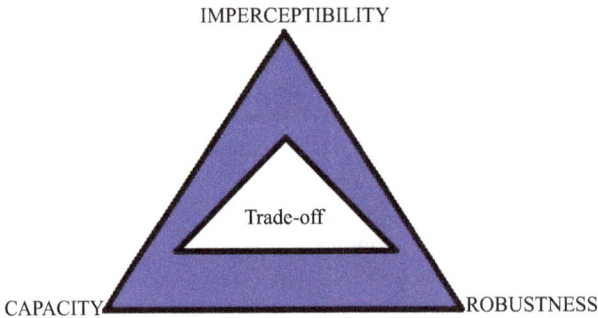

Figure 11.4   Trade-off among common evaluation metrics

11.5.2   Robustness against attacks in steganography

Steganographic methods aim to hide information in digital media (e.g., images, audio, video), but their effectiveness is often compromised by various attacks that alter the media. These attacks can *distort, remove, or reveal hidden data*, making robustness a crucial challenge in steganography.

11.5.2.1   Compression and noise addition

It is observed that many steganographic methods embed data in the LSBs or other delicate areas of media files. However, when these files undergo compression or noise addition, the hidden data can be lost or corrupted.

JPEG compression attack

- *JPEG compression* [35,53] reduces file size by discarding less important information, but in doing so, it also alters pixel values, potentially *destroying or modifying embedded data.*
- *Transform-based steganography* [23,24,27–30] *(e.g., DCT, DWT-based methods)* can help improve resistance to JPEG compression by embedding information in *frequency domains* rather than directly in pixel values.

Noise addition and filtering attacks

- *Random noise addition* (e.g., Gaussian noise, salt-and-pepper noise) alters pixel values, making the hidden message harder to extract.
- *Blurring or sharpening filters* modify pixel intensities, affecting the integrity of embedded data.
- *Robust steganography techniques* use *error correction codes (ECCs)* and *redundant embedding* to recover lost data after such attacks.

11.5.2.2   Geometric attacks

Conventional geometric transformations, such as *resizing, rotation, and cropping*, alter the spatial structure of an image, which can distort or completely remove embedded data.

Resizing attack

- Resizing (scaling up/down) changes the number of pixels in an image, which can cause *loss of steganographic data* or misalignment in extraction.
- Solutions include *feature-based embedding* techniques that use *keypoints (e.g., SIFT, SURF)* to ensure robustness against scaling.

Rotation attack

- Rotating an image can *misalign* the hidden data, making extraction difficult.
- *Rotation-invariant embedding* techniques use *circular embedding strategies* or *angle-adaptive watermarking* to improve resistance.

Cropping attack

- Cropping removes part of the image, potentially deleting *hidden data embedded in the cropped section.*

- A possible solution is *distributed embedding*, where the message is spread across the entire image instead of a single region, reducing the impact of cropping.

11.5.3   Detection and steganalysis resistance

Steganalysis is the process of detecting and analyzing hidden messages within digital media such as images, audio, video, or text. The goal is to uncover the presence of concealed data and potentially extract it, even without prior knowledge of the specific embedding technique. Steganalysis methods can be classified into four primary categories:

- *Passive steganalysis* [70,123]: Aims to determine if hidden data is present without retrieving it, such as detecting modifications in an image.
- *Active steganalysis* [70]: Focuses on extracting, altering, or eliminating hidden data, often employed by law enforcement agencies.
- *Targeted steganalysis* [71–74]: Identifies data embedded through specific techniques, utilizing knowledge of particular steganographic tools.
- *Blind steganalysis* [75–80]: A generalized detection method that does not rely on prior knowledge of the embedding technique and often leverages ML [81].

Traditional steganalysis techniques [1,2,7,71,74–76,79,80,82–97] detect hidden data in digital media through statistical, visual, noise, histogram, and algorithm-specific methods. Statistical analysis identifies embedding irregularities using RS analysis [82], Chi-square test [83], and fusion method [84]. Visual and noise analysis detect artifacts and distortions through edge detection, pixel flipping, and wavelet-based analysis. Algorithm-specific attacks exploit encryption flaws, making these techniques crucial for countering covert data transmission [90].

These days, steganalysis based on ML [81] is used to detect anomalies in data that may indicate the presence of hidden content. This technique relies on trained models to differentiate between clean and steganographically modified data. The methodology involves creating a dataset comprising both unaltered and steganographically altered samples for training purposes. Algorithms such as neural networks and support vector machines are commonly employed to analyze the data and identify subtle patterns indicative of hidden information. For instance, a classifier can be trained to distinguish between normal files and those altered through steganography, enabling automated and efficient detection of concealed content. This approach is particularly valuable in scenarios requiring high accuracy and scalability.

Some of the popular ML-based [81] analysis methods are briefly discussed:

- *SR-Net*
  With the introduction of SR-Net, a deep convolutional neural network (CNN) model [98], deep learning [67–69] has greatly enhanced steganalysis. In order to identify both visible and invisible artifacts, SR-Net employs a residual learning architecture to identify small differences observed between cover and stego images. It performs better than conventional techniques and is especially

effective at extracting hierarchical characteristics when embedded information causes less perceptual distortion. When it comes to automatic feature extraction for steganalysis, this model has proven to perform better, especially when embedded information causes little perceptual distortion. The model can identify even complex steganographic techniques because it can learn hierarchical aspects from the images [86].

- *YeNet*
  YeNet is a CNN made specifically for steganalysis tasks; it uses convolutional [98] and dense layers to learn to distinguish between stego and cover images. In terms of accuracy and generalization, it has surpassed certain previous deep learning models [67–69] and traditional techniques, providing strong performance across a range of distortion types. YeNet is a flexible steganalysis tool that uses a number of convolutional and fully connected layers to collect both low-level and high-level characteristics. It has been demonstrated to perform better than several previous deep learning models and conventional techniques. The benefit of YeNet is that it can generalize well to different forms of distortion and embedding techniques, maintaining good accuracy even when confronted with complex steganography [87].

- *SPAM*
  A deep learning technique called steganalysis with patch attention mechanism (SPAM) identifies tiny areas in images where embedding has probably taken place. To increase detection accuracy, it employs deep convolutional networks [98] for feature extraction and attention layers to concentrate on important areas of the image. When it comes to identifying intricate steganographic techniques that target particular regions of the image, SPAM is especially successful. By concentrating on crucial patches that are probably hiding information, it improves CNN performance in steganalysis tasks. This method increases detection accuracy, particularly for methods that focus on particular regions of the image. The model is especially good at identifying nuanced steganographic techniques because of this attention mechanism, which makes it more sensitive to tiny, localized changes [88].

- *DeepSteg*
  DeepSteg is a deep learning [52,99] architecture used to identify steganographic material in multimedia files. Convolutional and fully linked layers are used to extract features from noise patterns and unprocessed image data. This model is a top option for contemporary steganalysis tasks due to its adaptability and ability to generalize across different steganographic techniques. It is an effective tool for steganographic analysis since it can learn both local and global properties of the image. DeepSteg has proven to be highly accurate in identifying both contemporary and conventional steganographic techniques, providing notable advancements over previous methods that mostly depended on manually created features [89].

- *XceptionNet and ResNet for steganalysis*
  To enhance steganalysis performance, researchers have created models that employ depth wise separable convolutions [98] and residual learning. By

identifying subtle and complicated patterns caused by embedding, these models make it possible to extract more sophisticated features from images. They have proven to function competitively with traditional techniques, especially in intricate steganographic situations. Advanced CNN architectures like XceptionNet and ResNet have justified their effectiveness in detecting minute alterations brought about by steganographic methods in steganalysis tasks. Their capacity to extract deep hierarchical features from cover and stego pictures is responsible for their remarkable performance, especially in deep learning-based steganalysis [100,101].

11.6 Applications of steganography

Steganography is a powerful technique used across multiple industries due to its ability to conceal information within digital media, ensuring secure and covert communication. The following are some of the key applications of steganography:

- *Secure communication:* Steganography is widely used to facilitate confidential communication by embedding sensitive data within various media formats, including images, videos, and audio files [102]. This method ensures that the hidden information remains undetectable to unintended recipients, making it a valuable tool for secure correspondence, particularly in situations where secrecy is essential.
- *Copyright protection:* Intellectual property protection is a crucial concern in the digital era. Steganography enables the embedding of ownership details or digital watermarks within multimedia files to help track unauthorized distribution and establish proof of authorship. This application [103] is essential in copyright enforcement and legal disputes.
- *Data integrity verification:* To maintain the data integrity and authenticity of digital data, steganography can be used to embed verification codes or cryptographic hashes within files [31]. This allows for discreet verification of data integrity without visibly modifying the content. Any alterations to the file can be detected, ensuring data remains uncorrupted and unmodified.
- *Covert surveillance and intelligence:* Government agencies and intelligence organizations utilize steganographic techniques to transmit classified information covertly. By embedding messages within seemingly ordinary files, they can communicate sensitive data securely without drawing attention, ensuring the confidentiality of intelligence operations [104].
- *Authentication systems:* Steganography enhances authentication mechanisms by embedding security credentials such as passwords, encryption keys, or biometric data within digital files. This provides an additional layer of security [105], preventing unauthorized access to sensitive systems and ensuring robust identity verification.
- *Media fingerprinting:* Digital content providers leverage steganography to embed unique identifiers into media files, allowing for tracking and monitoring

of copyrighted material. This helps identify the source of leaks and unauthorized distribution, facilitating anti-piracy measures [106].

- *Healthcare security:* In the medical field, protecting patient data is of paramount importance. Steganography allows confidential health records, such as diagnostic images and reports, to be securely embedded within medical imaging files [107]. This ensures privacy while maintaining compliance with regulatory standards.

- *Military and defense communication:* Military operations rely on secure channels for communication. Steganography is used to conceal critical strategic information within digital media files, reducing the risk of interception and ensuring that messages reach only the intended recipients [108].

- *Secure online voting:* Electronic voting systems use steganography to embed voter information within digital ballots, providing an additional security layer that prevents tampering and ensures the authenticity of votes [109].

- *Cloud security:* As data is increasingly stored in the cloud, steganography offers a means of hiding sensitive information within seemingly innocuous files. This technique enhances data protection by concealing critical details from potential cyber threats [110].

- *Banking and financial security:* Financial institutions use steganography to secure electronic transactions by embedding authentication details within statements, receipts, and other digital financial documents [111]. This reduces the risk of fraud and unauthorized alterations.

- *Digital forensics:* Law enforcement agencies and forensic investigators utilize steganography to embed forensic markers within digital evidence, ensuring the authenticity of information and maintaining a reliable chain of custody [112].

- *Software protection:* To combat software piracy and unauthorized distribution, developers embed licensing details and serial numbers within software executables using steganographic methods [113]. This makes it difficult for cybercriminals to bypass licensing restrictions.

- *Anti-piracy measures in broadcasting:* Broadcasters implement steganographic techniques to embed watermarks or hidden identifiers in audio and video content. This allows content owners to trace unauthorized copies and enforce copyright protection more effectively [114].

- *Internet of Things (IoT) security:* Steganography plays a role in securing IoT devices by embedding encryption keys and sensor data within digital transmissions [115]. This adds an extra layer of security, ensuring the confidentiality of device communications.

- *Steganography in social media:* Social media platforms and users employ steganographic techniques to hide confidential messages within shared images or videos, allowing for secure and private communication while avoiding censorship or monitoring systems [116].

- *Historical and cultural preservation:* Digitized historical documents and artifacts can contain embedded metadata that provides information about their origins, ownership, and restoration history [117]. This ensures the preservation and authenticity of cultural heritage.

- *Blockchain integration:* Blockchain technology is combined with steganography to embed transactional data within digital assets, ensuring secure and traceable exchanges while maintaining data integrity and confidentiality [118–120].
- *Multilingual data transmission:* Steganography facilitates multilingual communication by embedding translations and annotations within multimedia files [121]. This technique enables seamless and secure international data exchange without altering the primary content.
- *Augmented and virtual reality (AR/VR):* Within AR and VR environments, steganography is used to embed interactive elements, metadata, or hidden instructions that enhance user engagement and functionality. This application improves immersive experiences while maintaining data security [122].

The applications of steganography continue to expand, playing a pivotal role in modern digital security, forensic investigations, intellectual property protection, and secure communications. As technology advances, steganography will remain a vital tool for safeguarding sensitive data and enhancing cybersecurity measures.

11.7   Conclusion

This chapter provides an extensive study of different steganographic techniques, with a particular emphasis on spatial and transform domain methods. It discusses various approaches, outlining their strengths, drawbacks, and areas of application. Spatial domain techniques, such as LSB substitution, are widely used due to their simplicity and high data embedding capacity. However, they are often susceptible to detection and attacks. In contrast, transform domain methods, including DCT, DWT-based techniques, offer greater security and resistance to steganalysis but typically involve higher computational demands.

Recent advancements in AI have paved the way for deep learning-based steganography, which introduces adaptive and intelligent approaches to data hiding. Additionally, hybrid models combining multiple techniques have gained attention for their ability to enhance security, imperceptibility, and efficiency. Despite these developments, challenges persist, particularly in ensuring resilience against advanced detection techniques, maintaining high payload capacity, and improving computational efficiency.

Future studies should prioritize the development of steganographic methods that can withstand evolving steganalysis techniques while remaining efficient for practical implementation. Furthermore, integrating blockchain and cryptographic techniques may strengthen data security and traceability. As digital communication continues to expand, the demand for secure and undetectable steganographic approaches will become increasingly critical, reinforcing the need for ongoing research and innovation in this field.

Funding statement: The author(s) received no specific funding for this study.

Conflicts of interest: The authors declare that they have no conflicts of interest to report regarding the present study.

References

[1] Bender, W., Gruhl, D., Morimoto, N., and Lu, A. (1996). Techniques for data hiding. *IBM Systems Journal*, *35*(3–4), 313–336.

[2] Abood, O., and Guirguis, S. (2018). A survey on cryptography algorithms. *International Journal of Scientific and Research Publications*, *8*(7), 495–516. https://doi.org/10.29322/IJSRP.8.7.2018.p7978.

[3] Subramanian, N., Elharrouss, O., Al-Maadeed, S., and Bouridane, A. (2021). Image steganography: A review of the recent advances. *IEEE Access*, *9*, 23409–23423. https://doi.org/10.1109/ACCESS.2021.3053998.

[4] Kazm, A. (2018). A survey of spatial domain techniques in image Steganography. *Journal of Education, College Wasit University*, *1*, 497–510. https://doi.org/10.31185/eduj.Vol1.Iss26.105.

[5] Nazari, M., Sharif, A., and Mollaeefar, M. (2017). An improved method for digital image fragile watermarking based on chaotic maps. *Multimedia Tools and Applications*, *76*, 16107–16123. https://doi.org/10.1007/s11042-016-3897-x.

[6] Bolourian Haghighi, B., Taherinia, A., and Monsefi, R. (2020). An effective semi-fragile watermarking method for image authentication based on lifting wavelet transform and feed-forward neural network. *Cognitive Computation*, *12*, 863–890. https://doi.org/10.1007/s12559-019-09700-9.

[7] Liu, H., Chen, Y., Shen, G., Guo, C., and Cui, Y. (2024). Robust image watermarking based on hybrid transform and position-adaptive selection. *Circuits, Systems, and Signal Processing*, *44*, 2802–2829. https://doi.org/10.1007/s00034-024-02946-1.

[8] Subhedar, M. S. (2021). Cover selection technique for secure transform domain image steganography. *Iran Journal of Computer Science*, *4*, 241–252. https://doi.org/10.1007/s42044-020-00077-9.

[9] Aslam, M., Rashid, M., Azam, F., Abbas, M., Rasheed, Y., Alotaibi, S., and Anwar, M. (2022). Image steganography using least significant bit (LSB) – A systematic literature review. *Proceedings of the IEEE International Conference on Computer and Information Technology (ICCIT)*, pp. 32–38. https://doi.org/10.1109/ICCIT52419.2022.9711628.

[10] Bhuiyan, T., Sarower, A. H., and Hassan, M. M. (2019). An image Steganography algorithm using LSB replacement through XOR substitution. *Proceedings of the IEEE International Conference on Information and Communications Technology (ICOIACT)*. https://doi.org/10.1109/ICOIACT46704.2019.8938486.

[11] Hiary, H., Sabri, K. E., Mohammed, M., and Al-Dhamari, A. (2016). A hybrid Steganography system based on LSB matching and replacement. *International Journal of Advanced Computer Science and Applications*, *7*(9), 374–380. https://doi.org/10.14569/IJACSA.2016.070951.

[12] Majumder, J., and Pradhan, C. (2019). High-capacity image steganography using pixel value differencing method with data compression using neural

network. *International Journal of Innovative Technology and Exploring Engineering*, *8*, 1800–1804. https://doi.org/10.35940/ijitee.L2839.1081219.

[13] Zhang, X., and Wang, S. (2006). Efficient steganographic embedding by exploiting modification direction. *IEEE Communications Letters*, *10*(11), 781–783. https://doi.org/10.1109/LCOMM.2006.060863.

[14] Cheng, L. M., and Chan, C.-K. (2003). Hiding data in images by simple LSB substitution. *Pattern Recognition*, *37*(3), 470–474.

[15] Grajeda-Marín, I. R., Montes-Venegas, H. A., Marcial-Romero, J. R., Hernández-Servín, J., and De Ita, G. (2016). An optimization approach to the TWPVD method for digital image steganography. In *Mexican Conference on Pattern Recognition* (pp. 125–134).

[16] Hajizadeh, H., Ayatollahi, A., and Mirzakuchaki, S. (2013). A new high-capacity and EMD-based image steganography scheme in the spatial domain. *2013 21st Iranian Conference on Electrical Engineering (ICEE), Mashhad, Iran.*

[17] Gaurav, K., and Ghanekar, U. (2018). Image steganography based on Canny edge detection, dilation operator, and hybrid coding. *Journal of Information Security and Applications*, *41*, 41–51. https://doi.org/10.1016/j.jisa.2018.05.001.

[18] Ghosal, S. K., Chatterjee, A., and Sarkar, R. (2021). Image steganography based on Kirsch edge detection. *Multimedia Systems*, *27*, 73–87. https://doi.org/10.1007/s00530-020-00703-3.

[19] Sultana, H., Kamal, A. H. M., Apon, T. S., and Alam, M. G. R. (2024). Increasing embedding capacity of stego images by exploiting edge pixels in prediction error space. *Cyber Security and Applications*, *2*, 100028. https://doi.org/10.1016/j.csa.2023.100028.

[20] Jan, A., Parah, S., Hussan, M., and Malik, B. (2022). Realization of efficient steganographic scheme using hybrid edge detection and chaos. *Arabian Journal for Science and Engineering*, *48*, 1–14. https://doi.org/10.1007/s13369-022-06960-w.

[21] Mandal, J. K. (2020). Discrete Fourier transform-based steganography. In J. K. Mandal (Ed.), *Reversible steganography and authentication via transform encoding* (Vol. 901, pp. 63–98). Springer, Singapore. https://doi.org/10.1007/978-981-15-4397-5_4.

[22] Goel, S., Rana, A., and Kaur, M. (2013). A DCT-based robust methodology for image steganography. *International Journal of Image, Graphics and Signal Processing (IJIGSP)*, *5*(11), 23–34. https://doi.org/10.5815/ijigsp.2013.11.03.

[23] Shehada, D., and Bouridane, A. (2023). A comparative analysis of image steganography techniques in spatial and transform domains. *2023 IEEE International Conference on Dependable, Autonomic and Secure Computing, International Conference on Pervasive Intelligence and Computing, International Conference on Cloud and Big Data Computing, International Conference on Cyber Science and Technology Congress (DASC/PiCom/*

CBDCom/CyberSciTech), pp. 155–160. https://doi.org/10.1109/DASC/
PiCom/CBDCom/Cy59711.2023.10361384.

[24] Atawneh, S., Almomani, A., Al Bazar, H., Sumari, P., and Gupta, B. (2017).
      Secure and imperceptible digital image steganographic algorithm based on
      diamond encoding in DWT domain. *Multimedia Tools and Applications*, *76*,
      18451–18472. https://doi.org/10.1007/s11042-016-3930-0.

[25] Falih, Z., and Karim, A. (2021). Digital watermark technique: A review.
      Journal of Physics: Conference Series, *1999*, 012118. https://doi.org/10.
      1088/1742-6596/1999/1/012118.

[26] Kim, H., Kim, C., Choi, Y., Wang, S., and Zhang, X. (2010). Improved
      modification direction methods. *Computers and Mathematics with Applica-
      tions*, *60*(2), 319–325. https://doi.org/10.1016/j.camwa.2010.01.006.

[27] Seyyedi, A., and Ivanov, N. (2014). High payload and secure steganography
      method based on block partitioning and integer wavelet transform. *Interna-
      tional Journal of Security and Its Applications*, *8*(4), 183–194. https://doi.
      org/10.14257/ijsia.2014.8.4.17.

[28] Mandal, J. K., and Ghosal, S. K. (2012). Separable discrete Hartley
      transform-based invisible watermarking for color image authentication
      (SDHTIWCIA). In *Proceedings of the Second International Conference on
      Advances in Computing and Information Technology (ACITY 2012)I* (pp.
      767–776).

[29] Subhedar, M. S., and Mankar, V. H. (2019). Image steganography using
      contourlet transform and matrix decomposition techniques. *Multimedia Tools
      and Applications*. *78*(16), 22155–22181. https://doi.org/10.1007/s11042-019-
      7512-9.

[30] Ghosal, S. K., and Mandal, J. K. (2014). Binomial transform-based fragile
      watermarking for image authentication. *Journal of Information Security and
      Applications*, *19*, 272–281. https://doi.org/10.1016/j.jisa.2014.07.004.

[31] Hambouz, A., Shaheen, Y., Manna, A., Al-Fayoumi, M., and Tedmori, S.
      (2019). Achieving data integrity and confidentiality using image stegano-
      graphy and hashing techniques. *Proceedings of ICTCS 2019*, pp. 1–6. https://
      doi.org/10.1109/ICTCS.2019.8923060.

[32] Khan, M., and Rasheed, A. (2023). A high-capacity and robust stegano-
      graphy algorithm for quantum images. *Chinese Journal of Physics*, *85*, 89–
      103. https://doi.org/10.1016/j.cjph.2023.06.016.

[33] Taremwa, M., Anaedevha, R. N., and Trofimov, A. G. (2024). *Robust audio-
      image Steganography using cross-modal based transformer models* (Version 1)
      [Preprint]. Research Square. https://doi.org/10.21203/rs.3.rs-5463235/v1.

[34] Shafi, I., Raza, M., Haq, I. U., Tariq, M., Qureshi, R. J., and Javid , M. A.
      (2018). An adaptive hybrid fuzzy-wavelet approach for image steganography
      using bit reduction and pixel adjustment. *Soft Computing*, *22*(5), 1555–1567.
      https://doi.org/10.1007/s00500-017-2944-5.

[35] Laishram, D., and Tuithung, T. (2023). A secure adaptive Hidden Markov
      model-based JPEG steganography method. *Multimedia Tools and Applica-
      tions*, *83*(13), 38883–38908. https://doi.org/10.1007/s11042-023-17152-5.

[36]   Setiadi, D. R. I. M., Rustad, S., Andono, P. N., and Shidik, G. F. (2024). Maximizing complex features to minimize the detectability of content-adaptive Steganography. *Multimedia Tools and Applications*, *84*, 23813–23831. https://doi.org/10.1007/s11042-024-20056-7.

[37]   Sondas, A., and Erturk, N. B. (2024). Dynamic data hiding capacity enhancement for the hybrid near maximum histogram image steganography based on multi-pixel-pair approach. *Multimedia Tools and Applications*, *84*(7), 3683–3699. https://doi.org/10.1007/s11042-024-19059-1.

[38]   Cao, Y., and Zhang, L. (2019). A comprehensive survey on image steganalysis techniques. *International Journal of Computer Science and Network Security*, *19*(8), 1–10.

[39]   Dong, F., and Liu, J. (2020). Advanced steganalysis for video content based on deep learning. *IEEE Access*, *8*, 119460–119468. https://doi.org/10.1109/ACCESS.2020.3005144.

[40]   Yang, A., Bai, Y., Xue, T., Li, Y., and Li, J. (2023). A novel image Steganography algorithm based on hybrid machine learning and its application in cyberspace security. *Future Generation Computer Systems*, *145*, 293–302. https://doi.org/10.1016/j.future.2023.03.035.

[41]   Dhawan, S., Agarwal, R., Singh, R., Patil, S., and Gupta, S. (2024). Secure and resilient improved image Steganography using hybrid fuzzy neural network with fuzzy logic. *Journal of Safety Science and Resilience*, *5*(1), 91–101. https://doi.org/10.1016/j.jnlssr.2023.12.003.

[42]   Li, X., Chen, L., Lai, J., Fu, Z., and Liu, S. (2024). GAN-based image steganography by exploiting transform domain knowledge with deep networks. *Multimedia Systems*, *30*(4), 224. https://doi.org/10.1007/s00530-024-01427-4.

[43]   Li, Z., Wang, Y., Zhang, Y., Zhu, Y., and Zhang, M. (2023). Adversarial feature hybrid framework for steganography with shifted window local loss. *Neural Networks*, *165*, 358–369. https://doi.org/10.1016/j.neunet.2023.05.053.

[44]   Wang, Z., Chen, M., Yang, Y., Lei, M., and Dong, Z. (2020). Joint multidomain feature learning for image steganalysis based on CNN. *EURASIP Journal on Image and Video Processing*, *2020*(1), 28. https://doi.org/10.1186/s13640-020-00513-7.

[45]   Chen, M., He, P., and Liu, J. (2022). HLTD-CSA: Cover selection algorithm based on hybrid local texture descriptor for color image Steganography. *Journal of Visual Communication and Image Representation*, *89*, 103646. https://doi.org/10.1016/j.jvcir.2022.103646.

[46]   Li, X., Guo, D., and Qin, C. (2023). Diversified cover selection for image steganography. *Symmetry*, *15*(11), 2024. https://doi.org/10.3390/sym15112024.

[47]   Jiang, N., Zhao, N., and Wang, L. (2016). LSB based quantum image Steganography algorithm. *International Journal of Theoretical Physics*, *55*(1), 107–123. https://doi.org/10.1007/s10773-015-2640-0.

[48]   Joshi, R., Gupta, A., Thapliyal, K., Srikanth, R., and Pathak, A. (2022). Hide and seek with quantum resources: New and modified protocols for quantum

Steganography. *Quantum Information Processing*, *21*(5), 164. https://doi.org/10.1007/s11128-022-03514-9.

[49] Khan, M., and Rasheed, A. (2023). A secure controlled quantum image Steganography scheme based on the multi-channel effective quantum image representation model. *Quantum Information Processing*, *22*(7), 268. https://doi.org/10.1007/s11128-023-04022-0.

[50] Maurya, S., Nandu, N., Patel, T., Reddy, V. D., Tiwari, S., and Morampudi, M. K. (2023). A discrete cosine transform-based intelligent image Steganography scheme using quantum substitution box. *Quantum Information Processing*, *22*(5), 206. https://doi.org/10.1007/s11128-023-03914-5.

[51] Rijati, N., Sahu, A. K., Setiadi, D. R. I. M., and Sambas, A. (2025). DWT-DCT image watermarking with quantum-inspired optimization. *International Journal of Intelligent Engineering Systems*, *18*(1), 1034–1044. https://doi.org/10.22266/ijies2025.0229.74.

[52] Chen, Y., and Wu, X. (2021). A hybrid model for image-based steganalysis using deep learning. *Computational Intelligence and Neuroscience*, *2021*, 5590145. https://doi.org/10.1155/2021/5590145.

[53] Jia, J., Luo, M., Liu, J., Ren, W., and Wang, L. (2022). Multiperspective progressive structure adaptation for JPEG Steganography detection across domains. *IEEE Transactions on Neural Networks and Learning Systems*, *33* (8), 3660–3674. https://doi.org/10.1109/TNNLS.2021.3054045.

[54] Eid, W. M., Alotaibi, S. S., Alqahtani, H. M., and Saleh, S. Q. (2022). Digital image steganalysis: Current methodologies and future challenges. *IEEE Access*, *10*, 92321–92336. https://doi.org/10.1109/ACCESS.2022.3202905.

[55] Song, S., Yang, S., Yoo, C. D., and Kim, J. (2023). Implicit steganography beyond the constraints of modality, *15144*, pp. 289–304. https://doi.org/10.1007/978-3-031-73016-0_17.

[56] Wang, X., and Sun, Z. (2020). Detecting steganography in images using deep neural networks: A review. *Pattern Recognition Letters*, *133*, 180–187. https://doi.org/10.1016/j.patrec.2020.01.005.

[57] Zhao, L., Li, H., Ning, X., and Jiang, X. (2024, January). THInImg: Cross-modal steganography for presenting talking heads in images. In *2024 IEEE/CVF Winter Conference on Applications of Computer Vision (WACV)* (pp. 5541–5550). https://doi.org/10.1109/WACV57701.2024.00546.

[58] Cui, W., Liu, S., Jiang, F., Liu, Y., and Zhao, D. (2020). Multi-stage residual hiding for image-into-audio Steganography. In *ICASSP 2020–2020 IEEE International Conference on Acoustics, Speech and Signal Processing (ICASSP)* (Vol. 2020, May, pp. 2832–2836). https://doi.org/10.1109/ICASSP40776.2020.9054033.

[59] Yi, B., Wu, H., Feng, G., and Zhang, X. (2022). ALiSa: Acrostic linguistic steganography based on BERT and Gibbs sampling. *IEEE Signal Processing Letters*, *29*, 687–691. https://doi.org/10.1109/LSP.2022.3152126.

[60] Yang, T., Wu, H., Yi, B., Feng, G., and Zhang, X. (2024). Semantic-preserving linguistic steganography by pivot translation and semantic-aware

bins coding. *IEEE Transactions on Dependable and Secure Computing, 21* (1), 139–152. https://doi.org/10.1109/TDSC.2023.3247493.

[61]   Namitha, M. V., Manjula, G. R., and Belavagi, M. C. (2024). StegAbb: A cover-generating text steganographic tool using GPT-3 language modeling for covert communication across SDRs. *IEEE Access, 12,* 82057–82067. https://doi.org/10.1109/ACCESS.2024.3411288.

[62]   Ziegler, Z., Deng, Y., and Rush, A. (2019). Neural linguistic steganography. In *Proceedings of the 2019 Conference on Empirical Methods in Natural Language Processing and the 9th International Joint Conference on Natural Language Processing (EMNLP-IJCNLP)* (pp. 1210–1215). https://doi.org/10.18653/v1/D19-1115.

[63]   Wen, J., Zhou, X., Li, M., Zhong, P., and Xue, Y. (2019). A novel natural language steganographic framework based on image description neural network. *Journal of Visual Communication and Image Representation, 61,* 157–169. https://doi.org/10.1016/j.jvcir.2019.03.016.

[64]   Gupta Banik, B., and Bandyopadhyay, S. K. (2020). Novel text Steganography using natural language processing and part-of-speech tagging. *IETE Journal of Research, 66*(3), 384–395. https://doi.org/10.1080/03772063.2018.1491807.

[65]   Gurunath, R., Alahmadi, A. H., Samanta, D., Khan, M. Z., and Alahmadi, A. (2021). A novel approach for linguistic steganography evaluation based on artificial neural networks. *IEEE Access, 9,* 120869–120879. https://doi.org/10.1109/ACCESS.2021.3108183.

[66]   Steinebach, M. (2024). Natural language steganography by ChatGPT. In *Proceedings of the 19th International Conference on Availability, Reliability and Security* (pp. 1–9). https://doi.org/10.1145/3664476.3670930.

[67]   Bayar, B., and Stamm, M. C. (2020). Deep learning for steganalysis of digital images: A survey. *IEEE Transactions on Information Forensics and Security, 15,* 2985–2999. https://doi.org/10.1109/TIFS.2020.2981450.

[68]   Zhang, Y., and Zhang, H. (2020). Passive and active steganalysis of image data using deep learning models. *Journal of Electronic Imaging, 29*(3), 033010. https://doi.org/10.1117/1.JEI.29.3.033010.

[69]   Wu, X., and Zhang, J. (2020). Active steganalysis of audio signals using deep learning. *IEEE Transactions on Information Forensics and Security, 15,* 1697–1710. https://doi.org/10.1109/TIFS.2019.2953290.

[70]   Miao, X., and Wang, X. (2021). A novel approach for passive and active steganalysis in digital images. *Digital Signal Processing, 114,* 102856. https://doi.org/10.1016/j.dsp.2021.102856.

[71]   Singh, A., and Agarwal, S. (2019). An effective targeted steganalysis approach using deep neural networks. *Neural Computing and Applications, 31*(9), 4815–4828. https://doi.org/10.1007/s00542-018-4075-5.

[72]   Wei, M., and Duan, X. (2020). Targeted steganalysis using deep learning: A detailed review. *Journal of Information Security and Applications, 53,* 102481. https://doi.org/10.1016/j.jisa.2020.102481.

[73] Yao, D., and Wu, Q. (2020). Targeted steganalysis of audio and video data using deep neural networks. *IEEE Transactions on Information Forensics and Security, 15*, 2898–2912. https://doi.org/10.1109/TIFS.2020.2979028.

[74] Fu, Y., and Zhang, J. (2021). Machine learning for targeted steganalysis of digital media. *Journal of Information Science, 47*(4), 468–478. https://doi.org/10.1177/0165551519896253.

[75] Zhang, C., Wang, X., and Liu, Y. (2021). A novel deep learning approach for blind steganalysis. *Journal of Visual Communication and Image Representation, 74*, 102909. https://doi.org/10.1016/j.jvcir.2021.102909.

[76] Zou, X., and Li, J. (2021). Advanced blind steganalysis for image data using convolutional neural networks. *Journal of Ambient Intelligence and Humanized Computing, 12*(10), 10085–10100. https://doi.org/10.1007/s12652-020-02525-x.

[77] Zhou, Y., and Yang, J. (2020). Blind steganalysis of digital images via deep learning. *Soft Computing, 24*(19), 14243–14253. https://doi.org/10.1007/s00542-020-05698-1.

[78] Guo, Y., and Li, L. (2021). Advanced techniques in blind steganalysis of images. *Multimedia Tools and Applications, 80*(6), 9605–9621. https://doi.org/10.1007/s11042-020-10165-3.

[79] Zhao, Z., and Wei, Z. (2020). Convolutional neural networks for blind steganalysis of images. *International Journal of Wavelets, Multiresolution and Information Processing, 18*(7), 2043001. https://doi.org/10.1142/S0219691320430013.

[80] Zhang, S., and Li, T. (2021). A deep learning-based approach for blind steganalysis of encrypted images. *Signal Processing, 183*, 107967. https://doi.org/10.1016/j.sigpro.2021.107967.

[81] Zhang, D., and Zhang, F. (2020). Advanced machine learning techniques for passive steganalysis of images. *Neural Computing and Applications, 32*(3), 877–889. https://doi.org/10.1007/s00542-019-04544-4.

[82] Fridrich, J., and Goljan, M. (2002). RS analysis for steganalysis. *Proceedings of the International Workshop on Information Hiding*, pp. 111–123.

[83] Zhang, Y., and Wang, Z. (2013). Primary set and feature-based steganalysis. *IEEE Transactions on Image Processing, 22*(10), 3456–3468.

[84] Deguillaume, F., and Lemoine, D. (2005). Chi-square test for steganalysis in digital images. *Journal of Electronic Imaging, 14*(4), 487–499..

[85] Mielikainen, J., and Kittler, J. (2010). Fusion of steganalysis methods for robust detection. *IEEE Transactions on Information Forensics and Security, 5*(4), 919–928.

[86] Hu, J., and Yang, Z. (2018). SRNet: A residual learning network for steganalysis. *Proceedings of the IEEE Conference on Computer Vision and Pattern Recognition*, pp. 4587–4596.

[87] Ye, Z., and Hu, M. (2019). YeNet: A deep neural network for steganalysis. *IEEE Transactions on Multimedia, 21*(6), 1589–1602.

[88] Ma, L., Zhang, X., and Li, W. (2020). SPAM: Steganalysis with patch attention mechanism. *IEEE Access, 8*, 55288–55301.

[89]   Xu, T., and Li, Q. (2021). DeepSteg: A convolutional network for stega-
       nalysis. *IEEE Transactions on Circuits and Systems for Video Technology*,
       31(1), 45–59.

[90]   Liu, F., and Yang, Z. (2020). A review of passive steganalysis techniques
       for image-based steganography. *Journal of Electronic Imaging*, *29*(5),
       053015. https://doi.org/10.1117/1.JEI.29.5.053015.

[91]   Xie, L., and Cheng, H. (2021). Survey on image steganalysis using deep
       learning. *Journal of Information Security and Applications*, *59*, 102789.
       https://doi.org/10.1016/j.jisa.2020.102789.

[92]   Chen, W., and Guo, Y. (2021). A survey on machine learning-based image
       steganalysis techniques. *Soft Computing*, *25*(4), 3087–3103. https://doi.org/
       10.1007/s00542-020-05716-2.

[93]   Yang, W., and Li, B. (2020). Blind image steganalysis with deep convolu-
       tional neural networks. *Multimedia Tools and Applications*, *79*(7), 4731–
       4753. https://doi.org/10.1007/s11042-019-08353-0.

[94]   Li, C., and Li, Z. (2021). Passive steganalysis using residual neural net-
       works. *International Journal of Computer Applications*, *175*(4), 20–25.
       https://doi.org/10.5120/ijca2021921685.

[95]   Zhang, Y., and Chen, H. (2020). Active steganalysis using deep learning for
       video data. *IEEE Access*, *8*, 32499–32508. https://doi.org/10.1109/
       ACCESS.2020.2978017.

[96]   Zheng, Z., and Liu, Y. (2021). Blind image steganalysis using residual con-
       volutional neural networks. *Journal of Visual Communication and Image
       Representation*, *73*, 102866. https://doi.org/10.1016/j.jvcir.2020.102866.

[97]   Kong, Y., and Hu, W. (2021). A survey on steganalysis methods and chal-
       lenges. *Multimedia Tools and Applications*, *80*(11), 16245–16272. https://
       doi.org/10.1007/s11042-021-10755-w.

[98]   Zhan, S., and Wang, X. (2020). Steganalysis based on convolutional neural
       networks with image data. *Computers, Materials and Continua*, *64*(1), 157–
       174. https://doi.org/10.32604/cmc.2020.017557.

[99]   Li, F., and Cheng, M. (2021). Deep learning approaches for detecting Ste-
       ganography in multimedia data. *Signal Processing: Image Communication*,
       89, 116016. https://doi.org/10.1016/j.image.2020.116016.

[100] Chollet, F. (2017). Xception: Deep learning with depthwise separable con-
       volutions. *Proceedings of the IEEE Conference on Computer Vision and
       Pattern Recognition*, pp. 1251–1258.

[101] He, K., Zhang, X., Ren, S., and Sun, J. (2016). Deep residual learning for
       image recognition. *Proceedings of the IEEE Conference on Computer
       Vision and Pattern Recognition*, pp. 770–778.

[102] Panigrahi, R., and Padhy, N. (2025). An effective steganographic technique
       for hiding the image data using the LSB technique. *Cyber Security and
       Applications*, *3*, 100069. https://doi.org/10.1016/j.csa.2024.100069.

[103] Araujo, I. (2024). *Information hiding and copyrights*. IntechOpen. https://
       doi.org/10.5772/intechopen.1004651.

[104]   Nag, A. (2019). Low-tech Steganography for covert operations. *IJCNIS, 2* (1), 21–27.

[105]   Al-Shqeerat, K. H. A. (2022). An enhanced graphical authentication scheme using multiple-image steganography. *Computer Systems Science and Engineering, 44*(3), 2095–2107. https://doi.org/10.32604/csse.2023.028975.

[106]   Cao, P., He, X., Zhao, X., and Zhang, J. (2019). Approaches to obtaining fingerprints of Steganography tools which embed messages in fixed positions. *Forensic Science International: Reports, 1,* 100019. https://doi.org/10.1016/j.fsir.2019.100019.

[107]   AlEisa, H. N. (2022). Data confidentiality in healthcare monitoring systems based on image Steganography to improve the exchange of patient information using the Internet of Things. *Journal of Healthcare Engineering, 2022,* 7528583. https://doi.org/10.1155/2022/7528583.

[108]   Dangi, K. O. P., Tandon, S., Deorari, S., and Kumar, R. (2024). *Steganography: Unveiling techniques and research agenda.* IntechOpen. https://doi.org/10.5772/intechopen.1005052.

[109]   Rura, L., Issac, B., and Haldar, M. (2016). Implementation and evaluation of steganography-based online voting system. *International Journal of Electronic Government Research, 12,* 71–93. https://doi.org/10.4018/IJEGR.2016070105.

[110]   Murakami, K., Hanyu, R., Zhao, Q., and Kaneda, Y. (2013). Improvement of security in cloud systems based on steganography. *2013 International Joint Conference on Awareness Science and Technology & Ubi-Media Computing (iCAST 2013 & UMEDIA 2013),* Aizu-Wakamatsu, Japan, pp. 503–508. https://doi.org/10.1109/ICAwST.2013.6765492.

[111]   Mangayarkarasi, S., Suganya, K., and Kala, T. S. (2019). Secure M-Banking using steganography. *International Journal of Innovative Technology and Exploring Engineering, 8*(11), 2769–2772. https://doi.org/10.35940/ijitee.K2253.0981119.

[112]   Dalal, M., and Juneja, M. (2021). Steganography and steganalysis (in digital forensics): A cybersecurity guide. *Multimedia Tools and Applications, 80,* 5723–5771. https://doi.org/10.1007/s11042-020-09929-9.

[113]   Warkentin, M., Bekkering, E., and Schmidt, M. B. (2008). Steganography: Forensic, security, and legal issues. *Journal of Digital Forensics, Security and Law, 3*(2), Article 2. https://doi.org/10.15394/jdfsl.2008.1039.

[114]   Prasad, N. A. (2023). Tackling movie piracy enigma employing automated infrared transmitter screen system and steganoanalysis techniques. *International Journal of Advanced Research in Computer and Communication Engineering,12*(2). https://doi.org/10.17148/IJARCCE.2023.12210.

[115]   Koptyra, K., and Ogiela, M. R. (2022). Steganography in IoT: Information hiding with APDS-9960 proximity and gestures sensor. *Sensors, 22*(7), 2612. https://doi.org/10.3390/s22072612.

[116]   Ramachandra, G., Klaib, M., Samanta, D., and Khan, M. (2021). Social media and steganography: Use, risks and current status. *IEEE Access, 9,* 153656–153665. https://doi.org/10.1109/ACCESS.2021.3125128.

[117] Chaczko, Z., Wazirali, R., Carrion Gordon, L., and Bozejko, W. (2018). Steganographic data heritage preservation using sharing images app. In *Advances in Intelligent Systems and Computing* (pp. 225–235). Springer. https://doi.org/10.1007/978-3-319-74718-7_18.

[118] Cao, Y., Li, J., Chao, K., et al. (2024). Blockchain meets generative behavior steganography: A novel covert communication framework for secure IoT edge computing. *Chinese Journal of Electronics*, *33*(4), 886–898. https://doi.org/10.23919/cje.2023.00.382.

[119] Wang, M., Zhang, Z., He, J., Gao, F., Li, M., Xu, S., and Zhu, L. (2022). Practical blockchain-based steganographic communication via adversarial AI: A case study in Bitcoin. *The Computer Journal*, *65*(11), 2926–2938. https://doi.org/10.1093/comjnl/bxac090.

[120] Partala, J. (2018). Provably secure covert communication on blockchain. *Cryptography*, *2*(3), 18. https://doi.org/10.3390/cryptography2030018.

[121] Shazzad-Ur-Rahman, M., Hosen Ornob, M. M., Singha, A., Kaiser, M. S., and Akhter, N. I. (2021). An effective text steganographic scheme based on multilingual approach for secure data communication. *2021 Joint 10th International Conference on Informatics, Electronics & Vision (ICIEV) and 2021 5th International Conference on Imaging, Vision & Pattern Recognition (ICIVPR)*, Kitakyushu, Japan, pp. 1–8. https://doi.org/10.1109/ICIEVicIVPR52578.2021.9564231.

[122] Li, C., Sun, X., and Li, Y. (2019). Information hiding based on augmented reality. *Mathematical Biosciences and Engineering*, *16*, 4777–4787. https://doi.org/10.3934/mbe.2019240.

[123] Yang, M., and Hu, H. (2021). An improved passive steganalysis method for detecting hidden data in multimedia. *Journal of Electrical Engineering & Technology*, *16*(2), 824–832. https://doi.org/10.1007/s42835-020-00203-z.

Chapter 12

Video tampering detection: deep learning techniques for temporal and motion artifacts

Ajantha Devi Vairamani[1]

Abstract

Video manipulation techniques, such as deepfake synthesis, frame interpolation, and frame deletion, pose a significant threat to both the cybersecurity field and misinformation prevention efforts. The difference between manipulating images and machining video footage is that in the latter case, spatiotemporally inconsistent errors may be introduced; special methods are required as tools to identify and analyze these types of gentler transformations over a sequence of serial motion. This chapter discusses deep learning techniques for video tampering detection, analyzing temporal and motion data. We study recurrent neural networks (RNNs), long-short-term memory (LSTM) networks, and temporal convolution networks (TCNs) with respect to their nature as models that can selectively encapsulate sequence dependencies in video media. The role of vision transformers (ViTs) and attention mechanisms in frame-wise anomaly detection could also be explored. We lay out methods for detecting sudden frame transitions, optical flow inconsistency, and the residual artifacts that indicate tampering by GANs. This chapter also introduces large-scale datasets such as FaceForensics++, DeepFake detection challenge (DFDC), and Celeb-DF, which are used to train and test deep learning models for video forensics. Issues including adversarial attacks on detection models, the generalization gap in deepfake detection, and real-time processing constraints are all covered. Finally, we also discuss countermeasures, such as adversarial training, multimodal analysis (audiovisual consistency analysis), and blockchain-supported video attestation. This is necessary in an era where people put trust in digital video as evidence that common knowledge must be the only way.

Keywords: Video manipulation; deepfake detection; temporal anomaly analysis; recurrent neural networks (RNNs); vision transformers (ViTs); adversarial attacks

[1]AP3 Solutions, Chennai, Tamil Nadu, India

12.1   Introduction

Video content in the digital age is the medium of communication, journalism, and entertainment, and even upholds safety. With the advancement of artificial intelligence and deep learning, video manipulation technologies have been keeping pace; contesting works of misinformation, security threats, and privacy violations. It is thereby becoming increasingly difficult to distinguish genuine content from fake videos manipulated through deepfakes and other video manipulation technologies, making it increasingly important to develop efficient detection and security systems to combat this menace.

12.1.1   Rise of video manipulation technologies

The growth of video manipulation technologies witnessed a surge as advancements in deep learning and computer vision progressed. Methods like generative adversarial networks (GANs) [1] and autoencoders are used to create hyper-realistic synthetic videos referred to as deepfakes. These techniques are capable of high-end video manipulations mimicking human facial expressions, voices, and body movements with astonishing accuracy on large datasets. Originally for the purposes of entertainment and research, deepfake technology has also been leveraged for filmmaking, digital avatars, and AI assistants that learn personalized information from an individual. However, unduly application has raised ethical concerns and security fears about its actual use. For instance, deepfake videos can impersonate people, disinform the public about political matters, and create fraudulent content that can lead to social damage [2].

12.1.2   Threats posed by deepfakes and video tampering

With the emergence of deepfake technology comes myriad threats that cut across the legal, ethical, and security divide. The most immediate concerns are:

1. Misinformation and disinformation
   Deepfake videos are increasingly being used to disseminate wrong information, capable of shifting public opinion and disrupting democratic processes. A distinct instance happened in the 2020 U.S. Presidential Elections, where, in fact, AI-generated videos were suspected to have been used for politically motivated misinformation. Some of the primary targets for such artful deception are social media platforms, especially when manipulated videos can reach millions within a matter of minutes [3].
2. Identity theft and fraud
   By mimicking people's appearance through deepfake technology, cybercriminals engage in identity fraud, financial scams, and social engineering attacks. A 2019 fraud case in the United Kingdom showcased how such deepfakes were employed by criminals who impersonated the voice of an executive by AI means, misleading an employee to transfer €220000 to a rogue account. Such

cases exemplify the increasing risk of AI impersonation in the financial and business world.

3. Political and social manipulation

    Deepfakes are video form of world leaders and politicians. Such geopolitical affairs, social discourses can also be manipulated by such deepfakes. Such a viral manipulated video of former U.S. President Barack Obama surfaced in 2018, stating things that he never said. The research by the University of Washington demonstrates the capability and danger of the technology for producing deepfakes [4].

4. Legal and ethical dilemmas

    The incorporation of deepfake technology has created legal and ethical concerns relating to privacy rights, defamation, and intellectual property. Most jurisdictions have no specific laws defining deepfake crimes, which renders their enforcement nearly impossible [5]. Indeed, deepfake pornography has arisen as a potential problem in some jurisdictions where victims of non-consensual explicit videos were left oblivious to their existence [6].

12.1.3 Role of AI and deep learning in video security

To solve the security issue posed by deepfake technology, AI-based solutions are being researched and developed by researchers and security experts. They are coming up with different methods of attesting the authenticity and security of media:

1. Deepfake detection algorithms

    The new collective research movement for AI detection models addresses deepfake video recognition based on the differences found in manipulated videos. These include convolutional neural networks (CNNs)-related investigation of facial expressions, as well as recurrent neural networks (RNNs)-based analysis of lip movements to find differences in blinking and lip movement when comparing deepfake videos to real footage. The deepfake detection challenge (DFDC), organized by Facebook and academic institutions, has greatly contributed to the improvement of deepfake detection models [7].

2. Blockchain for video authentication

    Current investigations about blockchain technology are focusing on tamper-evident digital signatures that will make it viable to attach anything to a video file. By its very nature, the blockchain keeps a record that cannot alter the origin of a video and all modifications. In so doing, it goes a long way toward establishing authenticity and discovers tampering [8]. Truepic, as well as Project Origin by Microsoft, is working on such a combination of blockchain with AI for anti-video manipulation purposes.

3. Real-time video analysis and forensics

    AI forensics software can analyze videos for pixel inconsistencies, frame artifacts, and lighting anomalies. Research shows that GAN-generated deepfakes often leave behind detectable artifacts, usually unnatural lighting reflections or irregular pupil shapes, which forensic tools can then identify. Integrating these tools into journalism and law enforcement through cyber-space will fight misinformation.

4. Ethical AI development and policy regulations
    Technology firms and governments have been exerting efforts to put in place legal frameworks for the regulation of deepfake technologies. The European Union's Digital Services Act (2022) and China's Deepfake Regulation (2023) mandate the labeling of content generated by artificial intelligence to allow better transparency. On a similar note, social media platforms such as Twitter and Facebook have been working on deepfake detection and policy measures to counter misinformation.

12.2   Types of video tampering

Video forgery has changed winsomely over the years, owing much to deep learning approaches, GAN techniques, and advanced image editing methods, making it easy to generate highly manipulated, convincing content. It is indeed a challenging task to look after some aspects of the modifications in a video, either with respect to its visual appeal or the time at which the changes were made; forensic experts, cybersecurity professionals, and AI researchers have much to deal with this fact.

Figure 12.1 illustrates the synopsis of the various types of video forgery classified as spatial forgery (intra frame), temporal forgery (inter frame), and spatio-temporal forgery. The spatial forgery is the integrity of individual frames compromised at the pixel level through copy-move, splicing, and morphing. Temporal forgery refers to manipulations carried out across frames using tech-niques such as frame duplication, insertion, and deletion. Spatio-temporal forgery ambiance blends both spatial and temporal manipulations into realistic video realizations.

12.2.1   *Spatial forgery through intra-frame manipulations*

Spatial forgery refers to manipulating the image content within single frames without affecting the timeline of the video. The main operations include:

-   *Copy-move forgery*: This technique involves creating an identical zone in the same frame and pasting it into a different area of that very frame. This is basically to hide or mask some information, such as the removal of an object from a shot or the removal of a face from a shot. The challenge is to make the pasted area blend seamlessly into the background such that a person would not know that any alteration was made.
-   *Splicing*: The technique consists of inserting many components from an external image or video into a frame. For example, the target frame can contain an object's face from a different source. An activity typical in deepfake synthesis is the swapping or alteration of a person's identity.

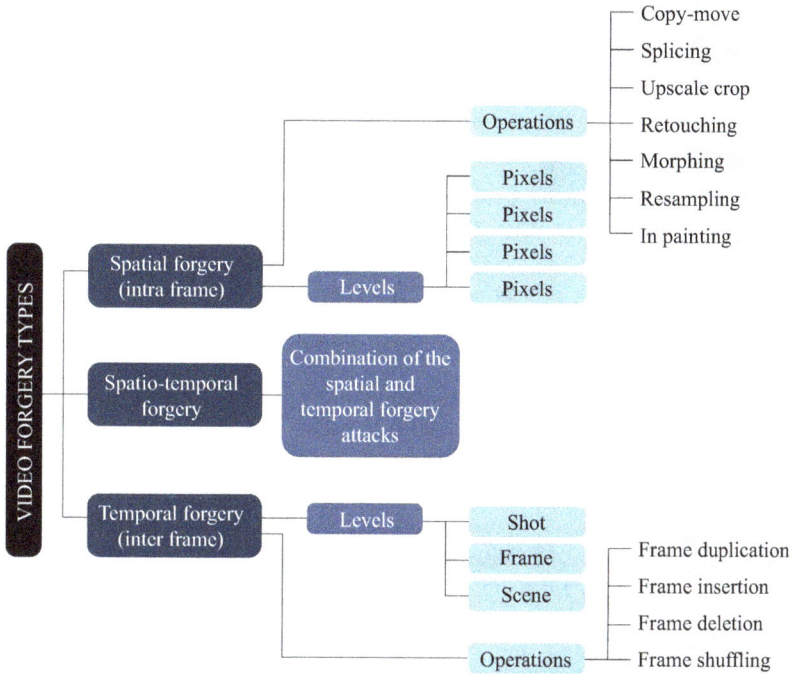

Figure 12.1   Types of video forgery: spatial, temporal, and spatio-temporal manipulations

- *Inpainting and object removal*: This operation uses advanced AI, like Contextual Attention GANs, to remove unwanted elements from a frame by filling the area from which they have been removed with synthesized pixels that match the texture and color of the surrounding area. This method functions extremely well for the elimination of objects when the outlines of the objects cannot be seen.
- *Retouching and morphing*: The art of retouching is commonly used for refinement, while morphing applies those tweaks to blending. By collapsing skin textures, facial features, or any two images together, an image could either be retouched for slight modifications to look better or morphed, merging faces to create a totally different identity. Manipulation at the pixel level is essential for both approaches.
- *Super-resolution and upscaling*: Upscaling can enhance or reduce the quality of a given area, and it's often used for subterfuge. Forgers can thus vary pixel density or resolution for the purpose of obscuring traces of tampering.

Figure 12.2 shows an original version of the video (*Vo*) and a tampered version of the video (*VT*). The original video maintains spatial and temporal coherence between frames {F(1), F(2), F(3), F(4), F(5), and F(6)}. Contrarily, the tampered

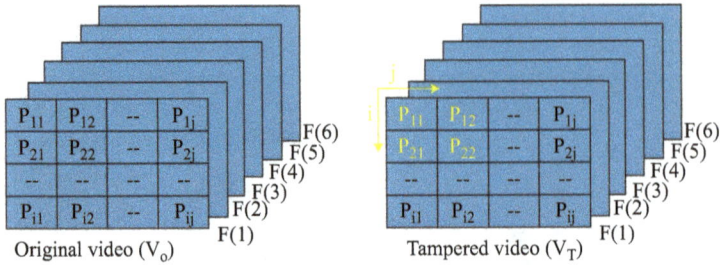

Figure 12.2   Comparison of original and tampered videos for video forgery detection

video gives away manipulations demonstrated with altered pixel blocks {Pij} and a dissimilarity in motion flow resulting in visual artifacts. This kind of visualization is used for video forgery detection for the identification of inconsistencies at the frame level.

- *Detection methods*: Detecting spatial forgery makes use of different forensic devices capable of scrutinizing images at a pixel scale, establishing the presence of the forgery.
- *Pixel-level inconsistency analysis*: Such methods include noise, color inconsistencies, and local texture examination, which are able to identify anomalies that were introduced by either copy-move or splicing processes.
- *Deep learning based image forensics*: CNNs are trained to identify artifacts or small discrepancies introduced by inpainting, retouching, or morphing; these networks' objective was training on detecting unnatural patterns that contrast real video frames [9].
- *Block-matching and texture analysis*: For copy-move forgery, block-matching algorithms detect areas of duplication by comparing overlapping blocks of pixels, while texture analysis techniques.

12.2.2   Temporal forgery (inter-frame manipulation)

Temporal forgery employs the modifications inter-frame through manipulative operations, whereby the changes are being wrought on the sequence or duration of frames instead of the content of individual frames. This type of forgery aims to break the temporal continuity of a video, and therefore, it becomes less detectable under conventional spatial analysis methods. In these methods, frame insertion, deletion, duplication, and shuffling are used in ways that tamper with the timeline and either distort events or hide specific actions. In this manner, attackers can easily manipulate an entire sequence of frames, causing them to deviate from the true interpretation or serve to eliminate incriminating evidence from video forensic analysis.

Different types of temporal video forgery methods are illustrated in Figure 12.3. The "Original Sequence" is that typical video frame order from 1 to

12. Extra frames were added for unnatural elongation of the video. Specific frames removed through "Frame Deletion" create abrupt motion or jumps in the scene. The "FD without Replacement" adds duplicates of the frame but does not substitute any of the frames, which results in visible stuttering. Any replacement of frames would be "FD with Replacement," and since these frames have copies instead, the motion appears seamless even though it has been manipulated. The video is made to look out of context and distorted story-wise by shuffling frames in what is called "Frame Shuffling." These methods are used in video forgery to fake events.

The category encompasses:

- *Frame deletion*: Deletion of key frames from the video seems to distort the chronological sequence of events. For example, one could remove frames that capture a critical moment in surveillance footage to either eliminate evidence of an event or misrepresent the timing of the event.
- *Frame duplication*: Frames are copied for exaggerations of the actions or other works to mask a missing portion. Frame duplication could thereby create an impression of continuity or slow down the action with no evident justification.
- *Frame insertion*: Synthetic frames, possibly generated by GANs, will then be inserted into the video. This could be used to fabricate events that never existed, like inserting an extra moment in favor of a false narrative.
- *Frame reordering (shuffling)*: Actions that change the contextual frame of view of an event by changing the ordering of frames may thus mislead the audience regarding the true order of events.

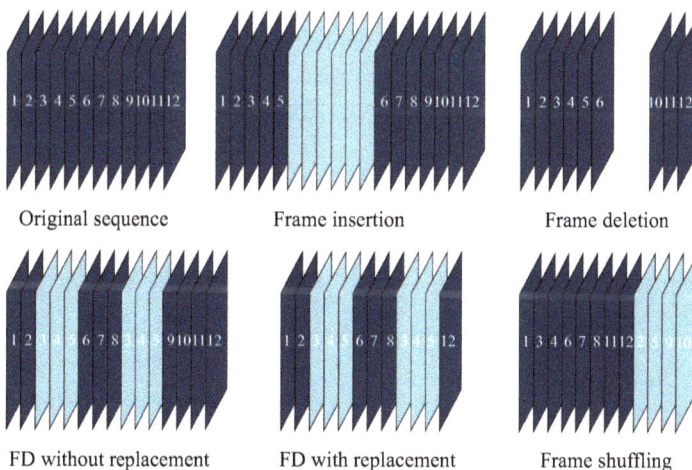

Original sequence   Frame insertion   Frame deletion

FD without replacement   FD with replacement   Frame shuffling

Figure 12.3   Types of temporal video forgery techniques

Detection methods: Temporal forgery detection mainly depends upon motion and continuity analysis across frames:

- Optical flow analysis: This method does the motion calculus of two consecutive frames. An earlier indication could be sudden changes, unusual stops, or irregular flow of motion vectors pointing to frame deletion, insertion, or reordering, etc.
- Temporal consistency checks: RNNs and long short-term memory (LSTM) networks are suitable for modeling temporal dependencies. These inconsistency models analyze motion and content directed across frames that indicate temporal tampering.
- Frame-level metadata comparison: This observation of the analysis of all the metadata (timestamps, frame rates, etc.) would again help with identifying inconsistencies brought about through manipulation of the frame. Surprise timing changes or peculiarities in the frame encoding might suggest frame tampering.

12.2.3 Spatio-temporal forgery (combination of spatial and temporal manipulation)

An extremely crafty hybrid form of forgery in video manipulation comprises the spatial and temporal ravages. This hybridization renders it among the most challenging cases of video manipulation. While spatial forgeries are content manipulations within individual frames and temporal forgeries alter the frame sequences, spatio-temporal forgeries comprise both approaches: the object can be added, modified, or even removed from within the scene, and at the same time, an alteration may take place in the order of its frames so as to falsify movements or events. Such forgeries would require specialized algorithms to analyze for visual and motion anomalies using deep learning models, optical flow analysis, and attention-based video transformers.

Frame duplication is a video tampering technique, and is shown in Figure 12.4. The original video sequence on the left consists of F(1)–F(6) frames. The tampered one on the right contains duplicated frames; here, F(4) and F(5) are duplicates of

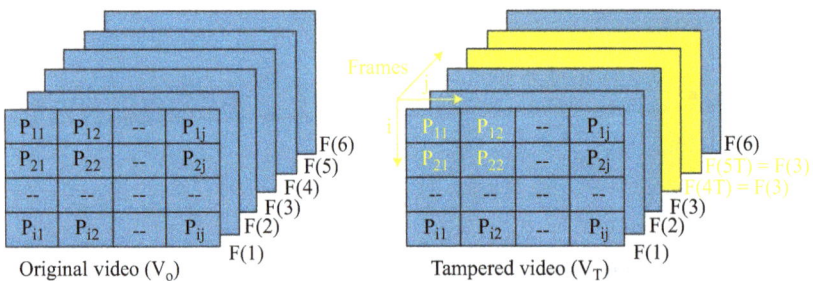

Figure 12.4 Frame duplication in video tampering

F(3). Such tampering is mostly done to mislead a viewer or hide some events. Frame-level analysis coupled with methods that could detect temporal inconsistencies can help in the detection of such alterations.

Operations include:

- Removal of objects gradually over time: This is the process wherein any object is gradually removed from the scene over several frames. The said object might exist at the first frame and gradually disappears, thus becoming a very natural change in the conscious ticking of the timeline of the video.
- Gradual adding object: In contrast, objects could be added to the video over more than one frame. For example, a person could gradually be made to appear in a scene, thus changing with the narrative of that event.
- Deepfake face swapping via temporal smoothing: Here, deepfake technology is used with algorithms that vary in the different frames to swap faces with those of other subjects. This provides a very different appearance in terms of the format of facial movements and expressions from that of the other content of the video.

Detection methods:

Because spatio-temporal forgeries involve alterations that may generate or distort complex manipulations in space and time, methods of detection mostly tend to combine various techniques for analysis:

- Hybrid CNN-RNN models: Bring together the spatial functionalities of CNNs that are used in analyzing frames to identify spatial artifacts while maintaining coherence along the sequence of frames. Because of the temporal tracking by RNNs, the hybrid models could reveal inconsistency between these dimensions.
- Spatio-temporal transformers: Recently, transformer models engineered to cater for video analysis have paled in front of their ancestors in the challenge of identifying spatio-temporal manipulations. The self-attention mechanisms used in these models help give rise to intricate relationships between spatial features and employ temporal dynamics.
- Optical flow plus object tracking: Optical flow used in conjunction with object tracking algorithms enables a comprehensive analysis of the appearance or disappearance of objects over time. This would be helpful toward identifying gradual object removal/addition.

12.2.4   *More manipulation techniques and their detection*

Additional types of manipulations: Many other very complicated manipulations lie apart from these primary ones:

- *Entirely AI-generated synthetic video*: The real challenge lies here in entirely synthetic videos generated from GANs, which do not follow the typical forge patterns (e.g., "This Person Does Not Exist"-like outputs).

- *Audiovisual synchronization forgery*: This involves manipulating the audio track and being independent of the video. Such manipulations often result in subtle discrepant things, such as unnatural voice modulation or lip movements that do not match the video and can be detected multimodally.
- *Exploitation of compression artifacts*: They may also deliberately misuse compression artifacts to camouflage the forgery. Thus, the methods of detection must also provide lines between artifacts as produced through legitimate compression and that caused by manipulation.
- *Adversarial attacks against detection models*: The detection algorithms would be specifically fooled through video manipulation by designing the manipulations in such a manner by attackers. These adversarial examples can be highly optimized for evading detection and thus need robust detection techniques along with adversarial training strategies.

Methods for detection of additional manipulations:

- *Multimodal AI approaches*: Such systems can find discrepancies between spoken content and lip movement that typically accompany manipulated videos by analyzing both the audio and visual signals together [2].
- *Blockchain-based verification*: Recording video metadata and cryptographic hashes using blockchain at the time of capture creates a forgery of an unalterable record, making it easier to verify the integrity of the video later [10].
- *Adversarial training*: Detection models adapt and learn from adversarial examples mostly during training, enhancing the robustness of the systems against unknown forgery techniques developed in the future. Learning could involve repeated retraining using an adversarial example generated from a method such as the fast gradient sign method (FGSM).

Maintaining an edge in the face of greater sophistication, the field of video forgery detection is developing evermore. Operations in spatial, temporal, and spatio-temporal domains present unique challenges requiring specialized detection methodologies. With pixel-level analysis, deep learning models (both CNNs and RNNs), and a hybrid approach, researchers have been working toward developing systems that are able to detect even the most subtle and high-quality forgeries. Also helping advance the cause of digital media integrity are techniques such as multimodal analysis, blockchain-based authentication, and adversarial training.

12.3   Detection of video tampering in deep learning

Traditional forensic methods fail to keep pro pace with new state-of-the-art video manipulation approaches when it comes to detecting the tampering of videos. The deeper learning-based models provide a powerful means to detect the inconsistencies in a tampered video. The four major approaches include CNNs at the frame level, RNNs for temporal consistency checking, vision transformers (ViTs) for motion-based analysis, and hybrid models for spatiotemporal analysis.

12.3.1 Convolutional neural networks (CNNs) for frame-level analysis

True to their name, CNNs, due to spatial modeling capabilities, have been in use for the analysis of images and videos. Abnormalities are reported on pixel levels as the texture meets and other physical effects of light, along with compression artifacts – the telltale signs of tampering.

- It is under video tampering detection, primarily dealing with the above-discussed CNNs about deepfake detection, such as frame insertion and deletion, manipulation, and object removal. In the most basic of terms, it can be said that CNNs work on subtle artifacts left in the process of generation by GANs, such as unnatural-looking skin textures or inconsistencies with reflections [11]. When frames are inserted or deleted, CNNs are able to locally identify the mismatch at the level of the frame by monitoring the pixel distribution and encoding artifacts [12]. They also identify unnatural gaps and missing shadows, along with, to some extent, interpolated pixels caused by concealment due to their presence in the scene but removed from the inputs during processing [13].
- Common architectures these days used generally for the joy that comes with the use of CNNs in video forensics include the much-coveted excellence of XceptionNet, which has been going toward depths of effects in detecting deepfakes because of enhanced performance in image classification and forgery detection [9]. EfficientNet applies scaling and computational efficiency to its top-performing video forensic applications. On the other hand, the ResNet-based architectures have been primarily known for forensic classification and audio forgery detection from videos due to the deep feature extraction they provide in increasing the chances of being manipulated. Despite their several advantages, CNNs are limited by many aspects, some of which include making the system vulnerable to adversarial attacks and the other is the proper generalization over varied video forgeries.

12.3.2 Recurrent neural networks (RNNs) for temporal consistency verification

RNNs, specifically LSTM and GRU, are best suited for the detection of temporal inconsistencies in video sequences. These analyze motion patterns and frame continuity, hence finding a suitable application in the detection of tampering techniques such as frame reordering and duplication.

- In video forgery detection, RNNs are useful in analyzing temporary inconsistencies and sequence anomalies. Their first main application is for the detection of temporal reordering and duplication, wherein motion discontinuities and scene transition discrepancies indicate probable tampering. The second important area of application is the synchronization of audio with video, whereby discrepancies between speech and facial expressions in either manipulated audio or video are dubbed potential forgery [10]. In addition, RNNs excel at recognizing spatiotemporal inconsistencies, especially when talking about unnatural transitions between the movements of objects between consecutive frames [14].

- Many architectures based on the RNN are usually applied in the detection of video tampering. LSTM-based video forensics models specialize in capturing long-term dependencies from video sequences. This makes them quite powerful in unnatural frame transition detections and duplicate detections. Conversely, the GRU-based deepfake detection models are an alternative to the LSTMs that emphasize long-range dependencies present in manipulated videos while relieving the computational intensity. These architectures realize robust video forensics by embracing the trade-off between efficient performance in manipulated content detection and computational efficiency.

12.3.3   Vision transformers (ViTs) for motion-based analysis

ViTs utilize a self-attention mechanism to model long-range dependencies in the video frames. Contrastingly with CNNs, focused on local feature extraction, ViTs view the global pixel-to-pixel and frame-to-frame relationships, thus are apt for motion-based detection of tampering.

- ViTs are considered strong systems for video forensics as they capture complex motion patterns, long-range dependencies, and the motion pattern in the scene. A major application of ViTs has been in the deepfake motion analysis, particularly from the capabilities of ViTs to find unnatural motion discontinuities in synthesized videos, which are usually devoid of the natural dynamics that characterize footage shot in reality [15]. Next comes another major locus for ViTs: the ability to detect face or object swapping through unnatural movements such as mismatched facial expressions or even misalignments in the objects, which give clues to manipulations. Also, using ViTs, motion-based forgery detection can be enhanced, where they learn closer details from optical flow analysis and thus can detect unnatural-looking movements and even inconsistencies within a scene.
- Many ViT-based architectures are commonly used for video tampering detection. TimeSformer is also a transformer-based model specifically designed for video analysis, which uses self-attention mechanisms for deepfake and video manipulation detection [16]. At the same time, the Swin Transformer is created in a hierarchical structure that captures the local and global dependency within a video; thus, it can be highly efficacious in those forensic applications needing fine temporal and spatial analysis. These transformer-based models are state-of-the-art for video tampering detection and portray unmatched accuracies on deepfake-generated content identification and motion-based inconsistencies.

12.3.4   *Hybrid models for spatiotemporal analysis*

Utilizes deep learning techniques that involve the hybridization of CNNs, RNNs, and ViTs, maximizing the benefits from both the spatial and temporal features for satisfactory evidence of tampering. These models can use the CNNs in enhancing detection verification, RNNs for accuracy in temporal coherent checks, and the transformers for proper conduct in motion-forensic analyses.

* Hybrid deep learning models use several architectures to their advantage to improve the detection of manipulated videos. The area of major application is in the detection of deepfakes and synthetic videos, whereby hybrid CNN-RNN frameworks combine CNNs, working with pixel-level artifacts, with RNNs for detecting motion inconsistencies [11]. In another important area, the detection of frame manipulations sees ViT–CNN hybrids detecting insertions, deletions, and reordering of frames by combining spatial feature extraction in CNNs with temporal tracking through ViTs' self-attention mechanism [17]. Also, hybrid models greatly enhance forensic analysis of real-time videos, much more than individual ones, which tend to be heavy on either computational cost or feature generalization.
* Widely used in video forensics are several hybrid frameworks. CNN+LSTM networks utilize CNNs for feature extraction and LSTMs for sequence modeling, allowing them to be effectively used to detect frame inconsistencies and spatio-temporal anomalies. ViT–RNN hybrids are known for exploiting the self-attention mechanisms of transformers with recurrent models, hence facilitating enhanced temporal analysis and robust detection of sophisticated video forgeries [18]. Moreover, multi-modal fusion models combine features from varied sources like audio, video, and text, giving a comprehensive avenue for deepfake detection by capturing inconsistencies across the different modalities. This hybrid model thus enhances the detection capability for many forms of video manipulation and is therefore an indispensable tool for obtaining forensic analysis and security solutions.

12.4   Feature extraction for forged video detection

Feature extraction is an important component of detecting forgery in videos, as it enables the deep learning models to identify manipulations through spatial and temporal analysis of consistencies. Feature extraction techniques include optical flow and motion anomalies, frame transition anomalies, and time attention distributions. They will be used to detect the presence of deepfake videos, frame insertions, object removals, and several other video tampering methods.

The pipeline, as described in Figure 12.5, is an approach to typical deepfake detection. It starts by preprocessing input video frames, which are formatted uniformly, and those processed frames are further subjected to the Feature Extractor, where the spatial and even temporal features are captured. Further

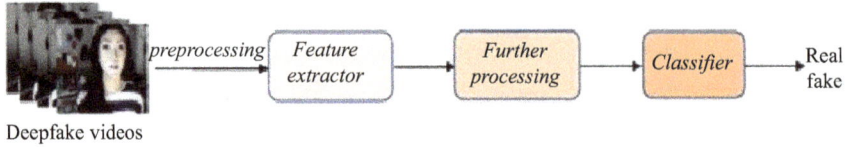

Deepfake videos

Figure 12.5   Deepfake detection pipeline overview

Processing may consist of noise reduction, frame alignment, or optical flow analysis, etc., and in the last stage, a Classifier behaves like a machine learning algorithm to decide whether the video under examination is fake or real. This entire system is incorporated well in various deepfake detectors for identifying manipulated videos.

12.4.1   Optical flow and anomalies of motion

Optical flow measures the motion of objects between consecutive frames with the help of pixel displacements [19]. As a result, it helps in detecting unnatural motion patterns introduced by deepfake synthesis, face swapping, and scene manipulation applications.

- In fact, optical flow methods have become one of the biggest applications of identifying video forgery by analyzing the motion inconsistency patterns induced by them. One of the most widely used applications is for revealing inconsistencies in motion in deepfakes, where unnatural head motion, mismatched facial expressions, and irregular gaze movements are evidence of tampering. An equally important application will be the identification of object removal or scene manipulation; in this two-way process, those absent synthesized frames will create discontinuities within the motion during the frame, which one way, could be discovered by analyzing the frames individually [13]. Additionally, fake audio–video synchronization detection can take the help of speech-lip movement analysis in identifying mismatches between spoken words and their lip movements, thereby exposing deepfakes with altered or synthesized speech [10].
- Various optical flow-based mechanisms are widely applied to detect video tampering. Farneback optical flow is designed for dense motion fields, which is of great help in revealing differences between consecutive frames. In addition, the Gunnar–Farneback algorithm enhances real-time tampering detection by analyzing the pattern of motion flow frame by frame and thus enables quick detection of manipulation. More advanced, deep learning-based models like FlowNet and PWC-Net leverage convolutional architectures to improve motion anomaly detection, providing much higher accuracy as compared to traditional methods in identifying video forgeries involving motion synthesis or distortion [20]. These and other optical flow techniques form the core basis

for detecting deepfake manipulations, frame insertions, and other forms of video tampering through the analysis of the underlying motion patterns.

12.4.2   Frame transition inconsistencies

Frame transition analysis focuses on unnatural scene transitions, frame duplication, or re-ordering within a video. There are certain tampering techniques like frame insertion, deletion, and duplication that disrupt natural transitions and leave detectable artifacts.

- Frame transition analysis is an important method of abnormality detection of alterations made in a video by focusing on inconsistencies in frame sequences, abrupt scene changes, and duplicate frames. The most vital application is frame duplication detection, where, by repetition of frames, an artificial slow-motion effect is created in video evidence, commonly used to alter evidence manipulation [12]. Another essential technique is temporal reordering analysis, wherein temporally shuffled frames in forged video surveillance footage indicate manipulations and misrepresentations of events. The change of figures is often marked by a very sudden, abrupt change in brightness, contrast, or background continuity, mostly brought about during tampering during scene changes [17].
- There are many methods of frame transition-based detection. Histogram difference analysis examines inter-frame intensity variations to pin down sequential frame changes from the abnormal. This makes it very effective at detecting inserted or duplicated frames [18]. The structural similarity index (SSIM), on the other hand, unlike almost all other methods, focuses on structural similarities across consecutive frames, proving useful for detecting frame duplications or missing frames by evaluating the frame's visual coherence. More advanced CNN-based frame consistency models that use deep learning to evaluate transition anomalies at the pixel level with very high precision in the detection of modifications [9] form additional room for enhancement regarding video forensics through the better detection of manipulated sequences, apart from dealing with robustness and generalization across different video formats.

12.4.3   Temporal attention mechanisms

Temporal attention mechanisms facilitate video analysis of tampering through spatial focus in many frames. In contrast to CNN-based processing approaches, attention models learn to weigh important frames together with spatio-temporal features to detect inconsistencies.

- Temporal attention mechanisms are vital in exposing video forgeries by framing themselves around the inconsistencies between consecutive frames and maintaining long-range dependencies. One major application is the detection of deepfakes and face-swaps, as discrepancies such as unnatural blending of frames or mismatching identities reveal synthetic manipulations.

Also, by identifying motion discontinuities, GAN-generated moving images that are commonly known to have abrupt velocity changes or unusually moving objects can be flagged for tampering. Tracking facial motions over time in these situations, where the real faces and AI-generated faces significantly deviate, can help identify GAN-generated videos, as most GANs would lose natural temporal coherence [14].

• Advanced mechanisms for video forgery detection based on temporal attention. Videos are analyzed via a transformer-based analysis, which includes self-attention mechanisms, to build long-term dependencies of videos and locate these subtle inconsistencies [15]. Prominent examples of this include TimeSformer, a ViT that allows modeling of temporal relationships of video content for detecting deepfake frames [16]. The LSTM-based attention network softens these temporal patterns with recurrent structures and attention mechanisms for even higher efficiency in deepfake detection. Attention-based mechanisms can significantly enhance the spatiotemporal pattern learning in manipulated video detection, thus assisting in getting robust forensic analyses.

12.5   Datasets from video-tampering detection

Datasets contribute significantly to training and evaluating deep learning models for video tampering detection. Labeled authentic and manipulated videos facilitate the efforts of researchers to develop robust techniques for detection. In this section, four major datasets are presented wherein deepfake videos, face-swapping manipulations, surveillance video forgeries, and general video integrity verification are detected.

12.5.1   FaceForensics++

FaceForensics++ [9] is one of the most used benchmark datasets in the world for deepfake detection and face manipulation. The dataset contains 1000 original videos and 500,000 manipulated videos produced by the most recent face-swapping methods, including Deepfake, Face2Face, FaceSwap, and NeuralTextures, as illustrated in Figure 12.6.

The dataset is hosted at three distinct compression levels – raw, lightly compressed, and heavily compressed – so that researchers can understand the models' performance when using deepfake models over real-world compression artifacts. Although FaceForensics++ is very effective in training CNN-based deepfake detection models and testing techniques for evaluating frame-level and temporal consistency, it is restricted to face manipulation and does not include wider video tampering techniques like the modification of a scene or even the removal of objects. This constraint may result in overfitting of detection models trained on the dataset with specific dataset artifacts.

Figure 12.6   FaceForensics++ dataset

12.5.2   Deepfake detection challenge (DFDC)

The DFDC dataset is one of the largest and most diverse datasets for deepfake detection, created in 2019 by Facebook AI and Kaggle. It features well over 100,000 deepfake videos with a variety of actors, ethnicities, lighting conditions, and backgrounds.

DFDC combines different senses of face-swapping models and is very useful for training deep learning frameworks like CNNs, RNNs, and ViTs; it tests deepfake detection models for low-quality and high-compressed videos, hence applicable in real-world cases. The massive size of it, however, requires high computational resources for training, and the impact made by certain deepfake manipulations made it difficult, as most videos look almost as if they have been recorded naturally.

12.5.3   Celebrity deepfake dataset (Celeb-DF)

Celeb-DF is a highly qualified deepfake dataset that surpasses all previous datasets, including FaceForensics++. It consists of original videos, approximately 590 in number, with another armada of 5639 deepfake videos characterized by fewer compression artifacts. Celeb-DF excels in realistic head movement, diversity in facial expressions, and varied lighting conditions when compared to earlier datasets, which makes it different from its predecessors and difficult to develop detection models to classify manipulated content as being manipulated only by visual artifacts. Hence, it is most utilized by training models that detect manipulation in deepfake videos closely resembling real videos. They do cover facial deepfakes, but not forgery of videos in terms of object removal or background changes.

12.5.4   UCF-crime dataset

The UCF-crime dataset is one of the most popular datasets used for video tampering detection and anomaly detection in surveillance. It contains 1900 long-duration CCTV videos; 1200 of which are normal activity, and the remaining 700 are suspicious or anomalous behavior. The dataset includes all types of manipulations, including frame tampering, scene insertions, and frame duplication, which are valuable for training models to discover inconsistencies based on real security footage. Normally, UCF-Crime is more interesting for crime detection and forensic video analysis. However, since it mainly comprises exterior surveillance footage, its application scope may not deal with a wide variety of video manipulations, such as deepfake distribution on social media and entertainment-forged videos.

12.5.5   Custom dataset creation for video integrity

Creation of customized datasets is the most common approach in cases where existing datasets are not sufficiently comprehensive in terms of a particular type of video forgery. This begins with collecting real video footage from public domains or newly shot videos with diverse representations regarding scenes, lighting conditions or subjects; and follows this up with manipulation using deep learning techniques such as GANs, face-swapping algorithms, or traditional video editing software. Each video is strictly annotated with forgery types such as frame deletion, deepfake, object removal, and others to ensure supervised learning. Finally, the dataset is validated by testing it against the state-of-the-art models for video tampering detection to ensure its effectiveness against practical applications. DeepFaceLab is one of the common tools used for dataset creation, as it allows deepfake generation. FFmpeg can also facilitate tampering at a frame level without getting stuck. OpenCV can detect and modify different elements of the video.

12.6   Hands-on: detecting video deepfakes using a pretrained transformer model

As deepfake videos become a more serious threat in digital media, developing dependable detection mechanisms has become imperative. This hands-on tutorial will deal with detecting video deepfakes using a pretrained Transformer-based model. The following flowchart, shown in Figure 12.7, describes a deepfake video detection pipeline, where optical flow analysis and convolutional LSTM

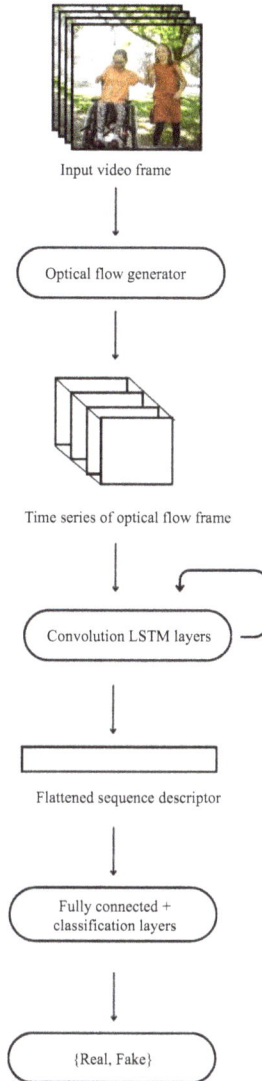

Figure 12.7 Optical flow-based deepfake video detection using convolutional LSTMs

networks are combined. A stepwise description of the working of the process is given below:

- *Input video frame*: The pipeline receives a sequence of video frames from the input video. These frames act as the source for assessing motion patterns and visual consistency.

- *Optical flow generator*: An optical flow generator calculates the motion of pixels from one video frame to the next frame. Optical flow is an estimation of direction and measurement of object motion, with the attribute of capturing temporal dependencies.
- *Time-series optical flow frames*: The optical flow is output as a time series of motion frames, effectively coding for the dynamic characteristics of the video. This time series will capture all the subtle motion artifacts created by the deepfake.
- *Convolution LSTM layers*: The motion data are processed through Convolutional LSTMs, which fuse the feature-extraction power of CNNs with the sequence model power of LSTMs. This means that Convolutional LSTMs can learn spatial features across frames while giving memory to the temporal dimension, which is very important for detecting temporal inconsistencies.
- *Flattened sequence descriptor*: The features extracted are then processed through LSTMs and deposited into a flat vector called the sequence descriptor. This puts together all the temporal and spatial features in a format friendly to classification.
- *Fully connected + classification layers*: The flattened sequence descriptor is fed to fully connected layers, followed by classification layers. These layers apply a nonlinear transformation and generate predictions.
- *Output (real or fake)*: Eventually, the model provides the output, predicting an authentic or manipulated video. This constitutes a binary classification to determine whether a video is genuine or modified.

This method becomes particularly beneficial for detecting deepfake videos whose manipulation affects motion patterns or creates inconsistencies in temporal dynamics.

12.6.1   Loading the pre-trained video deepfake detector

Deepfake detection involves using pretrained models trained on large-scale datasets of real and manipulated videos. The most commonly used state-of-the-art models are

- *DeepfakeTIMIT*: This is a dataset and detection model specifically trained on synthetic talking head videos generated from real speech recordings.
- *XceptionNet [9]*: A CNN-based model designed to detect forgery in faces, noted to be quite accurate in detecting pixel-level artifacts.
- *ViT*: A transformer-based model that performs self-attention mechanisms on motion-based deepfake analysis.

Here, we will be using a pretrained ViT model that is fine-tuned for deepfake detection.

```
import torch
from transformers import ViTFeatureExtractor, ViTForImageClassification
import cv2
import numpy as np

# Load pretrained Vision Transformer (ViT) model
model_name = "facebook/deepfake-vit"
feature_extractor = ViTFeatureExtractor.from_pretrained(model_name)
model = ViTForImageClassification.from_pretrained(model_name)

# Set model to evaluation mode
model.eval()
```

Installing dependencies

To effectively infer deepfakes, we shall utilize a ViT-based pretrained model on a deepfake dataset-trained deep learning models. ViTs are geared toward the analysis of images and videos because they account for self-attention mechanisms in both global and local feature extraction. Instead of the common CNNs that create an understanding of complex patterns by analyzing very long-range images as well as temporal consistency for their videos, ViTs accomplish that exactly by analyzing very long-range images as well as temporal consistency for their videos. Also, pretrained ViT models generalize quite well for different datasets and can also be fine-tuned to further improve their capacity to detect deepfake content. Conversely, the ViTs used previously have been trained on deepfake datasets, which is quite helpful in building an efficient ViT model for deepfake detection. In addition, ViTs are advantageous for image and video processing because they include a self-attention mechanism that captures not only local but also global features of objects. While conventional 2D CNNs analyze very long-range images and temporal consistency for their videos, ViTs accomplish that exactly by analyzing very long-range images and temporal consistency in their videos as well. Pre-trained ViT models generalize well across different datasets and can be fine-tuned on a task for better detection of deepfakes.

12.6.2 *Extracting frames and features*

One primordial step in video deepfake detection is frame extraction, which enables the model to analyze the spatial and temporal inconsistencies that are so characteristic of deepfake videos. Subtle artifacts like unnatural facial expressions, mismatched lighting, or texture inconsistencies that could easily escape the notice of an observer viewing the entire video may come to light when analyzed frame by frame. Subsequently, temporal analyses are preserved, whereby models are fed the sequence of frames to detect pixel-level anomalies.

```
def extract_frames(video_path, frame_interval=5):
    cap = cv2.VideoCapture(video_path)
    frames = []
    frame_count = 0

    while cap.isOpened():
        ret, frame = cap.read()
        if not ret:
            break
        if frame_count % frame_interval == 0:
            frame = cv2.cvtColor(frame, cv2.COLOR_BGR2RGB)  # Convert to RGB
            frames.append(frame)
        frame_count += 1

    cap.release()
    return frames

# Example usage
video_path = "sample_deepfake.mp4"
frames = extract_frames(video_path, frame_interval=5)
```

This is done using a frame capture function that extracts frames from a video at regular intervals (for example, every five frames), thus allowing for the reduction of redundancy while still capturing sufficient information. The function reads a video file frame by frame using OpenCV's VideoCapture, converts each selected frame into RGB color format (as is needed for Transformer-based models), and stores selectable frames in a list. Converting to RGB is necessary since pretrained deepfake detection models are typically trained on RGB instead of OpenCV's default BGR. This design ensures maximum computational efficiency and accuracy of the model, where it extracts frames relevant to deepfake detection.

12.6.3   *Using optical flow for temporal analysis*

The use of optical flow stands out as a powerful method to detect deepfakes by studying incoherencies in motion across video frames. Typically, deepfake videos do not contain motions that appear natural and authentic, making everything seemingly odd and giving birth to sudden spots where facial expressions change, peculiarities in the movements of irrelevant objects, or wrong positioning of the head. One can use optical flow to detect all of these by determining the movements of pixels between frames, revealing very small deformations that are barely noticeable by a human being.

Optical flow will be computed using the Farneback optical flow, which gives the estimation of dense motion fields between any two consecutive frames. The function takes into account the conversion of frames to grayscale and calculates the motion vectors that show patterns of pixel displacement. Analysis of such patterns will assist in identifying the abnormal movements that indicate

```python
def compute_optical_flow(prev_frame, next_frame):
    prev_gray = cv2.cvtColor(prev_frame, cv2.COLOR_RGB2GRAY)
    next_gray = cv2.cvtColor(next_frame, cv2.COLOR_RGB2GRAY)

    flow = cv2.calcOpticalFlowFarneback(prev_gray, next_gray, None, 0.5, 3, 15, 3, 5, 1.2, 0)
    return flow

# Compute optical flow between consecutive frames
optical_flows = [compute_optical_flow(frames[i], frames[i+1]) for i in range(len(frames)-1)]
```

Optical flow thus will improve deepfake detection in the following ways: It helps identify deepfake inconsistencies, for example, abnormal jumps in motion or unnatural distortions in facial features. It indicates whether movement patterns are realistic with respect to how movements of face and non-face objects should conform to the laws of natural physics. It also traces coherence from frame to frame, and indicates sections where deepfake models show deficiencies in their generation of smooth and believable-looking movement. Optical flow, therefore, sensibly enhances the degree of proficiency of deepfake detection models using these insights.

12.6.4   Running the model and interpreting predictions

First, you need to pre-process these extracted frames to be aligned to the input format of the pretrained ViT model for deepfake detection. These steps include resizing the images, normalizing pixel values, and manipulating them back into tensors. Since the ViT model needs a structure for the input, you will use a feature extractor to make sure each frame is well formatted for further analysis.

```python
def preprocess_frames(frames):
    inputs = feature_extractor(images=frames, return_tensors="pt")
    return inputs

inputs = preprocess_frames(frames)
```

Once the preprocessed frames are ready, they will be passed through the pre-trained model to generate predictions for those frames. The model evaluates each frame and assigns some probability scores to determine whether the frame is real or fake. The output will be a confidence score related to each class, where higher probabilities imply a greater likelihood of having manipulated a frame.

```python
with torch.no_grad():
    outputs = model(**inputs)
    predictions = torch.nn.functional.softmax(outputs.logits, dim=-1)

# Get deepfake probability for each frame
deepfake_probs = predictions[:, 1].tolist()  # Assuming class index 1 is "fake"

# Print results
for i, prob in enumerate(deepfake_probs):
    print(f"Frame {i}: Deepfake Probability = {prob:.4f}")
```

Interpreting these predictions will result in a further analysis of the probability scores. It will be highly likely that an image is manipulated if it has a deepfake probability around a frame. By checking for deepfakes across multiple frames of a video, we can check whether there are inconsistencies over time, which improve deepfake detection results. This method thus helps to differentiate real footage from AI-generated fakes with greater accuracy.

12.6.5   *Fine-tuning on a custom dataset*

Fine-tuning a pretrained model for deepfake detection is necessary when a user wishes to test it on new types of deepfake videos, which might not be well-represented in the data used for training pre-existing models. Since pre-trained models are trained based on some datasets, these biases may either not generalize to unseen manipulations or might lead a model to fail. By fine-tuning it on a custom dataset, we are able to improve the accuracy and robustness of the model in detecting deepfakes under different environments, lighting conditions, and facial manipulations.

Fine-tuning is a multi-stage process. To start with, we gathered a heterogeneous dataset that contains real and fake videos from a plethora of sources to ensure that the representation is as wide as possible. Next, we annotated the dataset with binary labels like 0 for real and 1 for fake. This allows for supervised learning. The most interesting and useful would be data augmentations like distortion, compression, noise addition, etc., mimicking true video conditions. Finally, we train on a lower learning rate so that it can adjust itself to new data without

```
from transformers import Trainer, TrainingArguments

# Define training arguments
training_args = TrainingArguments(
    output_dir="./deepfake_model",
    evaluation_strategy="epoch",
    save_strategy="epoch",
    per_device_train_batch_size=8,
    per_device_eval_batch_size=8,
    num_train_epochs=5,
    learning_rate=1e-5
)

# Create Trainer
trainer = Trainer(
    model=model,
    args=training_args,
    train_dataset=custom_train_dataset,
    eval_dataset=custom_eval_dataset
)

# Train the model
trainer.train()
```

forgetting the features learned previously. This is how we make better and more practical applications of the model's detection features.

The model identified the video frames 12 and 20 from Figure 12.8 as "Fake." The suspicious areas highlighted by the model are the regions in which faces were detected, enclosed in a red bounding box. Noticeable artifacts are observed all around the contours of the face, especially around the eyes and mouth, which are the most common areas of evidence of deepfake manipulation. The skin texture is not realistic relative to the natural features of a human and has clear lighting inconsistencies. They also lack smooth facial expressions, marked with slight distortions indicating synthetic tampering. Hence, these visual abnormalities have led the model to predict it confidently as fake. This frame 26 from Figure 12.8 is classified by the model as "Real." Thus, it has a green box around a detected face for better identification. Compared with frame 12, the facial features in this image seem natural and perfectly aligned. Even in texture on the skin, there is definitely no sign of visible manipulation. Facial expressions of the subject entirely cooperate with what might be expected from his human movements; there are no striking transitions or weird distortions that take place. Finally, with the eyes, mouth, and other facial components aligned, everything just affirms the frame's authenticity.

Input frame	Output classification frame
	Prediction: Fake
Frame 12	
	Prediction: Fake
Frame 20	
	Prediction: Real
Frame 26	

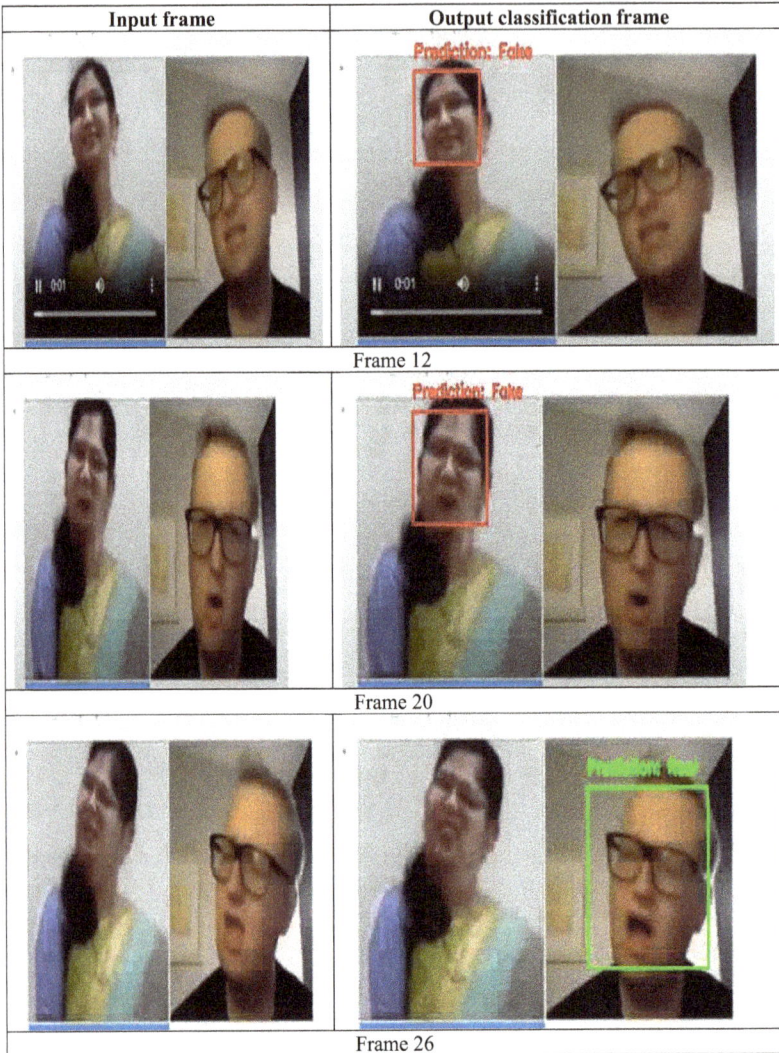

Figure 12.8   Deepfake detection: real vs. fake frame classification

Hence, it has been classified by the model as real at a confident level. In this hands-on session, we have successfully loaded a pretrained ViT model for deepfake detection, extracted frames from videos, and analyzed optical flow to detect motion inconsistencies. We also ran the model on video frames to interpret deepfake probabilities and explored fine-tuning the model on a custom dataset to improve accuracy. By following these steps, one can effortlessly create a very efficient deepfake detection application.

12.7 Conclusion

If there is a rapidly progressing domain, it is video manipulation technologies aimed at developing such unnatural events as deepfakes and other very advanced video forgeries, posing big digital threats to security and media integrity as well as an assault on misinformation. The availability of AI-based synthesis models has rendered it possible to create highly realistic tampered videos, thus making it even more complex to differentiate between original and manipulated content. This chapter discussed some of the deep-learning-based approaches to video forgery detection by means of CNNs, which analyze the frames, RNNs, which check for temporal consistency, and ViTs, which focus on motion-based detection. All these approaches take part in the detection of small inconsistencies, which may occur during the forgery of video. These include unnatural acceleration during transitions, frame duplication, and artifacts created by deepfakes. Forensic feature extraction procedures enhanced by the above-detection techniques include the like, optical flow analysis of motion irregularities, inconsistencies between frames in motion transitions, and temporal attention mechanisms. These help in recognizing abnormalities in video structure as well as movement dynamics, which serve as critical pointers toward manipulated content. Discussion also included large-scale datasets like FaceForensics++, DFDC, Celeb-DF, and UCF-Crime with regard to the aim of training robustly generalizable models for detection. Nonetheless, many challenges still lie ahead for deepfake detection, notwithstanding the aforementioned advancements. The very realistic forgeries created by ever-increasing sophistication due to advancement in generative AI models are evading most of the traditional means of detection. Real-time detection is computationally intensive and difficult to deploy on resource-constrained devices. Generalization gap is another important issue affecting almost all detection models. Such forms of detection models cannot maintain their accuracy in disjoint datasets and various video formats and techniques of manipulation.

To address the stated challenges, emerging paradigms that attracted attention included the multi-modal analysis, blockchain video authentication, and adversarial training. Multi-modal analysis is a combination of audiovisual modalities for improving detection accuracy by exposing discrepancies in lip-syncing, facial expressions, and voice modulation. Blockchain authentication utilizes decentralized ledgers to store and secure cryptographic hashes of videos, thus irrevocably preventing any unauthorized tampering through cryptographic means. Another view for adversarial training concerning the model's purpose, which is to increase robustness to evolving threats, is training detection systems on adversarially generated deepfakes. With widening borders of war in between deepfake generation and detection, future video security could depend on adaptive AI solutions designed for runtime implementations, which allow constant learning from the environment or with the environment.

References

[1]   Goodfellow, I., Pouget-Abadie, J., Mirza, M., *et al.* (2014). Generative Adversarial Nets. *Advances in Neural Information Processing Systems (NeurIPS)*, 27.

[2]   Chesney, R., and Citron, D. (2019). Deepfakes and the new disinformation war: The coming age of post-truth. *Foreign Affairs*, 98(1), 147–155.

[3]   Paris, B., and Donovan, J. (2019). *Deepfakes and cheap fakes: The manipulation of audio and visual evidence*. Data & Society Research Institute.

[4]   Suwajanakorn, S., Seitz, S. M., and Kemelmacher-Shlizerman, I. (2017). Synthesizing Obama: Learning lip sync from audio. *ACM Transactions on Graphics*, *36*(4), 1–13.

[5]   Kietzmann, J., Lee, L., McCarthy, I. P., and Kietzmann, T. C. (2020). Deepfakes: Trick or Treat? *Business Horizons*, 63(2), 135–146.

[6]   Ajder, H., Patrini, G., Cavalli, F., and Cullen, L. (2019). *The State of Deepfakes: Landscape, Threats, and Impact*. Deeptrace.

[7]   Dolhansky, B., Bitton, J., Pflaum, B., *et al.* (2020). The deepfake detection challenge (DFDC) dataset. arXiv preprint arXiv:2006.07397.

[8]   Wang, Z., Gu, Y., Xiong, H., Liu, H., and Wu, J. (2021). Blockchain-powered digital media authenticity verification framework. *IEEE Transactions on Multimedia*, *23*, 1010–1022.

[9]   Rossler, A., Cozzolino, D., Verdoliva, L., *et al.* (2019). *FaceForensics++: Learning to Detect Manipulated Facial Images*. ICCV.

[10]   Agarwal, S., Farid, H., Gu, Y., He, M., Nagano, K., and Li, H. (2020). Protecting world leaders against deep fakes. *IEEE Conference on Computer Vision and Pattern Recognition (CVPR)*.

[11]   Nguyen, H. H., Yamagishi, J., and Echizen, I. (2019). *Capsule-Forensics: Using Capsule Networks to Detect Forged Images and Videos*. ICASSP.

[12]   Hussain, M., Muhammad, K., Hussain, T., *et al.* (2021). Detecting video frame duplication forgery using convolutional neural networks. *Multimedia Tools and Applications*, 80(1), 1491–1509.

[13]   Zhou, P., Han, X., Morariu, V. I., *et al.* (2021). *Learning Rich Features for Fake Face Detection*. CVPR.

[14]   Haliassos, A., Doost, M. K., Tolosana, R., *et al.* (2022). Liveness detection for deepfake prevention. *IEEE Transactions on Biometrics, Behavior, and Identity Science*.

[15]   Dosovitskiy, A., Beyer, L., Kolesnikov, A., *et al.* (2021). *An Image is Worth 16x16 Words: Transformers for Image Recognition at Scale*. ICLR.

[16]   Bertasius, G., Wang, H., and Torresani, L. (2021). *TimeSformer: Is Space-Time Attention All You Need for Video Understanding? Proceedings of the International Conference on Machine Learning (ICML)*.

[17]   Guera, D., and Delp, E. J. (2018). *Deepfake Video Detection Using Recurrent Neural Networks. IEEE ICASSP*.

[18]   Saha, S., Sindagi, V., and Patel, V. (2021). *Adversarial Defense for Deepfake Detection with Variational Autoencoders*. CVPR Workshops.

[19]   Horn, B. K. P., and Schunck, B. G. (1981). Determining optical flow. *Artificial Intelligence Journal*, 17(1–3), 185–203.

[20]   Sun, D., Yang, X., Liu, M.-Y., *et al.* (2018). *PWC-Net: CNNs for Optical Flow Using Pyramid, Warping, and Cost Volume*. CVPR.

Conclusion

This volume has explored the evolving field of multimedia security by compiling research on assisting in the handling of images, video, audio, and physiological signals. As digital scope evolves and becomes paramount to communication, health care, and critical infrastructure, securing this data is no longer optional but a foundational requirement.

The chapters reveal the variousness and profoundness of current practices, ranging from perceptual hashing and reversible data hiding to quantum cryptography and blockchain-based authentication. Conjointly, they reminisce about a more general trend: one that incorporates signal processing, cryptography, artificial intelligence, and system-level thinking into multimedia protection.

A systematic theme across the chapters is the transition toward practical, scalable, and interdisciplinary solutions. In the case of securing medical data, embedding digital watermarks, or witnessing deepfake videos, the methods presented highlight both academic severity and real-world applicability.

Concurrently, the specialization persists in evolving. With quick advancements in deep learning, quantum computation, and synthetic media generation, future security challenges will need equally adaptive and intelligent countermeasures. This book proffers a foundation but also indicates the demand for persistent exploration, especially in explainability, automation, and cross-domain stability.

We thank all authors for their invaluable research and dedication. We expect this book to serve as both a reference and a call to act for researchers and practitioners dedicated to securing the digital world – one pixel, signal, and system at a time.

Index

access control mechanisms 145, 156–7, 159

accuracy (Acc) 107

active steganalysis 236

adaptive techniques 226–8

adoption barriers 160

advanced encryption standard (AES) 97, 145, 172

advanced persistent threats (APTs) 96

adversarial training 260

AI-based forensic techniques 152

AI-driven anomaly detection 146

AI-driven linguistic steganography 233

AI-driven optimization 150

AI-driven security models 147

AI-powered linguistic steganography 233

Anaconda3 130

anomaly detection 153

anonymity 82

anti-piracy measures 159

anti-steganalysis methods 233

area under the curve (AUC) 8

arithmetic coding 233

Arnold scrambling 34

artificial intelligence (AI) 142, 151–2
in steganography 228–9

artificial neural networks (ANNs) 17, 233

asymmetric encryption 32

attribute-based access control (ABAC) 145

attribute-based encryption (ABE) 95
for secure data storage 103–4

audio spectrogram transformers (AST) 231

audiovisual synchronization forgery 260

audio watermarking 37–8

augmented reality (AR) 240

Augustyniak watermarking scheme 18

authentication systems 145, 238

author citation linkages 66–8

author coauthor linkages 60–4

automated behavioral analytics 153

bandwidth (BW) 125

banking and financial security 239

bar charts 178

batch gradient descent technique 130

Bell's theorem 105

Bhalerao watermarking scheme 17

bibliometric analysis 172, 175

biometric data 199

Bitcoin 78–9

bit error rate (BER) 24, 42

blind steganalysis 236

blockchain
applications of 86
architecture 79–81
generations of 85

in image watermarking 86–8
integration 240
for tamper-proof multimedia security
   146
types of 84–5
for video authentication 253
blockchain-based content
   authentication 148
block-matching 256
box plots 178
byzantine fault tolerance (BFT) 83

capacity test 22, 224–5
celebrity deepfake dataset (Celeb-DF)
   267–8
center-symmetric local binary pattern
   (CSLBP) 2, 5
centralized systems 77–8
certificate-less proxy reencryption
   (CL-PRE) 98
chaos 173
chaos theory 174
chaotic encryption 34
chaotic maps 173
chaotic systems 201
ChatGPT 233
cheating text 200
Chebyshev polynomial approximation
   122
ciphertext-policy attribute-based
   encryption (CP-ABE) 98
Classic McEliece encryption algorithm
   200
cloud computing technology 140,
   155–6
cloud of things (CoT) 143
cloud security 96–7, 239
cloud service provider (CSP) 94
cloud services 96
CloudSim 96, 106

co-authorship analysis 60
   author coauthor linkages 60–4
   country coauthor linkages 65–6
   organizational coauthor linkages
      64–5
code division multiple access (CDMA)
   36
comma-separated values (CSV) 175
common image processing operations
   (CPOs) 3
comparison algorithm 131–2
computational complexity 35, 130, 143
computation power (CPU) 125
conflicts of interest 240
consensus mechanisms 82–4, 158
consortium-based blockchain 84
content analysis 177
content delivery networks (CDNs) 157
content-preserving manipulations 2
convolutional neural networks (CNNs)
   122, 199, 236, 253, 277
   for frame-level analysis 261
copy-move forgery 254
country citation linkages 68–9
country coauthor linkages 65–6
covert surveillance 238
cross-domain and cross-modal
   steganography 231–2
cross-modal transformer (CMT)
   models 231
crowd funding 86
cryptocurrency 78
cryptographic methods 32
cryptographic security 158
cryptography 16, 32, 173, 214
   in information hiding 33
cyberattacks 140, 156
cybercrimes 2
cyber-physical systems (CPS) 147

cybersecurity 172
cyclic convolution function 202

data analyses 176
  content analysis 177
  descriptive statistics 176–7
data augmentation 228
data availability 94
data breaches 155
data collection 175–6
data hiding (DH) 54
data integrity 94, 143, 156
  verification 238
data privacy 160
data security 97
datasets 266
data tampering 142
data visualization 177
  bar charts 178
  box plots 178
  line plots 177
  word clouds 178
data wrangling 176
Daubechies 4 wavelet 18
DCT-based steganography scheme 230
decentralization 82, 158
decentralized framework 141
decentralized identity management 151
decentralized system 77–8
decryption time (DT) 106
deepfake detection algorithms 152
deepfake detection challenge (DFDC)
    253, 267
deepfake technology 252
deepfake threats 151–2
DeepfakeTIMIT 270
deep learning based image forensics
    256
deep learning models 233

deep reinforcement learning (DRL)
    118
  analysis and evaluation of
      performance 130
    comparison algorithm 131–2
    result analysis 132–4
    setting parameter 130–1
  graph convolutional network 121–2
  machine learning-based virtual
      network embedding algorithms
      121
  network models and evaluation
      indicators 122
    evaluation metrics 124–5
    system model 122–4
  for policy-based network 125
    analysis of computational
        complexity 130
    feature extraction 125–6
    GCN–VNE algorithm 128–30
    graph convolutional network 127
    policy network 126–7
    training and testing 127–8
  security-based virtual network
      embedding algorithms 120
  virtual network embedding 119–20
DeepSteg 237
Deep Walk algorithm 121
delegated proof of stake (DPoS) 83
descriptive statistical analysis 175
descriptive statistics 176–7
detection methods 256, 259
differential privacy 148, 151
digital forensics 239
digital rights management (DRM) 141,
    157, 158–9
digital signatures 105
digital watermarking technique 27, 33,
    35, 214
  classification of 35–8

critical requirements for efficient
digital watermarking scheme
33–5
DCT transformation and DC
coefficient extraction 41
experimental outcomes 42
fragility analysis 43–9
imperceptibility analysis 43
timing analysis 49
image pre-processing and block
division 40–1
LSB embedding of bits 41
LSB modification 41
watermark extraction 42
discrete cosine transform (DCT) 3, 17,
36, 217, 222–33
discrete Fourier transform (DFT) 217,
222
discrete Gould transform (DGT) 39
discrete wavelet transform (DWT) 16,
19, 36, 88, 217, 222–3
disinformation 252
distortion measurements 16
distributed denial-of-service (DDoS)
attacks 142, 150
distributed embedding 236
distributed ledger technology 80, 88
distributed network 79, 82
distributed storage 159
distributed system 77–8
DNA-based cryptographic models 173–4
DNA computing 174
DNA encryption 34
doctor–patient interactions 16
document citation linkages 69–71
domain-based watermarking methods
35–6

eavesdropping 150
e-commerce 86

edge computing 96, 145
for secure and low-latency
multimedia processing 154–5
for secure multimedia processing
145–6
electrocardiogram (ECG) signals 16
experimental results and discussions
21
capacity test 22
imperceptibility test 23–5
robustness test 26
proposed watermarking scheme 18
RDWT-Schur watermark
extraction process 21
RDWT-Schur watermark
integration process 19–21
watermark generation 19
electronic health records (EHRs) 31
electronic voting systems 239
elliptic curve cryptography (ECC) 94,
145
embedding capacity (EC) 34
embedding process 134
encryption scheme 27, 32, 156
encryption time (ET) 106
end-to-end encryption (E2EE) 145, 200
entanglement-based uses pre-shared
entangled pair (EPR) 104
error correction codes (ECCs) 235
ethical AI development and policy
regulations 254
euclidean algorithm 204
evaluation metrics 124–5, 223
explainable AI (XAI) 153
exploiting modification direction
(EMD) technique 217, 219–21

F1-score 107
Face2Face 266
FaceForensics++ 266–7

FaceSwap 266
false positive rate (FPR) 7
Farneback optical flow 264
fast gradient sign method (FGSM) 260
fast Walsh Hadamard transform
    (FWHT) 88
feature extraction 125–6, 263
federated learning 148, 151
Feng Chia University 64, 71
fine-tuning 274
first-generation GCN formula 122
5G networks 96
flattened sequence descriptor 270
flexible representation of quantum
    images (IFRQI) 229
FlowNet 264
Floyd method 130
Floyd–Warshall method 125
forensics 253–4
fragile watermark 34
fragility analysis 43–9
frame deletion (FD) 257
frame duplication 257, 258
frame insertion 257
frame-level metadata comparison 258
frame reordering 257
Frame Shuffling 257
frame transition inconsistencies 265
frequency domain technique 16
Fudan University 71–2
fuzzy inference system 88
fuzzy logic 198, 227

Gaussian mask augmentation (GMA) 228
Gaussian noise (GAN) 8
GCN–VNE algorithm 128–30
generalization gap 277
generative adversarial networks
    (GANs) 151, 252

geometric attacks 235
government public key infrastructure
    (GPKI) 200
graph convolutional networks (GCNs)
    118, 121–2, 127
graph convolutional neural networks
    (GCNNs) 118
graph Laplacian techniques 122
Graph SAGE induction approach 122
Grover's algorithm 95
Gunnar–Farneback algorithm 264

hamming distance (HD) 4
hardware security modules (HSMs)
    150
hash-based signatures 152
healthcare digitization 197
healthcare security 239
hidden layers 127
hidden Markov model (HMM) 227
hierarchical attribute-based encryption
    (HABE) 98
high energy consumption 160
homomorphic encryption 147, 151
Huffman coding 233
hybrid CNN-RNN models 259
hybrid cryptographic models 97
hybrid fuzzy neural network (HFNN)
    228
hybrid near maximum histogram
    (HNMH) method 227
hybrid quantum-classical approaches
    230
hybrid techniques 226–8

identity-based encryption (IBE)
    function 97, 98
identity management 86
image encryption techniques 172
    background and motivation 173–4

evaluating research goals 190
  frequently cited papers and
    landmark research 190
  future directions 191
  interdisciplinary collaborations
    between multimedia security,
    cryptography, and machine
    learning communities 191
  keyword co-occurrence to reveal
    evolving research themes 191
  publication volume and growth
    rate over past five years 190
  technological shifts from
    traditional cryptography to
    AI- and blockchain-enhanced
    encryption 191
  top contributing authors, countries,
    and institutions 190
experimental setup 179–86
interpretation of findings 186–90
materials and methods 174
  data analyses 176–7
  data collection 175–6
  data visualization 177–8
  data wrangling 176
  research methodology 175
image pre-processing and block
  division 40–1
image quality-preserving technique 5
image watermarking 2, 37, 86–8
immutability 158
imperceptibility 33, 34, 225–6
  analysis 43
  test 23–5
implicit neural representations (INRs)
  231
industrial IoT (IIoT) systems 147
information encryption systems 32
information-hiding methods 32
input layer 127

INRSteg 231
integer wavelet transform (IWT) 39
intellectual property protection 238
intellectual property rights 156
Internet of Multimedia Things (IoMT)
  142
Internet of Things (IoT) 96, 140, 144,
  172, 200, 239
  blockchain and multimedia security
    157
    contribution of 158–9
    essence of 158
    obstacles and concerns in 160
    significance of 159–60
  challenges and considerations in IoT
    multimedia security 149
    AI and deepfake threats 151–2
    data privacy and confidentiality
      risks 151
    network vulnerabilities 150
    resource constraints 150
    safeguarding data integrity and
      authenticity 152
    scalability and security
      management issues 150–1
  cloud, and blockchain in multimedia
    security 160–2
  cloud computing and multimedia
    security 155–6
  emerging trends and research
    pathways in IoT multimedia
    security 152
    AI-driven threat intelligence and
      anomaly detection 153
    edge computing for secure and
      low-latency multimedia
      processing 154–5
    post-quantum cryptography for 152
    utilizing blockchain for ensuring
      multimedia data authenticity
      and integrity 154

zero-trust security models for 153–4

enhancing user experience 149

improved efficiency and automation 148

in multimedia security 144–6

real-time monitoring and secure communication 146–7

scalability and flexibility 148–9

significance of IoT in multimedia security 147–8

techniques for securing multimedia in cloud 156–7

interoperability 143

interplanetary file system (IPFS) 88, 159

intrusion detection systems (IDS) 150, 157

intrusion prevention systems (IPS) 157

inverse Schur decomposition 20

IoT-driven multimedia communication 146

IoT-powered multimedia systems 149

iQuantum 96, 106

Jupyter Notebook environment 175

Keras 176

key generation time (KGT) 96, 107

Kirchhoff's voltage law 18

Kullback–Leibler divergence (KLD) 227

Laplacian pyramids 4

large language models (LLMs) 233

lattice-based cryptography 152

LBP histogram Fourier (LBP-HF) 4

least significant bit (LSB) 35, 217

  embedding of bits 41

  modification 41

ledgers 79, 84

legal and regulatory uncertainties 160

Lempel–Ziv–Welch (LZW) lossless 38

lightweight encryption techniques 150

line plots 177

linguistic steganography 232–4

local binary pattern (LBP) 3, 228

local phase quantization (LPQ) 228

logistic map 39

log polar transform (LPT) 3

long short-term memory (LSTM) 258

LSB matching (LSBM) 218

LSB replacement (LSBR) 218

machine learning (ML) 118, 153, 173, 223, 228–9

machine learning-based virtual network embedding algorithms 121

man-in-the-middle (MITM) attacks 96, 150

MapReduce technique 4

Markov decision process (MDP) 121

Markov Random Walk framework 120

MATLAB tools 8

Matplotlib 175

MaxMiPOD 227

MD5 79

mean squared error (MSE) 225

measurement-device independent QKD (MDI–QKD) 97

media fingerprinting 238–9

medical imaging 31

memPool 81

Merkle root 81

Merkle tree 80

micro-segmentation 154

military and defense communication 239

misinformation 252

MIT-BIH Arrhythmia Database 16, 21
Monte Carlo search algorithm 121
Monte Carlo tree 120
Montgomery curves 200
morphing 255
most significant bits (MSB) 38
multi-authority attribute-based
    encryption (MAABE) 96, 109
multi-channel effective quantum image
    representation (MCEQI) model
    230
multi-factor authentication (MFA) 145
multilingual data transmission 240
multi-modal analysis 277
multi-pixel-pair (MPP) approach 227
multiple commodity flow algorithm
    121
multitenancy risks 156

natural language processing (NLP) 233
network vulnerabilities 150
neural networks 118
NeuralTextures 266
noise-resistant local binary pattern
    (NRLBP) 3
non-zero coefficients 18
normalized cross correlation (NCC) 26,
    42
normalized hamming distance
    (NHD) 7
novel enhanced quantum representa-
    tion (NEQR) 229
*N*th degree truncated polynomial ring
    unit (NTRU) 197
  motivation 198–9
  preliminaries 201
    cyclic convolution function 202
    *N*th degree truncation 202
    polynomial ring unit 202
  problem statement 198

proposed methodology 202
  key generation 202–4
  NTRU decryption 206
  NTRU encryption 204–6
research contributions 199
results and analysis 206
  performance evaluation 206–8
*N*th degree truncation 202
NumPy 175, 176

objective evaluation metrics 23
1D integer wavelet transform 17
OpenCV 268
optical flow 264
  analysis 258
  generator 270
  plus object tracking 259
  for temporal analysis 272–3
organizational citation linkages 71–2
organizational coauthor linkages 64–5
output layer 127

pairing-based provable multi-copy data
    possession (PB-PMDP) 96
Pandas 175
part-of-speech (POS) tagging 233
passive steganalysis 236
peak signal-to-noise ratio (PSNR) 23,
    218, 225
peer-to-peer network 79
People's Republic of China (PRC) 66
percentage residual difference (PRD) 24
perception-based watermarking
    techniques 36–8
perceptual image hashing (PIH)
    technique 2–3
  analysis of discrimination 10–11
  center-symmetric local binary
    pattern 5
  detailed design 5–7

distance measure and performance
    evaluation 7–8
  perceptual robustness 8–9
  performance comparison 11–12
perceptual robustness 8–9
performance assessments 112
performance comparison 11–12
performance evaluation 206–8
permissioned blockchains 85
permissionless blockchain 82
permissionless model 80, 84
pixel-level inconsistency analysis 256
pixel value differencing (PVD) 217
polynomial ring unit 202
post-quantum cryptography (PQC)
    algorithms 112, 198, 200
post-quantum encryption 202
practical BFT (PBFT) 83
precision (Pre) 107
predictive threat mitigation 153
preprocessing 5
pre-trained video deepfake detector
    270–1
principal component analysis (PCA) 3
privacy preservation 147–8
privacy-preserving techniques 152
private blockchain 84–5
probability distribution model 128
proof of authority (PoA) 83
proof of burn (PoB) 82
proof of elapsed time (PoET) 82
proof of importance (PoI) 83
proof of location (PoL) 83
proof of stake (PoS) 82, 158
proof of work (PoW) 82, 158
protecting critical infrastructure 147
provable data possession (PDP) 99,
    109
provenance tracking 159

publication structure analysis 56
  annual publications and type 56–7
  country trends 58–9
  productive researchers and
    organizations 57–8
  publishing sources 57
  WoS indices 59
public blockchain 84–5
public-key cryptography 32
public key infrastructure (PKI) 96
PWC-Net 264
PyCharm 130
Python 176
Python-based data analytics tools 173
Python libraries 175

Q-factor wavelet transformation 18
Q-learning algorithm 121
quantitative analysis 55
quantum attacks 198
quantum computing technology 95, 97
quantum cryptography 95–7, 172, 281
quantum cryptography-based cloud
    security model (QC-CSM) 93,
    96, 109, 112
  attribute-based encryption for secure
    data storage 103–4
  objective 99–100
  quantum authentication mechanism
    104
  quantum certificates and digital
    signatures for integrity and
    non-repudiation 105
  quantum key distribution 101–2
  quantum key distribution protocol
    process flow 106
  quantum no-cloning theorem 102,
    104
  secure key distribution using QKD
    102–3, 105–6

quantum digital signatures (QDS) 105
quantum hash function 105
quantum-inspired annealing (QIA) 230
quantum-inspired techniques 230
quantum key distribution (QKD) 95,
    97–8, 101–2, 152
quantum key distribution protocol
    (QKDP) 96, 106, 229
quantum-resistant cryptography 153
quantum secure direct communication
    (QSDC) 98
quantum steganography 229–30
quantum variational circuits (QVC)
    230
quaternion discrete Fourier transform
    (QDFT) 3
quaternion Fourier–Mellin transform
    (QFMT) 3
quick response (QR) decomposition 17

Radon transformation 4
random noise addition 235
RDWT-Schur watermark extraction
    process 21
RDWT-Schur watermark integration
    process 19–21
real-time anomaly detection 153
real-time monitoring and secure
    communication 146–7
real-time video analysis 253–4
receiver operating characteristics
    (ROC) curve 7, 11
recurrent neural networks (RNNs) 253,
    277
    for temporal consistency verification
        261–2
redundant discrete wavelet transform
    (RDWT) 16, 19
Reed–Solomon method 19, 38
reinforcement learning (RL) 118

remote data storage 94
research methodology 175
reserving room before encryption
    (RRBE) 55
ResNet 237–8
resource constraints 150
retouching 255
reversible data hiding (RDH) 54
reversible data hiding in encrypted
    images (RDHEI) 54
    block diagram of 54
    citation analysis 66
        author citation linkages 66–8
        country citation linkages 68–9
        document citation linkages 69–71
        organizational citation linkages
            71–2
    co-authorship analysis 60
        author coauthor linkages 60–4
        country coauthor linkages 65–6
        organizational coauthor linkages
            64–5
    publication trend and productivity
        analysis 55
        data collection source and strategy
            56
        publication structure analysis 56–9
Rivest–Shamir–Adleman (RSA) 94,
    145
RL-based VNE algorithms 118
robustness 33–4, 226
robustness tests 26, 27
role-based access control (RBAC) 145
rotation with crop (RCrp) 8

salient edge detection technique 4
salt and pepper noise (SPN) 8
Schur decomposition 16, 24
Schur transforms 38

Science Citation Index Expanded
(SCI-EXPANDED) 59
SciPy 176
Scopus database 175
Scopus-indexed information tools 173
second-degree Chebyshev polynomial
130
secure communication methods 214,
238
secure data transmission 145
secure multimedia streaming 148
secure multi-party computation
(SMPC) 146
secure real-time transport protocol
(SRTP) 148
secure socket layer (SSL) 150
secure telehealth services 146
security 34–5
security analysis 173
security-aware RL-based VNE method
119
security-based virtual network
embedding algorithms 120
self-attention mechanisms 262
semi-fragile watermarking technique 39
sensitivity (Sen) 107
sentence-by-sentence hiding (SSH)
scheme 233
sequence descriptor 270
setting parameter 130–1
SFrame 200
SHA256 79
sharpening techniques 227
shifted window local loss (SWLL)
mechanism 228
Shor's algorithm 95
shortest path technique 121
shortest vector problem (SVP) 198
singular value decomposition
(SVD) 3, 36

smart contracts 86, 141, 143, 160
smart grids 147
sniffing 150
software-defined networking (SDN)
150
sparse coding techniques 4
spatial domain techniques 121, 217,
240
exploiting modification direction
(EMD) technique 219–21
LSB technique 217–18
PVD technique 218–19
spatial domain watermarking 35–6
spatio-temporal forgery 258–9
spatio-temporal transformers 259
speckle noise (SPKN) 8
splicing 254
SR-Net 236–7
stacked denoising autoencoder
(SDAE) 88
steganalysis with patch attention
mechanism (SPAM) 237
steganography 33, 214
applications of 238–40
block diagram of 216
challenges and open issues 234
detection and steganalysis
resistance 236
robustness against attacks in
steganography 235–6
trade-off among evaluation
metrics 234
classification of information hiding
schemes 215
digital steganography 216
spatial domain techniques 217–21
transform domain techniques
221–3
evaluation metrics 223
capacity 224–5

imperceptibility 225–6
robustness 226
recent trends and advances in 226
adaptive and hybrid techniques 226–8
cross-domain and cross-modal steganography 231–2
linguistic steganography 232–4
machine learning and AI in steganography 228–9
quantum steganography 229–30
StegoFormer 228
stego-image (SI) 218, 225
Struc2Vec approach 122
structural similarity index measure (SSIM) 24, 42, 225, 265
subgraph isomorphic behavior identification method 120
substitution box (S-Box) 230
substrate node 128
Sun Yat-Sen University 64
symmetric encryption 32

tamper-proof digital content 146
tamper-proof verification 159
targeted steganalysis 236
telemedicine 16
temporal attention mechanisms 265–6
temporal consistency checks 258
temporal forgery 256–8
temporal video forgery techniques 257
TensorFlow 131, 176
texture analysis 256
text watermarking 37
THInImg 231
time-series optical flow frames 270
TimeSformer 262
timing analysis 49
Tompkins method 19

trade-off among evaluation metrics 234
transform-based steganography 235
transform domain techniques 221
discrete cosine transform 222–3
discrete Fourier transform 222
discrete wavelet transform 223
transform domain watermarking 36
transparency 82, 158
transportation networks 147
transport layer security (TLS) 145, 150
true positive rate (TPR) 7

UCF-crime dataset 268
unauthorized distribution 156
user authentication methods 104
U-shaped CNN network 228

vacating room after encryption (VRAE) 55
vector quantization (VQ) 39
video tampering detection 251
datasets from 266
celebrity deepfake dataset 267–8
custom dataset creation for video integrity 268
deepfake detection challenge 267
FaceForensics++ 266–7
UCF-crime dataset 268
in deep learning 260
convolutional neural networks for frame-level analysis 261
hybrid models for spatiotemporal analysis 263
recurrent neural networks for temporal consistency verification 261–2
vision transformers for motion-based analysis 262
detecting video deepfakes using pretrained transformer model 268

extracting frames and features
271–2

fine-tuning on custom dataset
274–6

loading the pre-trained video
deepfake detector 270–1

running model and interpreting
predictions 273–4

using optical flow for temporal
analysis 272–3

feature extraction for forged video
detection 263

frame transition inconsistencies
265

optical flow and anomalies of
motion 264–5

temporal attention mechanisms
265–6

rise of video manipulation
technologies 252

role of AI and deep learning in video
security 253–4

threats posed by deepfakes and
252–3

types of 254

manipulation techniques and their
detection 259–60

spatial forgery through intra-frame
manipulations 254–6

spatio-temporal forgery 258–9

temporal forgery 256–8

video watermarking 37

virtual network embedding (VNE) 117,
119–20

virtual network requests (VNRs) 118

virtual networks (VNs) 117, 120

virtual nodes 119, 128, 131

virtual reality (VR) 140, 240

vision transformers (ViTs) 231, 260,
271

for motion-based analysis 262

voice assistants 147

watermark generation process 19

watermarking approaches 16, 157

watermark integration process 20

wavelet transform 227

Web of Science (WoS) database 56

witnesses 83

word-by-word hiding (WWH) scheme
233

word clouds 178

XceptionNet 237–8, 261, 270

YeNet 237

zero trust architecture (ZTA) 145

zero-trust security models (ZTSMs)
153–4

zero-watermarking scheme 86, 88

zigzag pattern 230

Zoom 200

www.ingramcontent.com/pod-product-compliance
Lightning Source LLC
Chambersburg PA
CBHW050510190326
41458CB00005B/1490